HERSTAL group
FN HERSTAL - BROWNING

Distributed outside Benelux
by Yale University Press, New Haven and London

MERCATORFONDS

Distributed outside Benelux
by Yale University Press, New Haven and London

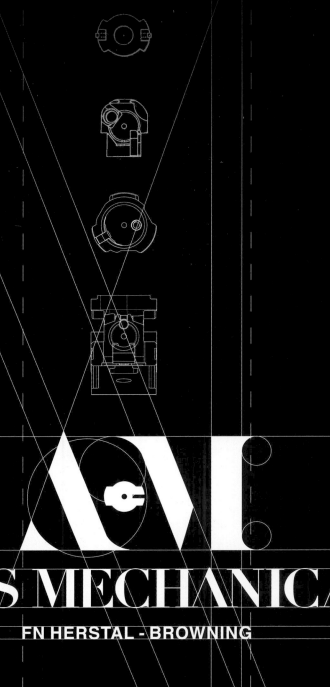

ARS MECHANICA

FN HERSTAL - BROWNING

DRIVING
INNOVATION

The Fabrique Nationale d'Armes de Guerre (National Factory of Weapons of War), also known as FN, is the pure product of its century and region and could not have emerged at any other time or in any other place. At the end of the 19th century, a key characteristic of Belgium, and particularly of the Liège basin, was economic power. The second industrial revolution, driven by electricity and steel, put the country at the top of continental Europe's pecking order. The armoury tradition of the Liège region, which has thrived and been famous since the 15th century, was the other key factor.

Contents

FABRIQUE NATIONALE
HERSTAL

Main entrance to the building of the Herstal Group's General Management, 2019
Completed in 1906, the building is still the headquarters for the company's strategic decisions.

Foreword

The Herstal Group is today one of the global leaders in small arms, thanks to its world-renowned marques – FN Herstal, Browning and Winchester. This success is rooted in the history of Liège and its industrial area, which was a fertile ground and inspiration for the Fabrique Nationale d'Armes de Guerre [National Factory of Weapons of War].

The company was created by gunsmiths in Liège, joining forces to meet the order requirements of the Belgian State. That was over 130 years ago... Five generations have since then followed one another and built FN – which has constantly adapted, navigated the twists and turns of history, and diversified by capitalising on its technological know-how. The company combined its arms expertise with its experience as an ammunition manufacturer – a unique phenomenon in this sector – before expanding to produce civilian arms. Subsequently it moved into the spheres of motor sports, leisure as well as aerospace and space. In so doing, the company underlined its capacity to work at the highest technological levels and showed there were no limits to what it could achieve.

Since the turn of the century, thanks to the support of the Walloon Region, the company has undertaken a major economic reorientation, based on its two business divisions: Hunting and Sports Shooting, plus Defence and Security. Backed by over a century of technological expertise, the Herstal Group is now engaged in diversification within its core business. Recently for example, it invested in a new assembly line making electronic cards for critical systems. The company is thus positioned as a European leader in innovation, design and production as well as the assembly of tomorrow's electronic equipment and systems. This sends out a clear message about the competitiveness of Walloon start-ups, SMEs and technology industries. In future, they will be able to benefit from the flexibility and agility of this new production tool. Because being agile is key in a context of globalisation and planet-wide competition.

This international scope is crucial for the company, since 99% of its turnover comes from either civilian or military exports. Global recognition of the Herstal Group's expertise enables the company to sign intergovernmental agreements and to join highly symbolic partnerships, notably the one constructing Europe's defence capabilities. The company also has industrial infrastructure located in several different countries, including Portugal, Finland, England, the US and Japan.

The Herstal Group can be proud of its 3,000 employees around the world, its state-of-the-art technological expertise, not to mention its outstanding skills in precision mechanics and technology. It is undoubtedly one of the best examples of Wallonia's boldness, creativity and innovation. A company that is always tenacious as well as eager to harness its tradition of excellence in support of the vital process of Europe's industrial renewal.

This book is an opportunity to retrace the evolution of this flagship of Walloon industry, a global leader in its business areas, from its origins through to the present day.

Willy Borsus
Vice President and Minister of Economy of the Walloon Region

1 The booming 19th century

The long march towards industrialisation

Up until the 19th century, the firearms produced in Liège, as was the case throughout Europe, were 'hand-made'. Workers copied whatever existed and there were no gauges or standards. A traditional rifle was made by forging and filing. Its components were finely adjusted to fit each other and they were not interchangeable.

Gun production was outsourced at the time. The 'manufacturer' responsible for finding and allocating orders would give craftsmen and home workers a part of the firearm to make. It would also provide the raw material to be processed and, sometimes but rarely, the necessary tools. The manufacturer would then pay the agreed sum of money for the work completed.

Each craftsman gunsmith would bring a specific element within the manufacturing process that took place in specific locations, which were selected because of their local stability down the centuries and which became highly specialised. For instance, the manufacturers of barrels were based in the Vesdre Valley, the stamping of break-action firearms would take place in Jupille and the wood assembly in Vottem. The gun being manufactured would thus travel frequently within the Liège region.

Manufacturing firearms had become a business for professionals, both men and women. They had many trades (e.g. barrel-maker, borer, woodworker, barrel polisher, engraver, chiseller, etc.) that followed on from each other in the manufacture of a firearm. This extreme division of labour clearly ensured excellence in each of the processes. However, it made adjustments necessary at almost every stage.

This way of doing things was already obsolete at the start of the 19th century. There was a mismatch between the very slow manufacturing processes and the high demand for weapons of war, since the 19th century was also a period of large-scale wars. From 1803 to 1815, the Napoleonic Wars resulted in much blood being shed on the continent and they were not the only ones. Others of a similar scale included the Russo-Turkish war of 1828-1829, the Crimean War of 1853-1856, the Austro-Prussian war of 1866, the Franco-Prussian war of 1870-1871, not to mention national uprisings and civil wars.

Constantin Meunier, *La coulée d'Ougrée*, around 1880
In 1817, John Cockerill arrived in Seraing and installed a coke oven there. Thirty years on, the factory had expanded considerably and this contributed to making Belgium a globally renowned industrial power. In 1863, the Cockerill blast furnaces were the country's first to use the Bessemer converter: this enabled the rapid, rational and flexible production of industrial iron and liquid steel. Steel production, along with electricity and the internal combustion engine, lay at the heart of the second industrial revolution in the late 19th century.

Luxury Liège rifle, 1866
This Lefaucheux system side-by-side shotgun was produced by Pierre-Joseph Lemille for the International Exhibition of Paris of 1867. It reflects the weaponry know-how of the Liège region. Using the modern pinfire cartridge system, the shotgun stood out because of its Damascus steel barrel, but above all for the carving and the embossed gold inlay work done by Joseph Boussart. It was done in the neo-Renaissance style and drew inspiration from hunting subjects: stylised dogs and decorative plants. Engraved in relief, the trigger guard represents a dog observing a couple of partridges, while the hammers are produced in the same hunting scene. This gun is listed among the Treasures of the Wallonia-Brussels Federation.

For instance, the gunsmiths from Liège Ancion & Cie, Pirlot frères, Renkin frères and Auguste Francotte supplied 20,000 rifles to the British Army during the Crimean War. There was so much geopolitical instability that countries that had regularly relied on foreign suppliers up until then – whether for weapons or other industries – would set up national factories capable of supplying them at any time, whatever was happening in the world.

The long supply chains of the past were no longer in operation and little by little, Liège would limit its production to hunting guns only. These were superbly made, according to methods passed down over the centuries.

Adaptation to modern warfare presupposes that weapons be produced rationally. In other words, the different parts of which they are composed should be defined with great precision and made meticulously, so that they can be produced in large quantities (and therefore possibly in different factories). Above all, they should be interchangeable. The efficiency of repairs depends on this, particularly on the battlefield.

This interchangeability has been sought after since the last quarter of the 18th century. In 1777, France experimented with a somewhat standardised rifle – the 'Model 1777' – whose parts could be assembled without modification or after minor alterations. The principles of machining and control were rigorously implemented. Thanks to standardisation, the Model 1777 could be manufactured – almost identically – by the thousands in Saint-Étienne, Tulle, Charleville, Maubeuge, Mutzig, Roanne, Versailles, Culembourg, Turin and Liège. Nearly two million of these rifles were manufactured up until the middle of the 19th century.

Even though the principle of interchangeability was well established, there was still a lack of technical capacity to implement it fully. The idea was there, but had not been perfected, which meant that products still had to be finished by hand. These constraints could only be overcome through mechanisation.

The challenge for the 19th century was therefore clear: mechanising production at a time when manufacturing was still firmly based on traditional processes and practices, while incorporating numerous recently invented processes (percussion lock, rifling of the barrel, breech loading and so on).

The craftsmen and women, who today we would call 'freelance', were working at home in the suburbs and outskirts of the cities. They were the masters of their tools, which they maintained themselves. They were specialists and responsible for a complete stage in the manufacturing process. Moreover, they had exceptional know-how, inherited from a long tradition of producing firearms.

These craftspeople would soon become highly specialised workers in the mechanical manufacture of parts. This evolution was driven by the interchangeability of parts, which had become essential for efficiency on the battlefield, as well as because of the mass production required to meet the growing demand from various countries. Now these specialists could call on advanced tooling, made possible by new sources of power like steam, which were owned and maintained by others. From now on, their activities would be coordinated and measured by foremen and inspectors and carried out in workshops, where the entire manufacturing chain was based.

Small arms were certainly not the only product to have been impacted by a series of technical revolutions. The buzz of industrial activity was such that, in the city of Liège alone, between 1825 and 1845, no fewer than 568 industrial plants were created after prior authorisation.[1] The Liège region was the country's leading industrial area. Coal mines, spinning mills, foundries, blast furnaces, steam engines and glass factories were dotted around the landscape and they shaped the lives of those working there: "Liège no longer has the enormous cathedral of the prince-bishops built in the year 1000 and demolished in 1795 by who knows who, but it does have Mr Cockerill's factory", noted Victor Hugo in 1842.[2] This modernisation, which would lead to the 1889 'model factory' concept, could be introduced because the 19th century brought innovations in many fields – including political, economic, industrial, social, energy and technical. These innovations provided solutions to many of the former contradictions between ideas and turning those ideas into practical reality. Take the example of the interchangeability of a weapon's parts. Although mechanisation seems the obvious solution here, this mechanisation is completely dependent on having advanced materials, high-precision machines that must first be invented and manufactured, a regular and reliable source of energy, a qualified workforce, and so on.

The 19th century enabled the firearms industry to flourish, together with its successes, as well as its damaging effects and excesses. Countries – especially the young nation of Belgium, born in 1830 – played a key role in creating an economic and political context that would inspire an entrepreneurial mindset. A royal decree in 1830, which was later extended, sought to 'meet the needs of industry', notably by encouraging the establishment of innovative industries in Belgium. Similarly, the construction of a national railway, decided by a law of 1834, created new and much larger markets by tackling the challenge of long distances.

2 The early years

Factory exit, 1912
At the time, 4,000 workers worked on the industrial site, which covered an area of 20 hectares and housed 5,000 machine tools.

3 July 1889: the birth of the Fabrique Nationale d'Armes de Guerre

Today, the Company for the manufacture of weapons for the Belgian army is established before Mr Biar, a notary in our town. This Company will have its registered office in Herstal. Its corporate name will be: 'Fabrique nationale d'armes de guerre', a limited company with a capital of three million Belgian francs. It is made up of the ten most important firms in our town.

Those were the words used by *La Meuse*, a local daily newspaper, to report the creation of the Fabrique Nationale d'Armements de Guerre (National Factory of Weapons of War). This was an event that had created a stir among journalists and members of parliament, the royal entourage and ministers, industrialists and banking companies in the preceding months. So one could have perhaps expected a more enthusiastic announcement...

Ten days later, when the company's deed of incorporation was published, the same newspaper came back to the subject. We learn from the newspaper that the company's purpose would be "mechanical manufacturing and especially that of weapons and parts of weapons" as well as their sale, and that the term of the company was set at 30 years, with the General Assembly having the power to extend this term.

So it was a limited term but one that could be extended. The company could also manufacture products other than those in the weapons sector. Its articles of association contain portents of the upheavals to come for the company in the years ahead.

The deed of incorporation lists those involved: Albert Simonis, arms manufacturer; Jules Ancion, arms manufacturer; Allard Bormans, representative of the arms manufacturers Dresse, Laloux et Cie; Léon Collinet, Managing Director of the Manufacture liégeoise d'Armes à feu; Auguste Dumoulin, arms manufacturer; Joseph Janssen, arms manufacturer; Henri Pieper, arms manufacturer; Gustave Pirlot, arms manufacturer; Nicolas Vivario (known simply as the 'owner', but who was also an arms manufacturer) and Alban Poulet, banker, Managing Director of the Crédit général liégeois.

All of them subscribed to part of the 6,000 preferred shares issued at a value of 500 Belgian francs each, forming a capital of three million francs.

Two thousand ordinary shares without a designated value were also issued. They represented the contribution of the aforementioned arms manufacturers (with the exception of Nicolas Vivario) together with two other arms manufacturers, Émile Nagant and Auguste Francotte.

The monumental entrance of the Fabrique Nationale d'Armes de Guerre [National Factory of Weapons of War] leading to the production workshops, 1920

AN AFFAIR OF STATE

This contribution from arms manufacturers was virtual at this stage. It constituted the promise of an order for 150,000 to 200,000 repeating rifles of a model that had yet to be chosen. This would then take the form of a definitive contract with the Belgian State, once the company had been created.

This provision was somewhat illogical, from a timing point of view. The State was committing itself to place a very significant but rather vague order with a manufacturing company that did not yet exist and this order was an integral part of the company's deed of incorporation. Let's step back a bit, for a better understanding of this apparently tortuous logic.

The years from 1886 to 1888 saw the large-scale implementation of an invention that would give the armies acquiring it a definite advantage in combat: the repeating rifle. It was a rifle capable of firing several shots in succession, without having to reload magazine between each shot. This system, which was developed in the mid-19th century, had now reached maturity. Several models were now being marketed and the technical progress they offered was significant enough to encourage armies to equip themselves with them.

Germany began to equip its soldiers with these rifles as early as 1886, after having greatly increased its manufacturing capacities. France began trials with the repeating rifle the same year. In the months that followed, the Ottoman Empire (encouraged by the German Empire),

Austria, Italy (for its troops going to Africa), Great Britain, Denmark and the Netherlands acquired this gun, which had become essential for future battles. Belgium had no choice but to upgrade its firearms.

Talk of an upcoming war was bubbling away in Europe. The newspaper *La Meuse*, a keen observer of the twists and turns in Europe, wrote on 9 February 1888:

For two years now, wherever you turn, you can hear the clatter of arms; there are manoeuvres at the borders and spies are being pursued; fort after fort is being built and oppressive military laws are being passed. At times alarmist pamphlets appear; at other times hundreds of more or less dirigible balloons are bought; and then a very long-range repeating rifle is invented [...]. One would have to be deaf and blind to deny that, while the Parisians are working on their future Exhibition,[3] in Russia, France, Germany and Austria, some gigantic firework display is being prepared [...]. Let a spark fly, and we will immediately have the grand finale.

Amid this heavy atmosphere, the dynamic Minister of War, General Charles Pontus, who was in office from 1884 to 1893, strove to enhance Belgium's defence capabilities. In particular, he had 12 forts built around Liège and nine around Namur. At the same time, he decided to equip the army with a new rifle, which was, naturally, a repeating rifle.

Placing orders for these new guns from abroad was impossible, as that would have made the Belgian Army dependent on another country. Worse still, ordering guns

View of the General Management building, 1912

from France or Germany, at a time of guaranteed neutrality, would be tantamount to siding with one or the other in a conflict that was considered imminent between the two countries.

The members of parliament from Liège convinced the minister – apparently without too much difficulty – not to acquire these new firearms abroad, but to have them manufactured on national territory and specifically in Liège. The city was the only arms manufacturing base in the country capable of mobilising its industry to meet such a large demand.

STRENGTH LIES IN UNITY

At that time, no single company in Liège was capable of successfully carrying out such a project on its own. Already, in 1870, just before the Franco-Prussian war, some arms manufacturers had understood the need to club together to obtain orders. This was the period, up until 1876, of the 'Petit Syndicat' ['Little Syndicate'], a group that supplied arms or components of arms in particular to France, the Ottoman Empire and Greece.

Subsequently, in 1886, other manufacturers joined forces in a general partnership called 'Les Fabricants d'Armes réunis' [Weapons Manufacturers United]. Their objectives were the same as those of the Petit Syndicat: to create sufficient clout to win contracts and then to divide up the manufacturing processes. This union brought together seven manufacturers who would all later participate in the creation of the National Factory

of Weapons of War: Ancion, Dumoulin, Dresse-Laloux, Janssen, Nagant, Pirlot, Simonis and, a little later, Pieper. Les Fabricants d'Armes réunis was not a great commercial success, coming up against competition on European markets from the Loewe group, which offered the most advanced weapons of the time – the Mauser and Mannlicher rifles.

However, this union made it possible to lay the foundations of the future FN. In March 1887, when the rumour of a forthcoming State order spread, Les Fabricants d'Armes réunis were quite naturally candidates. But their strongest rivals – the Manufacture liégeoise d'Armes à feu and the Établissements Francotte – were not ruled out by the Inspection des Armes de Guerre [the Inspectors of Weapons of War], which was responsible for the implementation of the order. All of them were invited, in August 1888, to submit a quotation for the manufacture of 150,000 repeating rifles of a model to be determined.

Instead of competing with each other, the Fabricants d'Armes réunis, the Manufacture liégeoise d'Armes à feu and Auguste Francotte decided to join forces and formalise this alliance in an agreement signed on 15 October 1888. This agreement specified that: "It has been agreed that as soon as the undersigned have obtained from the Belgian government the order for the infantry rifles for the manufacture of which they have joined together, they will constitute a limited company between them"

THE MAUSER MODEL 1889

The Belgian government was undecided as to which repeating rifle it would choose to equip its army with, as was the French government, incidentally. So it ran a succession of comparative tests over several months. Decision-makers hesitated between the Nagant, the Pieper-Mannlicher, the Schulhof and the Marga. Both the results of these tests and the rankings were regularly published in the newspapers, which sometimes contradicted one another. At the end of 1888, other rifles entered the fray, with variations being introduced. There was the Mannlicher by the Manufacture de l'État and the one manufactured in Steyr in Austria, the Pieper with a cylindrical-conical magazine and the Pieper with a modified Mannlicher magazine. New models also emerged: the Mauser with a Mannlicher or Engh type magazine from the Manufacture liégeoise d'Armes à feu and so on. New tests were carried out in July 1889 in Beverloo. The Mauser, the Nagant and the Mannlicher were still in the race.

On 23 October 1889, the official decision was finally taken. It would be the Mauser Model 1889, a 7.65 mm calibre rifle, firing a rimless grooved cartridge. A report by General Pontus justified the decision of the King, who, in Article 2 of his royal decree, announced that "all the necessary measures will be taken so that the standard rifle is promptly delivered to the National Factory of Weapons of War in Liège, which has been called on to manufacture the new armament".

The decision was widely reported in the press. It was no great surprise that the most hostile reaction to this decision came from *Le Peuple, Organe quotidien de la démocratie socialiste* [*Le Peuple,* the daily newspaper of the Socialist Democracy Party]: "The Mauser repeating rifle, of foreign invention, has been officially adopted by the Belgian government", it wrote, adding that

Mr Vandersmissen[4] is "a mercenary who will have carried out this business without worrying about comments and criticism".[5] *La Réforme, Organe quotidien de la Démocratie liberale* [*La Réforme,* the daily newspaper of the Liberal Democracy Party] also expressed concern, in an ironic tone, that a foreign firearm had been chosen: "Yesterday, newspapers from Antwerp said that the Mauser rifle, which has been adopted for our army, is a Belgian rifle... because the patent was bought by a Belgian."[6]

The choice of a foreign model was not just a source of concern for the country's independence. Other voices expressed their views too: the Cercle des Intérêts matériels de la Province de Liège [an association of small arms manufacturers that were rivals of FN] organised a meeting on Sunday 17 November 1889 at 11.30am to protest against this choice and the conditions under which it was made. The meeting, which was held in the Royal Hall of the Renommée, was designed mainly to protest against the favouritism that had been supposedly given to the arms manufacturers who had set up FN, but also against the fact that the factory was going to have to equip itself with German machines to manufacture the Mauser. They even considered sending a 'protest committee' to the King to call on him to withdraw the decree of 23 October. A press campaign was organised. The headline of the 17 November edition of *La Réforme* ran as follows: "A new scandal. The manufacture of Mauser rifles – The National Factory of Weapons of War – A twenty million Belgian franc deal in hard cash – Protest by the Liège arms manufacturers."[7] The body of the article regretted the supposed lack of fair competition.

These strong criticisms, often relayed to the Chamber of Representatives, continued to be made until at least February 1890. Then they were replaced by another concern: when would infantry soldiers be equipped with the new Mauser rifle?

TIGHT DEADLINES

It was not until September 1890 that the Mauser Model was handed over to FN[8] and the first three rifles were only assembled in Herstal in December of the following year. One of them was to be offered to the Minister of War, General Pontus, on 6 January 1892.

The rifles competing to equip the Belgian Army:
Pieper-Mannlicher repeating rifle, 8 × 50 mm calibre
Schulhof repeating rifle, 8 mm calibre
Marga repeating rifle, 7.5 mm calibre
Nagant repeating rifle, 8 mm calibre
Mauser 1889 repeating rifle, 7.65 × 53 mm calibre

Belgian carabinier-cyclists equipped with FN-Mauser 1889 rifles, greeting King Leopold II, c. 1900

FN quickly set up what was needed to manufacture 150,000 rifles. A factory had to be built, the machine tools sourced and purchased, the steel required had to be made, workers and managers hired, and so on. The sequence of events sheds light on FN's sheer drive and the risks taken by the company, in full knowledge of the facts:

- 3 July 1889: creation of Fabrique Nationale.
- 12 July 1889: signature of the contract under which FN undertook to supply the War Department with 150,000 repeating rifles of a model to be defined, at a price per unit of 79 Belgian francs. It was thus a contract worth 11,850,000 Belgian francs.
- End of August 1889: signature of an agreement with Ludwig Loewe & Co. of Berlin, directors of the Mauser Waffenwerke, for the acquisition of the machines needed for future production as well as technical assistance. The machines, tools and monitoring instruments were to cost a little more than 240,000 Belgian francs.[9]
- September 1889: FN's Board of Directors hired the engineer Léon Castermans as Director.

- 22 October 1889: the licence to manufacture the Mauser rifle was bought by the State from Alexandre Résimont, the Mauser Waffenwerke's agent for Belgium. This licence provided that the weapons could be manufactured by the private sector.
- 23 October 1889: publication of the royal decree establishing the choice of the Mauser 1889 and recalling that FN was entrusted with its manufacture.
- 1 February 1890: the first plot of land for the construction of the factory was bought from the Bonne Espérance et Batterie collieries: just over six hectares for 50,000 Belgian francs. The area was a field, known as Terre de l'Évêque, located in Herstal.
- April 1890: award of the main construction and development works.

FN had largely anticipated government decisions and orders, with all the events appearing to follow on from each other in a climate of confidence.

Manufacture of FN-Mauser rifle receivers, 1927

FN-Mauser Repeating Rifle 1889 Belgian Model

FUSIL A RÉPÉTITION

Système Mau

Fig. 1. Coupe longitudinale,

System
Bolt-action repeating rifle

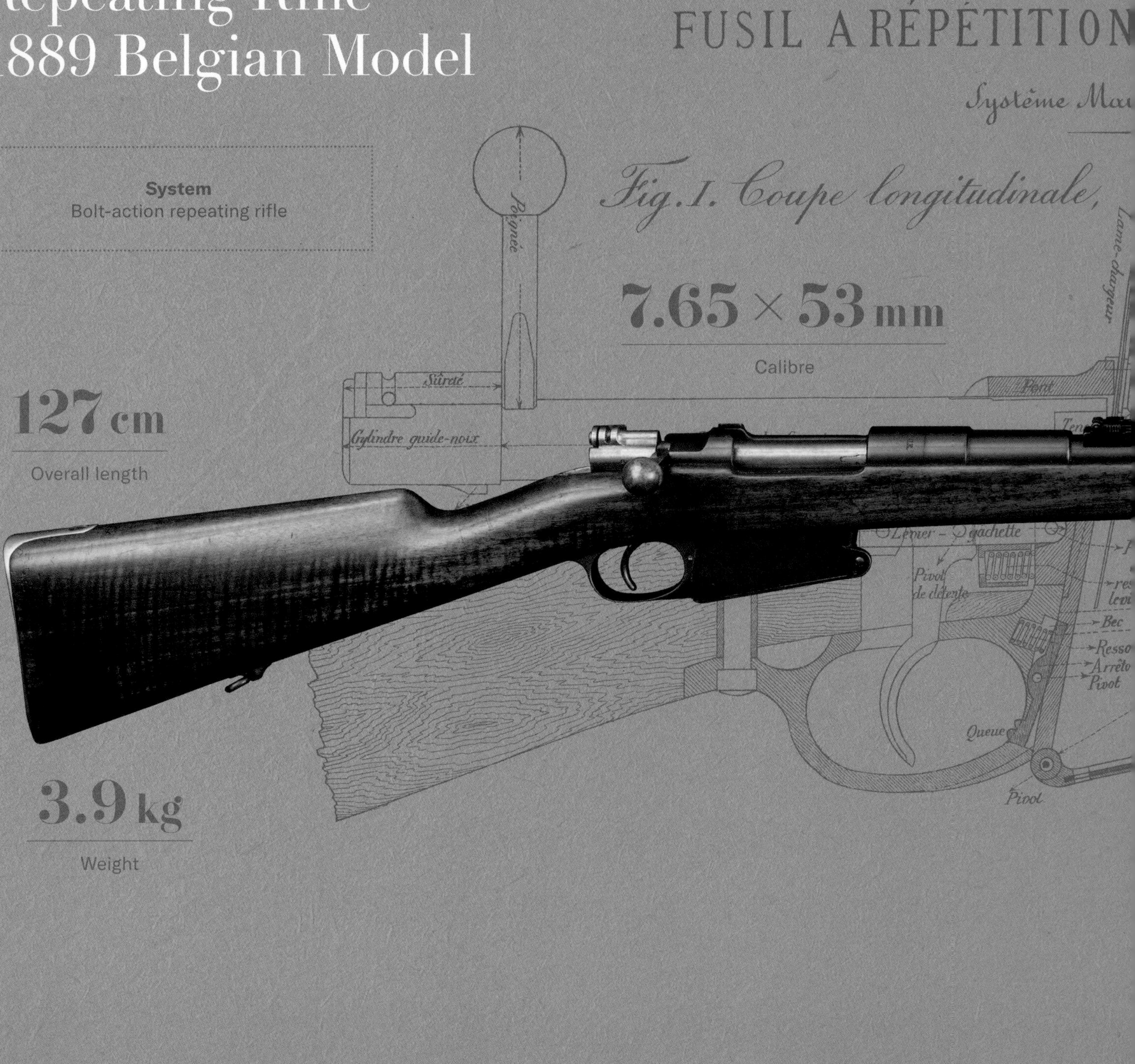

7.65 × 53 mm
Calibre

127 cm
Overall length

3.9 kg
Weight

The **FN-Mauser M1889** is based directly on the M1888, which was itself designed to replace the M1871, the German Imperial Army's first regulation rifle to use a rigid rimless brass case. The M1888 was developed to enable the use of smokeless gunpowder, which was more powerful than black gunpowder and reduced fouling. Besides the choke point for smokeless gunpowder, the barrel of the M1888 was fitted with a jacket, so it floated, in order to protect the user from any burns and to improve its efficiency; however rust could form over time and affect its longevity. The FN-Mauser M1889 was an adaptation of this rifle, to meet the demands of the Belgian Army. Firstly in terms of its calibre, which went from 7.92 × 57 mm to 7.65 × 53 mm. Secondly, a new type of semi-detachable magazine was installed to increase the firing rate. It would now be fed by a stripper clip for

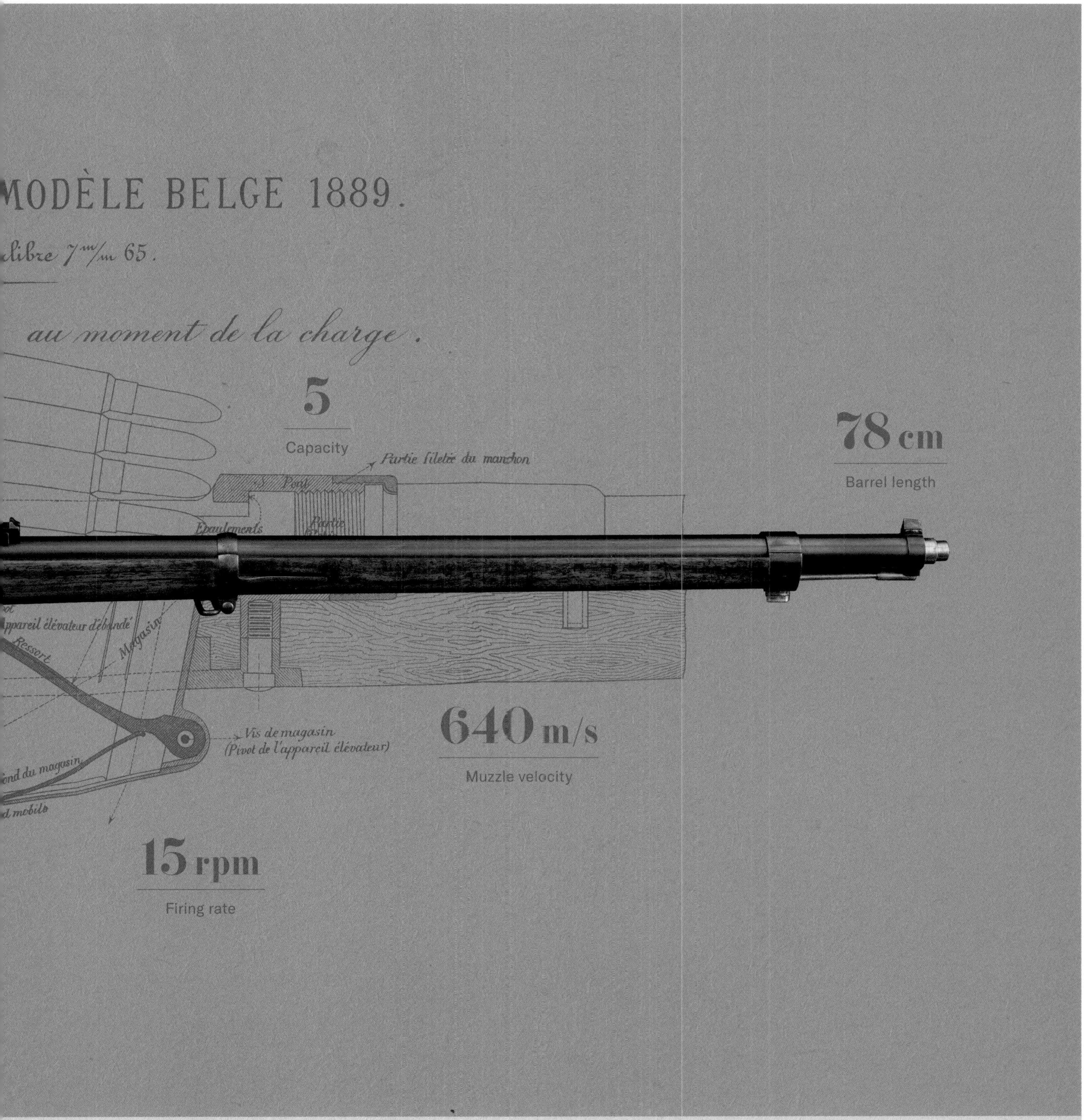

MODÈLE BELGE 1889.

calibre 7 m/m 65.

au moment de la charge.

5
Capacity

Partie filetée du manchon

Pont

Épaulements

Partie

pareil élévateur d'bande

Magasin

Ressort

Vis de magasin
(Pivot de l'appareil élévateur)

ond du magasin

d mobile

78 cm
Barrel length

640 m/s
Muzzle velocity

15 rpm
Firing rate

inserting cartridges into the upper opening of the breech block. A final key element was the improved design of the bolt head, which was no longer detachable. This made assembly easier and enhanced safety, by preventing dual feeding, which was a common problem in previous models.

Gas engines room, 1912
The Fabrique Nationale d'Armes de Guerre equipped itself with the period's most modern machines.

The model factory

From the very early days of its construction, FN captured the imagination of the public, who were mostly well informed about the work's progress. Construction work began with a new 11-metre wide access road being laid out. The road, called Rue Pépin d'Herstal, was inaugurated and paved from October 1890. Another road, which runs perpendicular to it and was known as the Voie de Liège, was widened. A galvanised steel fence both isolated and showed off the factory.

In October 1890, the construction work was already at an advanced stage: 6,000 m³ of excavated material had been removed, the structure that had been put in place weighed 645 tonnes and 18,500 m² of buildings had already been covered. An announcement was made that the electricity would be used not only for lighting, but also for the transmission of a 400 horsepower engine.

From that moment on, the idea that FN would be "the most beautiful arms factory in existence"[10] started doing the rounds. The press highlighted the huge size of the buildings, the sophistication of the machine tools, and

the thousand or so people who would be employed there. Outlandish calculations were made. If daily production was 250 Mauser rifles per day, as expected, then it would take an estimated four men to make one rifle in one day.[11] The machine tools were acquired from the firm *Loewe & Co.* of Berlin, but the forging machines came from the US, while the dynamos and engines were manufactured in Belgium.

On 21 June 1891, FN played host with great fanfare to the Minister of War, General Pontus. One can well imagine what the company looked like in its early days, from the many articles that reported on the event:

Although the installations have not yet all been completed, the general was able to get a good idea of the size of the new factory, the layout of the various departments and the way in which the work would be carried out. The factory has been set up on an eight-hectare site with 20,000 square metres of covered buildings. A railway line connected to Herstal station runs around the factory and serves the various departments.

The installations include a boiler room, a room for the engine, a hall for the machine tools for the iron parts of the mechanism and the rifle fittings, a hall for the machines for working and finishing the woods for the rifles, a high-precision mechanical workshop, forges, a case-hardening and annealing workshop, a room for the manufacture of the barrel jacket, workshops for drying wood, a cartridge factory, a wood store, a room for checking finished weapons, a room for testing rifle barrels, a stand, plus numerous and large offices, etc.

The boiler room contains four boilers supplied by the firm Jacques Piedbœuf de Jupille, each with 150 square metres of heating surface. There is also a Gaillet system purifier, which can process 100 cubic metres in ten hours.

Apart from its elegant ornamentation, the room with the engine is eye-catching in particular thanks to the remarkable installation of the 500 horsepower Corliss system compound steam engine, supplied by the highly reputed firm of Vanden Kerckhove of Ghent.

All the power of this major machine is transformed into electrical energy by means of a dynamo of 4 m 80 in diameter, and the fluid is transported via bronze cables to the different parts of the factory, where it drives 16 receiver dynamos with the force required for the tools that they control.

The regenerative dynamo, the largest ever built in Europe, was supplied by the Compagnie Internationale d'Electricité in Liège 'H. Pieper', which was also entrusted with the rest of the electrical installations.

Apart from long-distance transmission, to make use of the power from the rivers, this is, we believe, the first time that electricity has been used for installations as big as an arms factory. Therefore, it was only after submitting its project to the scrutiny of a Commission of special scientists and engineers that the new factory decided on the use of electricity.

Due to the lack of water at the factory site, special arrangements were made for the condensation. By means of a cooling device, the See system, a cooling pump from Bronne et Simon of Liège, and a 50 cubic metre tank at a height of 12 metres, the condensation water is continually circulated.

As for the boiler feed water, it is taken from the Liège-Maastricht canal and delivered by a pump driven by a 10 horsepower electric motor.

The machine-tools hall for the iron parts of the mechanism and fittings is surprisingly large, with an area of 1,000 square metres.[12] When the 900 machines that must be installed in it – and which are essential for the interchangeable manufacture of the model 1889 Belgian rifle, given the scenario, as it is in this case, of a level of daily production of 250 to 300 weapons – are installed and when this hall is lit by 76 arc lamps and the 100 16-candle incandescent lamps that it contains, it will offer a fairy-tale like spectacle.

The main transmissions attached to the columns, with a length of about 1,200 metres, are supplied by the firm Célestin Martin of Verviers. The high-precision mechanical workshop contains a series of extremely accurate machine tools, working, like the machine tools in the main hall, to the nearest 50th of a millimetre.

This workshop is intended to deal with the maintenance and renewal of clamps, cutters, drills and other tools.

All the machine tools, both for the iron and the wooden parts and for the high-precision workshop, come from Loewe et Cie, of Berlin, which delivers them at a price of 2,700,000 Belgian francs. Of this sum, 175,000 Belgian francs were used for the high-precision workshop alone.

The forges were built in America, at a cost of 300,000 Belgian francs, by the firm of Pratt Whitney et Cie, of Harsford.[13] They include 26 drop forge hammers of 1,000, 600 and 480 pounds sterling and six quick-impact hammers, Broadley system, of 60 and 40 pounds sterling. Each of these hammers has its own special operating temperature established according to a perfected system.

The parts will be case-hardened in a continuous gas furnace, fed by two Siemens gas generators, part of which will be used to heat the annealing furnaces.

The production of the barrel jacket of the rifle barrels requires 26 operations.

Strong presses and laminating machines will be used for this purpose.

The drying of the wood is carried out by the mechanical 'Dulzen' process. The process enables enough wood to be dried for a daily production of 500 rifles.

The cartridge factory, which will be installed shortly, will include the latest models of special machines required for the precision currently demanded for the manufacture of munitions.

The rifle barrels will be tested at 4,000 atmospheres.

The range for the firing test of the finished weapons is 225 metres long and will meet all safety requirements.

The metal part – 800 tonnes of iron – of all the buildings, columns, structures, tanks, etc., comes from the firm Vve Fredérix in Liège.

The transmission belts, 12 kilometres long, were supplied by Messrs Antoine Fetu-Defize et Cie, in Liège.[14]

FN brought together all the qualities that people liked to find in the industry at that time: strength and light, science and innovation, gigantic scale and precision. A rational approach thus made a sensational entry into the arms manufacturing process.

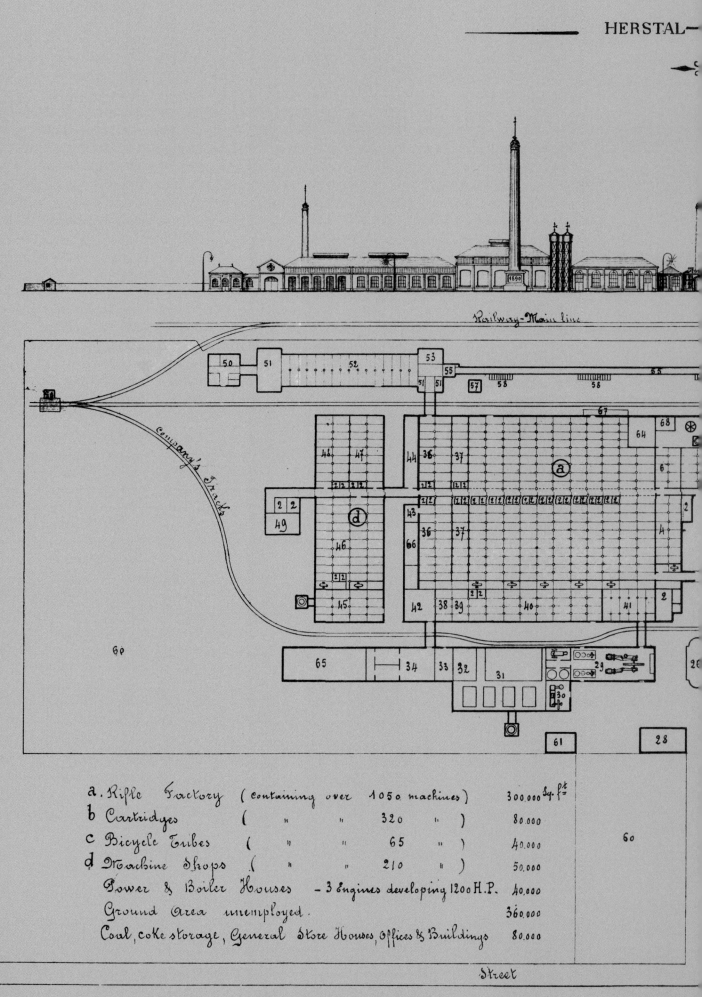

a. Rifle Factory (containing over 1050 machines) 300.000 sq.ft.
b. Cartridges (" " 320 ") 80.000
c. Bicycle Tubes (" " 65 ") 40.000
d. Machine Shops (" " 210 ") 50.000
Power & Boiler Houses — 3 Engines developing 1200 H.P. 40.000
Ground Area unemployed. 360.000
Coal, coke storage, General Store Houses, Offices & Buildings 80.000

Street

Scale

#		#	
1	Steel Room	35	Assembling Room
2	Offices	36	Revision
3	Bronzing Room	37	Revision of parts
4	Polishing Shop	38	Revision of Stocks
5	Hardening Shops	39	Oiling Stocks
6	Blueing & spring Hardening Room	40	Stocking Shop
7	Revision of forgings	41	General Store house
8	Smithy	42	Brazing shop
9	Annealing Ovens	43	Straightening barrels
10	Tumbling Drums	44	Tool Store Room
11	Cartridge shell shop.	45	Smithy for machine Shop.
12	Tool Room	46	Machine Shop
13	Cartridge testing & velocity Range	47	Model & gauge Shop.
14	Store house	48	Cutting Tools
15	Gauging Room	49	Draughtsmen
16	Clip Shops	50	Offices of military commission
17	Projectile Shop	51	Shipping Room
18	Cartridge Packing Room	52	Gauging of finished Rifles
19	Experimental Room	53	Proof firing at 60000 lbs. per sq. in.
20	Tool Room	54	Office of controllers
21	Revision of Bicycle Tubes	55	Range for sighting, 200 mètres
22	Tube Packing Room	56	Store-house for coke etc...
23	Stock Room for Ammunition Depot	57	Cartridge magazine
24	General Offices	58	W. C.
25	Eating Hall for women	59	Scales
26	Refrigerating Fountain	60	Ground Aera unemployed
27	Porter	61	Petroleum Store-house
28	Eating Halls for Engineers & officers of company	62	Charcoal Store-house
29	Engines	63	Proof firing
30	Pumps	64	Sub-superintendent's office.
31	Boilers	65	Store Room for green stocks
32	Repair shop	66	Model Room
33	Carpenter Shop	67	Rotary " Oil séparators "
34	Drying Room for stocks	68	Siemens Gasometer

Plan of the installations of Fabrique Nationale d'Armes de Guerre, 1890s

300   400   500 feet

Fabrique Nationale d'Armes de Guerre steam-powered electrical power station, 1912
The steam-powered electrical power station operated a dynamo generator, the biggest built in Europe up until then, from the Compagnie Internationale d'Électricité de Liège [Liège International Electricity Company]. This equipment powered the electric motors driving the transmission shafts of the machine tools, while providing lighting for the buildings and steam for heating. Extensive use of electricity in industry was considered very innovative at the time.

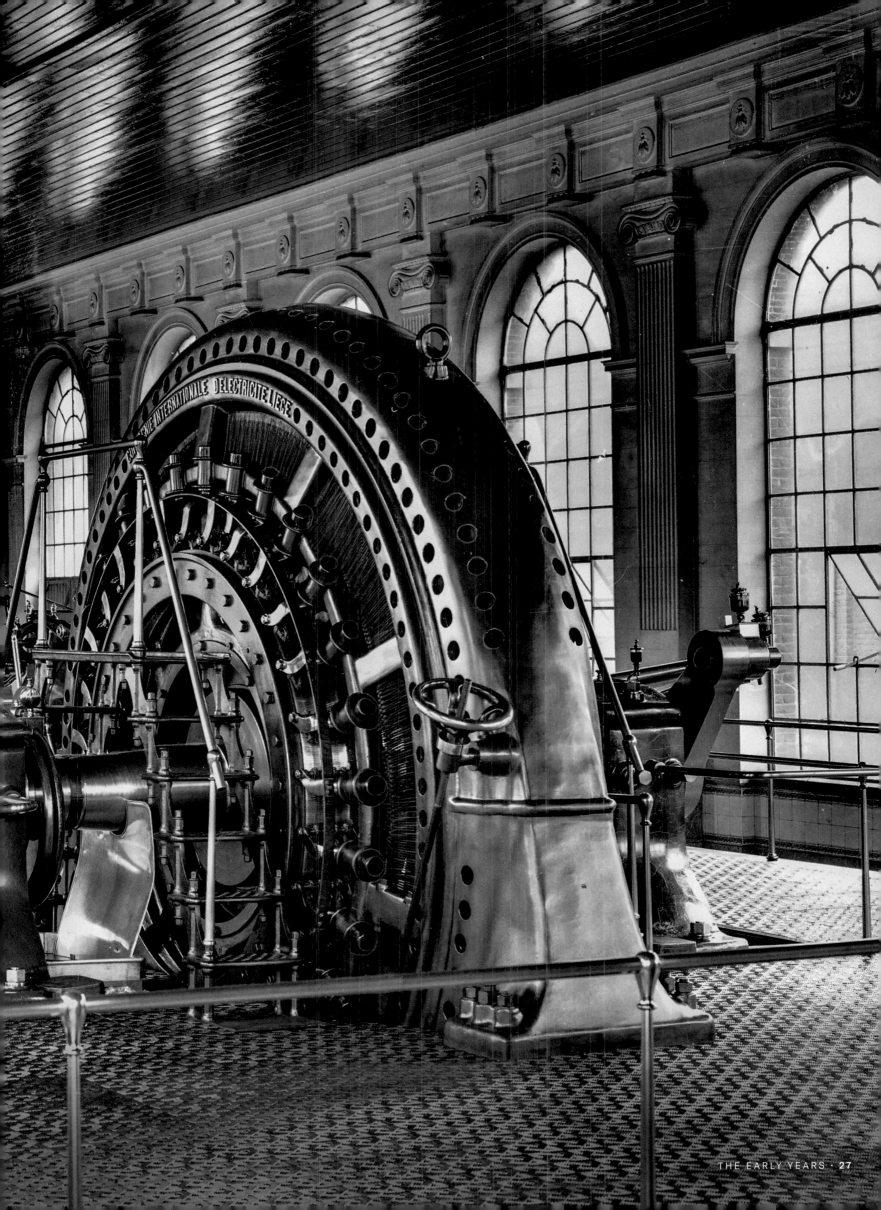

It took more than a century and considerable resources to move from the exceptional quality of craftsmanship, for which Liège had become famous, to efficient industrial production, which would make a fortune for the company.

The period of Businesses

Even before its establishment, FN was assured of a major order that would guarantee its stability for several years. With no pressure to establish a customer base, the company could have limited its industrial ambitions to executing this order. Because big investments, largely offset by the 11,850,000 Belgian francs expected from the Ministry of War, could have been enough to satisfy its directors. But that was far from being the case. Most likely the directors had differences of opinion on the central question of FN's future after the Belgian order. After all, for these gunsmiths, the company was a competitor to their own business. Yet right from the start, and until the 'restructuring of the company' in 1896,[15] the directors were apparently concerned about the company's survival after it had supplied some 150,000 Mauser 1889 rifles. Up until 1896, they were to participate – or would try to participate – in different business – with Russia, Serbia, China, Latin America, the Netherlands, Norway and, of course, with Belgium. Business cases were analysed, prices proposed, discounts granted, etc. Many visits were made to Brussels, Berlin, Paris, Vienna, Buda and Pest as well as St Petersburg, in order to meet intermediaries or government officials so as to lay the foundations for an agreement with similar companies and to avoid harmful competition.

THE CARTRIDGE BUSINESS

In 1891, the directors got wind of the Belgian government's wish to acquire cartridges for the Mauser rifle: "According to the information received, there is good reason to be in a position to supply thirty million cartridges for the Belgian War Department within a short time frame," they wrote on 19 February 1891. FN put itself forward for this contract, though it had no experience in the field of pyrotechnics. On 7 March, it offered to take on Lieutenant-Colonel Delmotte as an expert. In May that year, he travelled to Karlsruhe in Germany to study the operation of the facilities of the Deutsche Metallpatronenfabrik, an ammunition manufacturer working with Ludwig Loewe & Co.[16] Manufacturing equipment was ordered in a hurry, once again from the same Loewe company in Berlin. Terms of agreement were sought with specialised companies, such as the

Société d'Anderlecht. Finally, the order was secured and the cartridge factory was built on the Herstal site. The shareholders were informed as follows: "As a natural and necessary complement to the arms factory, during this financial year we have decided to set up a cartridge factory capable of producing 25,000 war cartridges in 10 working hours."[17]

This 'cartridge business' shows how FN operated to some extent, even in its early days. The construction of the factory was not yet finished, but the company was already seeking new outlets, which it achieved through a master stroke. To this very day, one of FN's strengths is combining the production of firearms and associated ammunition. Yet this impressive efficiency came with an unavoidable weakness, which was to tie, once again, its fate to the German supplier of manufacturing materials. Each 'business' thus came with its own specificities. This resulted in FN repositioning itself and, wherever possible, making improvements. Because, in spite of the considerable effort made by FN to create favourable commercial and industrial situations, structural problems kept emerging. Its senior managers' efforts were not enough to position FN in the new world taking shape in the late 19th century. In the last few years of that century, the pace of change accelerated. Economic and social markers were replacing traditional corporate values. For many years, these markers would be used to measure the efficiency of companies and thus how attractive and successful they were.

SALES AND MARKETING

Until its restructuring in 1896, FN would often use, and sometimes successfully, the empirical methods that it had learned. An example of this is the Russian contract in 1891, when the company eventually secured an order to do the conversion work for 400,000 rifles. In that case, it was the Nagant brothers, key figures at the court of St Petersburg, who successfully influenced the Russian Minister of War in his decision-making. Other transactions would turn out to be less successful, with the business entrusted to intermediaries who knew – or who claimed to know – the people likely to back an order, and who had to be paid for that. In addition to the costs these intermediaries incurred, they would receive a fiercely negotiated commission, calculated per rifle or per thousand cartridges sold.

The sales and marketing of industrial products had become a profession, in a highly competitive time and environment. In 1894, it dawned on FN that the networks that had proved effective in its early years were no longer sufficient. After some hesitation, the company decided to set up a new commercial organisation.

Hall for machine tools used for manufacturing parts of rifles, 1912
These machine tools, driven by a forest of conveyer belts, came from the firm Ludwig Loewe & Co.

The ammunition factory, 1928
FN's military ammunition factory was established in 1891 in partnership with the Deutsche Metallpatronenfabrik [German metal ammunition factory], which was also part of the Loewe Group. The factory was completed in 1929, with a workshop manufacturing hunting cartridges.

The Board made three decisions: to reorganise the agencies that had been set up in the capitals of the main European countries; to retain the services of Mr Burgers – who had previously acted as an intermediary in a few cases – for the Paris agency; and to instruct the management to recruit other agents, taking advice from the Cockerill Company. A few weeks later, an agent in Hamburg was appointed, and new business links were forged with companies or individuals who already represented Cockerill in Denmark, Spain and the United Kingdom. However, it was not until August 1895 that FN had a general agent capable of generating orders and negotiating them, in close cooperation with the Board.

FINANCIAL NEEDS

A problem emerged in the very first years of FN's operations. Orders from the Belgian State were paid without difficulty along with deliveries made. However, the same was not true for orders received from abroad. Brazil and Chile for instance requested, from the start of the negotiations, that their payments be spread over several accounting years. They wanted to reduce the cost pressure of the armaments on their national budgets. In 1895, China tried to demand that the first payment be deferred for a year. These demands made the situation pretty difficult for FN, as it did not have a sufficient financial base

at the time. The benevolent support it had – even within its Board of Directors – with the Crédit général liégeois was not enough. In March 1895, FN sought support from a financial consortium that might be able to advance the necessary sums.[18] The banks that were approached – the Crédit général liégeois and the Caisse commerciale de Bruxelles – turned down the proposal, while providing assurances that they were ready to help FN on a case-by-case basis. These assurances were insufficient for the company, which was rightly seeking to ensure regular and automatic financial facilities.

FACING COMPETITION

As early as 1893, FN was looking to reduce the impact of competition, which obviously leads to lower profits. Rather naively, it made this observation:
In view of the considerable sums spent in each country on negotiations for war weapons orders and the discounts which are being offered because of the presently acute nature of the competition, and given that FN has now proved itself and will most probably be able to produce for foreign countries, the Management asks the Board for permission to give special consideration to the desirability and necessity of being able to form a Syndicate with Steyr and Loewe. The Board is unanimously of this opinion and finds that the time has come

to try to reach this solution and instructs Mr Pieper to approach Steyr and Loewe and, as appropriate, to enter into negotiations with these companies.[19]

They clearly wanted to form a cartel, or at least an agreement among the biggest companies in their field, and thus to have more impact on the market. This proposed Syndicate was raised several times during FN Board meetings.[20]

Ludwig Loewe and August Schriever from the Steyr company, who were contacted by Henri Pieper, initially showed polite interest in the proposal but eventually stopped replying to FN's requests. Loewe probably already had another organisation in mind for the European war weapons industry.

The dreams of a syndicate of arms manufacturers came true when Ludwig Loewe & Co. took over the National Factory of Weapons of War in 1896. This was unlikely to have been a surprise for the Board of Directors.

An efficient commercial network, solid support from the banks and a more structured European arms manufacturing sector were three of the objectives pursued by FN. It pursued them doggedly but with mixed success over the seven years when it was the sole master of its destiny. These goals were perfectly in line with the existing industrial economy in the last years of the century.

The development of economic and financial structures was of key, and one might even say, essential importance in those years. Yet FN always remembered that the company relied on the quality of its production and therefore on the qualifications of its staff, workers and engineers.

NEW SOCIAL ASPIRATIONS

Despite the company's paternalistic image of caring for its workers, there could be no hiding what was already an outdated human resources management policy. The company did create a refectory for female workers in 1892, a facility that was separate from the refectory for male workers. Nonetheless, these women, who had been mostly employed in mechanical work in the factory's 'big hall' since September 1891, worked from 7am to 12 noon and from 1pm to 6pm (for a lower salary than the men, it goes without saying). That was a far cry from the eight-hour day regularly demanded by the workers' movement on May Day demonstrations on 1 May.[21]

On that very day of 1 May, in 1895, one retracted decision spoke volumes about the situation. On 27 April, the Board put up a poster in the factory with these words: "The management informs the staff that, as happened last year, work will take place regularly on Wednesday 1 May." It also made the decision to "purely and simply" apply Article 5 of the factory regulations to workers who did not show up for work on that day, i.e. probable dismissal. On 3 May, the Board could only note that "the vast majority of the special workers (setters, turners and fitters) did not come on 1 May". Almost all of them had asked in writing for permission not to show up for work. The Board therefore decided not to take any action on these absences, even if the factory's operations were greatly disrupted. It simply envisaged vague measures to be taken against the "ringleaders" in the future.

FN's attitude was typical of the era, when employment and welfare regulations were uncommon. This was due to what was called "freedom to work". In reality, this was only the freedom of the employers to control the way that work was organised, as they saw fit.

However, these certainties started to fade following a few developments. In March 1895,[22] the Board decided that elective functions for the Chamber, the province or the municipality and the function of worker or employee of FN were incompatible. Two directors – the Chairman, Joseph Janssen, and Auguste Dumoulin – called in vain for this proclamation to be tempered by wording that would limit its scope: "insofar as the exercise of these functions encroaches on the statutory working hours".[23] The staff's reaction to this decision was undoubtedly more than just lukewarm because, in October of the same year, the Board reversed its decision.[24]

In its early years then, FN assessed the social changes that were underway. With their roots in smaller companies, which were anchored in a more distant past with strong traditions, the firearms manufacturers/administrators gradually learned how to deal with the new aspirations emerging among their staff.

Marching towards diversification

THE JOURNEY UNDER GERMAN CONTROL

Completion of the manufacture of Mauser rifles to the satisfaction of the Belgian Army, the successful cartridge manufacturing immediately afterwards, as well as the industrial and commercial proactivity of its managers on the international markets could not fail to attract the attention and then whet the appetite of those who were becoming competitors of FN.

"The Germans in Herstal",[25] "German invasion",[26] "New meal for the ogre"[27] – the Belgian newspapers sounded the alarm at the end of 1895. FN came under the control of the Berlin-based arms manufacturer Ludwig Loewe & Co. which, after long negotiations conducted on the Belgian side by Henri Pieper, bought up the company's shares. For a few days, *Le Soir* spoke of the 'Fabrique nationale (?)', with a question mark after the adjective 'national'.

In 1890, Loewe had already tried unsuccessfully to become the main shareholder of FN. The attempt failed, with FN curtly refusing the proposal on the grounds that "the Belgian Government and public opinion would certainly be dissatisfied".[28] The events of 1896 underlined how relevant this analysis was, as we see it through the eyes of newspapers of the time. Loewe, however, was determined to create a European arms giant, of which FN would be one of the jewels.

Did the Loewe company – which was 'Prussian', as the press[29] liked to remind everyone – seek to destabilise FN a few years later, in 1894? Or did it only want to enforce its commercial rights, which was a legitimate aspiration? We can only point to a coincidence or an overlapping of dates between the end of the order for 150,000 Mauser rifles for the Belgian Army (December 1894) and the fact that Paul Mauser and Ludwig Loewe set the Chilean deal in motion (also in December 1894).

Chile had announced, in December 1894, that it wanted to buy 60,000 Mauser rifles (the Spanish 1893 Model) from FN. For FN, this model was a straightforward improvement of the Mauser 1889, for which it had acquired the licence. For the German side though, this rifle was a new model to which Belgium had no right. Paul Mauser wrote to FN on 11 December 1894: "You are manufacturing rifles for which you have no patents. Stop manufacturing within three days..." This was followed by recourse to arbitration, which would soon emerge as being unfavourable to FN. Although this arbitration's ruling was not delivered until January 1896, FN made a series of decisions in the first few months of 1895 to protect its existence as far as possible. An attempt at a settlement was made by the Belgian Minister of Industry in the summer of 1895. This ended in failure, as the claims made by Loewe – notably through its lawyer Victor Fris, a senator from Leuven – proved unacceptable to FN.

Rumours and denials, lawsuits and resignations, attempts to revoke patents and accusations with sometimes anti-Semitic[30] overtones against Ludwig Loewe followed one another. In early 1896, the German group took a majority stake in the capital of FN. The Managing Director, Jules Chantraine, was "called to other duties", resigned and a number of directors[31] and shareholders from the former factory made way for a new Board and a new General Assembly.

On 6 February 1896, the press announced the formation of the new Board of Directors. Baron Charles del Marmol, President of the Banque liégeoise, was appointed as its head. Léopold Vapart, industrialist; Georges de Laveleye, Director of the Moniteur des Intérêts matériels; while Jules Dallemagne, industrialist and Henri Pieper represented the Belgian side. Isidor Loewe and Alexis Riese, its Production Manager, represented the German side.

Émile Berchmans, *Fabrique Nationale d'Armes de Guerre Herstal-Liège*, lithography produced in Liège by Auguste Bénard, around 1900
Émile Berchmans called on mythology to highlight the quality of new mechanical productions. A mother-goddess, with her attributes – a sheaf of plants and a horn of plenty filled with spare parts – holds a bicycle fork in her right hand instead of her traditional scythe.

FN IN A CONSORTIUM

The newspapers of the time, whose hostility towards the Germans had subsided, strove to prove that FN had indeed remained a Belgian company on the pretext that there were more Belgian directors. While there were indeed few Germans on the Board, they nevertheless held a large share of the capital. In votes, they were assured of support from Victor Fris, their lawyer in the Mauser trial. They could possibly also count on the support of Henri Pieper, whose interests – both in the arms sector as well as in the tramway and electricity sectors, where he was very active – were partly, or almost totally, controlled by the Loewe group.[32]

It is useful to take a brief look at the 'takeover bid' though it was not yet called that. Whether hostile or friendly, nowadays this takeover bid would only merit a few adversarial debates as to its economic validity. It is likely that this 'internationalisation' of FN's capital, this European dimension that was imposed on the company, would not have aroused so much disapproval if a country other than Germany had been involved. Neutral Belgium carefully assessed its relations with its two big neighbours, France and Germany, taking care not to favour either of them or to offend anyone. The takeover of FN by Loewe – after Germans had taken many other stakes in Belgian companies – could have been seen as an attack on this balance. It is also possible, although no proof has been found, that French diplomats in Belgium whipped up the press to the point of unleashing some verbal attacks.[33]

The 'restructuring of the company' began, naturally enough, with the cessation of legal hostilities. The new FN had no further interest in pursuing a lawsuit against what had effectively become its parent company. Both the appeal against the arbitration award and the request for the lapsing of the Mauser patents were withdrawn. Once this formality was achieved, a new chapter could begin. It was to last 23 years, until the end of the First World War. This was a period that would contribute to building the company – through the diversity of the challenges it faced – much more than the short period beforehand when it was established.

FACING RESTRICTIONS

Membership of a cartel of arms manufacturers made it considerably less likely that FN would finalise contracts with foreign military administrations. The company was shut out from some markets and, everywhere else, any possible orders had to be shared with the other members of the cartel. In January 1897, a document fixed each of their shares in the manufacture of weapons: *DWM* in Berlin, would receive 32.5% of the orders; Mauser Waffenfabrik in Oberndorf, 20%; Steyr Waffenwerke in Steyr, 32.5%... and FN only 15%.[34] A similar agreement was reached for the supply of ammunition.[35]

In the years after, FN managed to supply significant quantities weapons of war to Uruguay, to the Belgian Army, plus cartridges to Spain and a large quantity of ammunition to Serbia.[36]

Given that the production of weapons of war was caught up in this cartel trap, FN's survival depended on the diversification of its activities, as provided for in Article 2 of its modified articles of association of 1896: "the manufacture and sale [...] of all objects that can be executed with its mechanical installations".[37] It should be noted, however, that diversification of production was not a new concept: other arms manufacturers have used or were to use it, in Belgium, as elsewhere.[38]

From October 1897, shareholders were informed that diversification of production was becoming a necessity: *Given that the manufacture of weapons and munitions of war is currently coming into a period of general lull, we busied ourselves seeking other branches of work in order to keep our staff occupied and to use our tools.*

Henri Pieper

Henri Pieper played a key role in the takeover of FN by Loewe. Born in 1840 in what is now Rhineland-Westphalia, he arrived in Belgium when he was barely 20. He settled in Liège around 1866, where he opened a mechanics and arms production workshop and, in 1870, a second factory in Nessonvaux, some 20 kilometres from Liège, specialising in Damascus steel rifle barrels. He soon abandoned handcrafted products in order to mechanise his business. From 1892 onwards, he marketed hunting guns under the Bayard brand. Less expensive than those of his competitors, they proved to be a success. Pieper was an inventor and registered 69 patents and created 24 trademarks between 1892 and 1898, the year of his death. His firearms were sold as far away as Mexico, for which he became honorary consul in Liège in 1894. With his son, he also founded the Compagnie Internationale d'Électricité in Liège in 1889 and was active in all the industrial sectors that were in fashion at the time: bicycles, electric cars, tramways, etc. He maintained links with Ludwig Loewe via several of his companies.

It was his son Nicolas who, in his father's name, proposed to the FN Board on 18 November 1895, to buy back all the shares – at a unit price of 700 Belgian francs! – and to redeem all the bonds.

Workshop producing and assembling the frames for motorcycles and bicycles, 1924

We have already succeeded in securing special products, such as bicycle parts and other items related to our industry, so that we have already managed to supply a large part of our workshops.

Restructuring of the company in 1896 had an immediate impact on production. Up until 1896, FN had only dealt with weapons and mostly with orders for Belgian or foreign governments. However on 12 October 1897,[39] new areas of business were opened: chainless bicycles, Aerator cartridges and even the authorisation to buy a motor car for testing purposes.

This diversification process continued until 1914 and was significantly helped, in terms of sporting firearms, by a chance meeting with John Moses Browning. There is still broad consensus today that the partnership was a major one and that it shook up the arms manufacturing world of the time, with a lasting impact that even shaped this sector differently.[40] Moreover, it was not the only opportunity that FN would grab to successfully diversify its production.

The chance meeting with John M. Browning

In its early years, FN had not sought to enter the hunting and sporting firearms market, because it did not want to compete with its own directors. The arrival of new managers and the departure of many of the historical shareholders allowed the company, once it had been reorganised, to envisage these new types of production. Thanks to similarities in the manufacturing processes of commercial and war weapons, FN set about producing, from November 1896,[41] 50,000 calibre .22 sporting rifles. Regular collaboration with the Syndicat des Pièces d'Armes Interchangeables[42] also allowed FN to deliver large quantities of shotgun parts. In 1903, these deliveries amounted to some 25,000 rifles.[43]

However, it was the meeting in 1897 between FN and John Moses Browning that determined the positive outcome of civilian arms production. The story – as it was later told – is that Hart Ostheimer Berg, FN's Commercial Director, met John Moses Browning by chance during a business trip to the United States in 1897. Berg wanted to explore the booming bicycle industry. One of the leaders in the sector, the Pope Manufacturing Company, was developing a new type of chainset in Hartford, Connecticut, which had caught the attention of FN. Another major company was located in Hartford: the Colt Manufacturing Company. Berg and Browning met there one day in April by a stroke of fortune that did not seem like chance at all.[44] The two men quickly realised that there was a convergence of interests between them. FN had skills and technical resources that had been underused since it had become part of the German cartel; Browning was looking for an astute and qualified entrepreneur who could put some of his many inventions into production.

What history fails to highlight is that John Moses Browning, while completely unknown to the general public in Europe, was already a major figure in the United States. His reputation as an inventor of firearms had spread from West to East and likely crossed the Atlantic, at least among the generally very knowledgeable arms-producing community. As an American citizen, Berg was very familiar with Browning and his inventions.

John Moses Browning, late 19th century
Today viewed as the Thomas Edison of firearms, John Moses Browning registered 128 patents, which laid the foundations for the functioning of the modern arms industry. A genius inventor, but not an industrialist, he joined forces with American firms that had a global reach such as Winchester, Colt and Remington, before turning to Europe and the Fabrique Nationale d'Armes de Guerre, a partnership with a bright future.

In 1879, upon the death of their father, John Moses and his brothers Matt (Matthew Sandefur) and Ed (Jonathan Edmund) inherited his small arms production shop in Ogden, Utah. It was here they repaired small arms, which were found widely throughout what was still the 'Wild West', while making a few of their own with very limited material means. The three brothers turned this modest business into a legend. The Browning Brothers – abbreviated to Browning Bros. in the American style – each brought his own skills to the table: Matt was a businessman, Ed a skilled mechanic and John was a genius inventor.

In the arms sector, the second half of the 19th century was an era of inventions: breech-loading, automatic repeating guns, smokeless powder, semi-automatic handguns and calibre diversification. The genius of John Moses was to consolidate and improve these new technologies in order to produce technically sophisticated guns that were bound to find their customers. Supported by regular advertisements for the Browning Bros. in local and then national newspapers, John quickly became well known. All the arms manufacturers in the country were now interested in his creations and his numerous patents: Winchester, Remington and Colt took over his inventions to put them into production, but also sometimes to prevent competitors from getting hold of them. In 1895, *The Salt Lake Herald* devoted a two-column article to 'John M. Browning, Inventor'. The tone was flattering: 'Many people in Utah have watched, with great interest, the successful career of the great Utah inventor, John M. Browning of Ogden...'[45]

More factual elements were also reported, contributing to the 'Browning legend': his birth in Ogden on 23 January 1855, his many brothers and half-brothers, his father, a gunsmith, from whom he likely inherited his fascination for guns at a very young age. While still a teenager, he repaired guns and tried to improve the way they worked, which gave him great knowledge of the mechanisms that he would later use in his inventions. This was Mormon territory, so the newspaper stressed that his successes were the result of sheer hard work, and not simply the fruit of a brilliant idea. After all, although John Moses had many successes, he also experienced countless failures. The article concluded by referring to Browning's wealth and the fact that sportsmen owed him a considerable debt.

The *Herald* recalled that his first patent had been taken out 16 years earlier for a breech-loading rifle of 'great simplicity and wonderful elegance', which was said to be the first successful application of this type of firearm.[46] The patent was taken over in 1884 by Winchester, which sold the gun as the 1885 Model. Eleven years later, the gun was still being manufactured and 75,000 had been sold. Inventions kept coming one after another, just like the patents (more than 40 had already been registered in 1895), notably for the Winchester 1886 and 1894 models of lever-action rifles.[47]

In 1895, 90% of the firearms manufactured by Winchester were based on Browning patents and tens of thousands of them were being sold. The *Herald* pointed out, however, that no pistol had yet been developed by Browning.[48] When Hart O. Berg returned from the United States with the promise of a first contract with the inventor for the manufacture of his 7.65 mm automatic pistol, this obviously generated great enthusiasm. On 17 July 1897, a contract was signed. The Browning brothers received 2,000 dollars and royalties fixed at two francs per gun sold.[49] Winchester had never offered to pay them a fee.

The armoury of the Browning brothers in Ogden, Utah
After opening their first armoury at 168 Main Street in Ogden in the late 1870s, the Browning brothers went on to construct another one in 1890 in the same town, at 2461 Washington Avenue. Yet they had no intention of stopping there. On 5 February 1890, a commercial advertisement in the local newspaper, *The Salt Lake Herald*, gave an indication of the new scale that the Brownings planned for their company: 'The Browning Bros. are moving to Salt Lake [at 155 S. Main Street]! We are going to make this old shop one of the best equipped sporting goods shops anywhere in the West. We will have a full range of weapons, rifles, pistols and ammunition. We will also have cutlery, boxing gloves, Indian clubs, dumbbells, playing cards, poker chips, waterproof boots and shoes and, in fact, a thousand and one things too numerous to mention.'

**The team of shooting champions from Utah, nicknamed
'The Four Bs' in the US, 1892**
From left to right: Gus I. Becker, John Moses Browning,
Archie P. Bigelow and Matthew Sandefur Browning.

BROWNING GOES TO BELGIUM

On 15 May 1898, *The Salt Lake Herald* announced that John M. Browning had left Ogden for Belgium to help the factory manufacture his automatic pistol, one of his most recent inventions. This appears to be one of the first accounts in the United States of the inventor's new contract with FN.

In 1900 Utah, which had become the 45th state of the United States in 1896, was well represented during the International Exhibition of Paris. Most of the products on display were Browning firearms.[50] The journalist added that Belgium was exhibiting a pistol manufactured in Herstal. It was prominently displayed among the Belgian products, even though it was under a Browning licence. Subsequently, each trip made by Browning to Belgium – and there were many of them – was carefully announced by the Utah press.

➻ **Patent filed by John Moses Browning for a semi-automatic pistol, 1897**
This pistol was based on the FN-Browning 1899 model. Thousands were produced before it was modified to become the FN-Browning Pistol, 1900 Model, which sold on a huge scale.

⚲ **Receipt signed by John Moses Browning and his brother Matthew for the payment of manufacturing rights by FN of a semi-automatic pistol, 1897**
Production of this firearm marked the beginning of relations between Fabrique Nationale d'Armes de Guerre and the famous American inventor and his family.

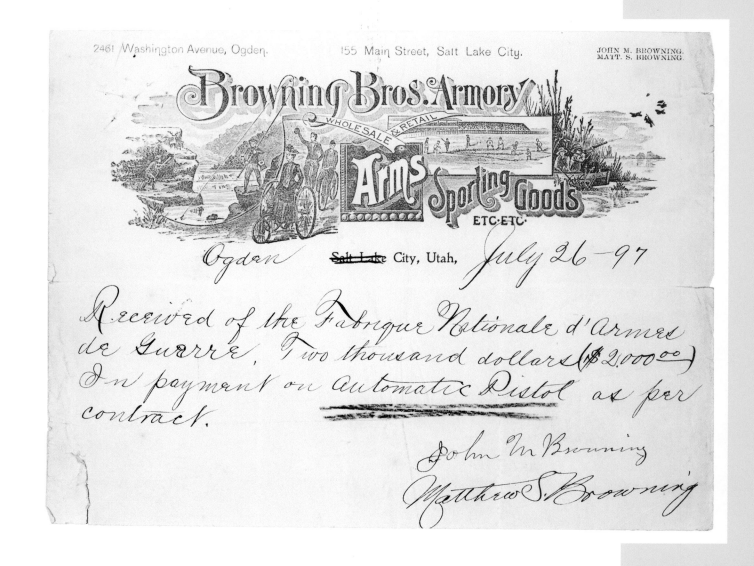

J. M. BROWNING.
GAS OPERATED FIREARM.
(Application filed Dec. 28, 1897.)
(No Model.)

2 Sheets—Sheet 1.

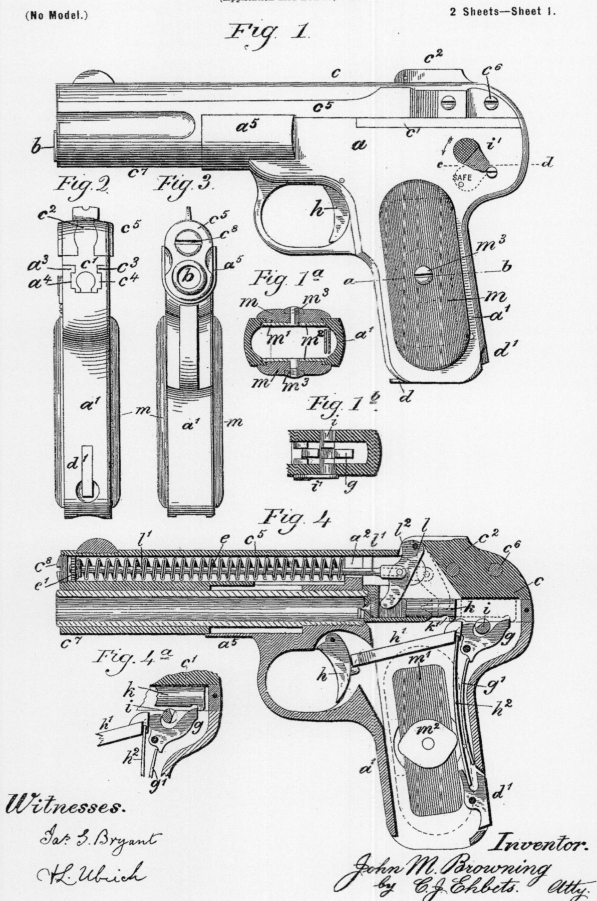

Witnesses.
Jas. S. Bryant
H. L. Ubrich

Inventor.
John M. Browning
by C. J. Ehbets. Atty.

The Salt Lake Tribune of 3 September 1909 gave more details: the inventor returned from Liège. He had looked after his interests at the factory in Belgium, where 3,500 workers, most of them women, were employed. FN was producing an average of 500 Browning pistols per day, or 175,000 per year.

Unsurprisingly, the partnership did not cause much of a stir in Belgium at first. After all, it was only an agreement with a firearms inventor. In May 1899, the Belgian press mentioned the name of the inventor, apparently for the first time. *La Meuse* reported on the gift received by Prince Albert of Belgium during his visit to the factory. Del Marmol and Frenay persuaded him to accept a Browning pistol in a case marked with its number.[51] On 6 July, the same newspaper devoted half a column to the qualities of the 'Browning automatic repeating pistol', which all officers of the Belgian Army would be equipped with from now on: 'The new gun is one of the most perfect that has been manufactured so far.'[52]

In Belgium, up until 1913, 'Browning' was a common name, except at FN. Prestigious visitors – the son of the King of Siam, Prince Tsaï Chen, a representative of the Qing dynasty and many others – were offered 'Browning' guns and the archives record courts cases in which a 'Browning' has played a lethal role, with the pistol being described as a 'fearsome weapon'.[53]

Hundreds of thousands of copies of the Model 1900 automatic pistol were sold. This would be the first of the benefits resulting from the association between a genius inventor and a factory open to innovation. However, the relationship between the factory and Browning had not yet taken the familiar and faithful turn that it would subsequently take.

In 1902, John Moses developed the world's first automatic shotgun. Initially, he did not plan to entrust FN with its manufacture. Instead, he approached American manufacturers. Winchester was the first and it refused, thus ending its fruitful collaboration with Browning. Browning immediately prepared to negotiate with Remington, but at the very moment when he was about to be received by its president, Marcellus Hartley died of a heart attack in his office. FN therefore acquired the licence via a contract on 24 March 1902. John M. Browning immediately ordered 10,000 of these firearms for the American

Semi-Automatic FN-Browning Pistol, 1910 Model
This Browning 7.65 × 17 mm calibre (.32 ACP) pistol operated with the blowback system.
The weapon became famous because it was used by Gavrilo Princip, a Bosnian Serb nationalist,
to assassinate the Austrian Archduke Franz Ferdinand, an event triggering the First World War.

market. The seeds of the relationship that FN would subsequently maintain with the Browning family were sown in this joint agreement.

In 1903, the inventor developed a 9 mm calibre pistol: the Swedish Army purchased large quantities of this weapon in 1907. In 1909, he developed the 'Model 1910', a pistol that, when the barrel was changed, could fire 7.65 and 9 mm cartridges alternately; and in 1913, a .22 calibre rifle that Browning ordered 50,000 of, again for the United States. He also developed a .22 calibre Trombone rifle, of which he bought 25,000, but which was never manufactured because of the war.

It's worth noting that, in 1907, the inventor granted FN the right to use the name Browning as a trademark. This clearly underlined the trust and team spirit between the company and Browning.

ADMIRATION ACROSS THE ATLANTIC

In 1914, a few months before the outbreak of war, Americans and Belgians alike shared their admiration for the inventor. FN was celebrating the production of the one millionth Browning pistol in its factories, albeit with a delay of a few months. *The Ogden Standard*, of 9 March 1914, wrote:

We have received, in Ogden, a copy of the newspaper 'La Meuse', from Liège, Belgium, dated 31 January. Nearly the whole of the front page and part of the second page of the newspaper are devoted to the celebration during which John M Browning was awarded the Knighthood of the Order of Leopold in the name of the King of Belgium.[54]

The Ogden Standard then quotes lengthy passages from the speech by Alfred Andri, FN's Managing Director.[55] Six front-page columns in *La Meuse* reported, on Monday 2 February 1914, on this lavish celebration intended to honour John Moses Browning, to thank the King and the Government, industrialists, the military, suppliers, customers and the company's heads of department. Fifty workers, chosen from among the oldest and most deserving, took part in the banquet. Some 500 people paid tribute to the inventor... and to FN.

Shouldering with the help of the FN-Browning Pistol, 1903 Model or the Grand Modèle, 1911
This semi-automatic pistol with an internal hammer operated with the blowback principle and was developed in a Browning Long 9 mm calibre version. Because Fabrique Nationale d'Armes de Guerre wanted to offer the military market a pistol with stopping power greater than that of the 1900 model and its 7.65 mm calibre.

FN-BROWNING
Semi-Automatic Pistol
1900 Model

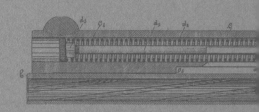

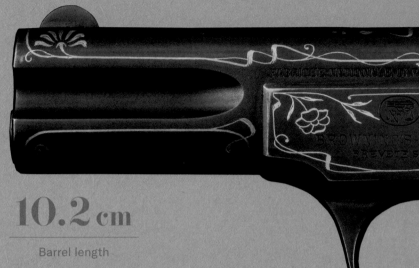

10.2 cm

Barrel length

Position, au départ, du mécanisme de la détente

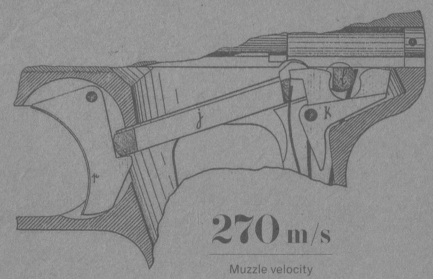

0.625 kg

Weight

270 m/s

Muzzle velocity

The **FN Browning 1900 Model** was not the first semi-automatic pistol to be designed. However, it stood out from predecessors such as the Borchardt C93 because of the simplicity and efficiency of its mechanism, lightness, refinement, its 7.65 mm BRG calibre ammunition (which was very powerful for the time) and especially because it could be produced easily in large quantities. It was in fact the first semi-automatic pistol to be produced on an industrial scale.

The 1900 Model was based on the 1899 Model, which was designed by John Moses Browning according to the patent that he filed in 1897. The gun uses recoil force (blowback) to produce different movements: pressure on the trigger moves the sear, which releases the firing pin and detonates the cartridge. The slide, which is not locked to the fixed barrel, recoils under the pressure of the gases generated by the propellant charge, thus extracting and ejecting the case. As it slides back, the

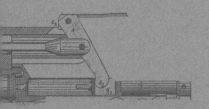

16.3 cm
Overall length

7.65 × 17 mm Browning (.32 ACP)
Calibre

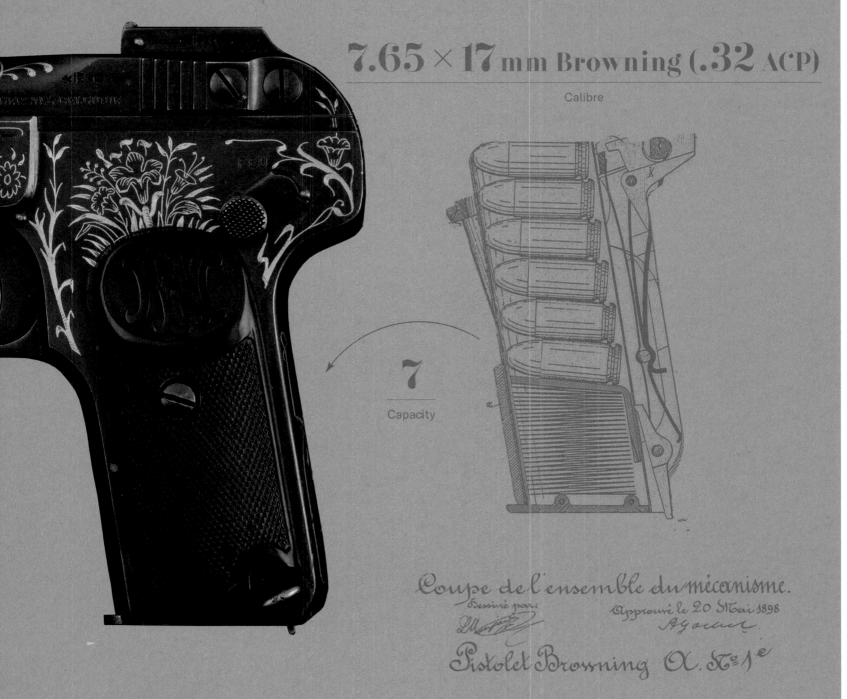

7

Capacity

Coupe de l'ensemble du mécanisme.
Dessiné par:
Approuvé le 20 Mai 1898

Pistolet Browning α. №1ᵉ

slide compresses a recoil spring, which sends it back to the front. During this forward movement, it picks up a new cartridge, which it takes from the magazine and inserts it into the barrel, fires and so on. The 1900 Model merely featured ergonomic adjustments, to match the Belgian Army's requirements: the grip-plates were reinforced and enlarged, a lanyard ring was incorporated and safety markings were reviewed. The 1900 Model also came in a luxury version, such as the grade V pistol illustrated here,

produced in 1908, blue bronzed with Modern Style gold inlays and standard vulcanite grip-plates.

The Liège daily newspaper's headline ran as follows: 'A Great Celebration at the National Factory. The Grande Manufacture de Herstal celebrates the 1,000,000th Browning and its Inventor with a huge banquet.' The speeches of the day – high on hyperbole as was typical of the era – came one after the other. After a toast to the King's health and the customary thanks, Andri took the floor to talk at length and in detail about the relationship that had bound FN to its brilliant inventor for 16 years:

Attracted by [the] reputation [of FN], Mr Browning had planned to enter into a union with it. The thing he counted on most to be approved was an automatic pistol [...] based on a new principle. It was like love at first sight for FN! And the agreement... sorry, the contract was soon signed. [...]. Let me tell you why and to what extent this partnership [...] has been so successful. [...] Both parties had all the qualities needed to give birth to the most perfect of products. [...] I could not possibly count all the qualities of the various weapons invented by Browning. [They are to be found in all his inventions] the ingenuity of the mechanisms, [the] simplicity and what I will call their aesthetics, i.e. their elegant form.

[...] The 'Fabrique Nationale', for its part, endowed the most modern and sophisticated of installations.

[...] In terms of his role in the cooperation, Browning remained the creative genius with all of his productive capacity: he invented various automatic pistols of different calibres, pocket pistols or pistols for combat, automatic hunting rifles, big game hunting rifles, and... we can announce it today, since FN has just given birth to it, its favourite child, its youngest child, the small automatic 10-shot rifle, of calibre 22. Each of these weapons was more beautiful and more ingenious than the others. FN was quick to adopt and to make a success of all of them, thanks to the finesse of their production. [...]

[It is this] partnership between Browning and FN

Photograph taken when John Moses Browning was departing for the US, 28 June 1910
From left to right: Mr Christiaens, FN representative in West Flanders; a member of a reception committee; A. Galopin, chief engineer; E. Michels, secretary to the management; Mr Chesney, permanent delegate and receiving clerk for Aerators Ltd (Sparklets); Mr Darrier, head of department for automobiles; Mr Englebert, FN Holland representative; Mr Israel, head of the technical weapons office service; T. Böll, chief inspector for the manufacture of weapons and cartridges; a member of the Russian committee; J.M. Browning; Mr Bräuning, inventor of a semi-automatic military rifle; A. Andri, managing director; Mr Hansgraeve, commercial director; an FN representative; J. Leroy, head of industrial accounting, future secretary general for the Manufacture d'Armes de Paris; G. Joassart, secretary and chief legal officer, future managing director; Mr Tritschel, head of the weapons manufacturing service, and Mr Stassart, head of the arms sales department.

[which is] the main architect of this marvellous development of our beautiful factory. FN had a dowry of nine hectares, three of which were built on; today it owns 20 hectares, three-fifths of which are covered with the most beautiful workshops. Its equipment has grown from 1,700 machine tools, worth 3.5 million, to more than 4,000 machines, worth nearly 10 million! The turnover, which only came to two million in 1898, has increased almost tenfold today and is worth about 20 million francs. [...] 730 workers were employed at FN in 1898; today there are more than 4,000 of them.

Only *La Meuse* gave a detailed account of this day, with the other newspapers being content to produce short and sometimes ironic reports. This detailed information was clearly provided to the Liège daily by someone who was in the FN camp.

Alfred Andri's explicit references to the union, engagement, marriage and children of the Browning-FN couple go beyond the notion of an industrial partnership, even if it was a major player in the firearms production sector. The Browning legend was fuelled by the warm international recognition enjoyed by FN. The subsequent, and undoubtedly embellished, account of their shared history given by Andri that day, focuses solely on Browning's contribution to FN. It says nothing about the other sectors of the company that also contributed greatly to its success. Nor does it reveal anything about the circumstances that led the war weapons factory to diversify its activities.

The war would put a strain on this relationship. From 3 August onwards, the press began to question the future of FN and, after the United States entered the conflict in April 1917, FN was only described as a "German factory". Browning, on the other hand, was hailed as a hero because he was supplying the American Army. Dozens of American newspapers enthusiastically praised his patriotism, including the *Romanian Daily News* in Cleveland, Ohio, the only Romanian language newspaper in the United States, which ran with the headline 'Patriotul J. Browning' ['Patriot J. Browning'] in April 1918.

The meeting with John Moses Browning and the phenomenal development and mass production of semi-automatic firearms were not the only fruitful venture of that period for the factory. Bicycles, automobiles and then motorcycles would also strongly contribute to its reputation.

The flamboyant Hart O. Berg

The engineer Hart Ostheimer Berg was born in Philadelphia in 1865. He was 30 years old when Henri Pieper recommended him to the FN Board of Directors.[56] At the time, the factory was looking for a sales representative for the Asian markets and particularly for the Chinese market, which it had been trying to enter for a year. The job profile was straightforward: 'Go and find the lay of the land, assess the positions and lead the negotiations.' In so doing, the company was embracing the newest sales techniques from around the end of the century. Hart O. Berg had no other commitments and he must have been convincing because, just a week later, FN offered him the position of general sales representative for all countries, with a comfortable salary (25,000 Belgian francs per year),[57] reimbursement of his travel expenses and of his 'high and exceptional expenses'... and a commission of one franc per rifle or per thousand cartridges sold, whether or not the order was obtained through him.[58] Apart from weapons of war, Berg's commission would be 1.5% for other items that FN would deal in. It was agreed that the contract could be terminated after three years, if the turnover from contracts brought in during this period by Berg was less than 10 million francs. When the company was reorganised in 1896, the contract was amended. The salary remained the same, but Berg's share of the net profits was increased to 3%, with a minimum of 15,000 francs per year. He also received an allowance of 25,000 francs in shares in the company, as well as the title of Director in charge of external affairs, while Fresnay stayed on as Director in charge of all internal services.

Berg made trips on behalf of FN to Serbia, Greece, the United States, Central America, South America and a number of other countries closer to home. In 1897, he used his social skills to persuade John Moses Browning to make a contract with FN and to license a chainset from the American firm Pope. This product helped the company to successfully launch its bicycle manufacturing business.

Berg suddenly resigned in April 1898. The reason for his departure is not known, but he later enjoyed a brilliant international career. He was the European agent for the Wright brothers, the American aviators, which led to him rubbing shoulders with high society like Crown Prince Wilhelm of Germany or the King of Spain. His wife was the first woman in the world to fly in a plane. He died in New York in 1941.

Celebration at the Fabrique Nationale d'Armes de Guerre, in the presence of John Moses Browning, of the millionth Browning pistol, 31 January 1914

FN-BROWNING
Semi-Automatic Shotgun
Auto-5 Model

Fig. 9

System
Semi-automatic shotgun, operating with a locked breech and a long barrel recoil

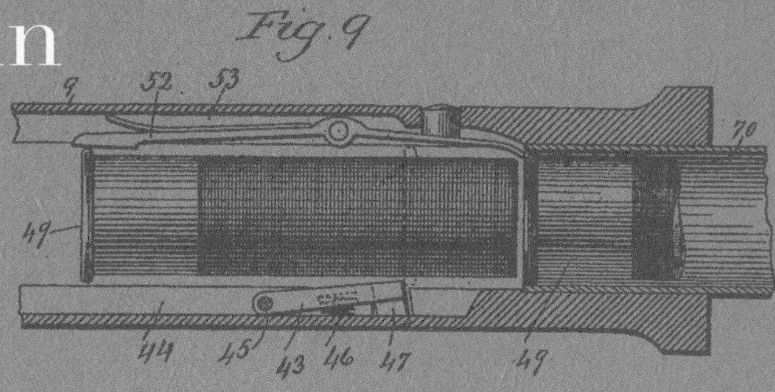

4 + 1
Capacity

± 120 cm
Overall length

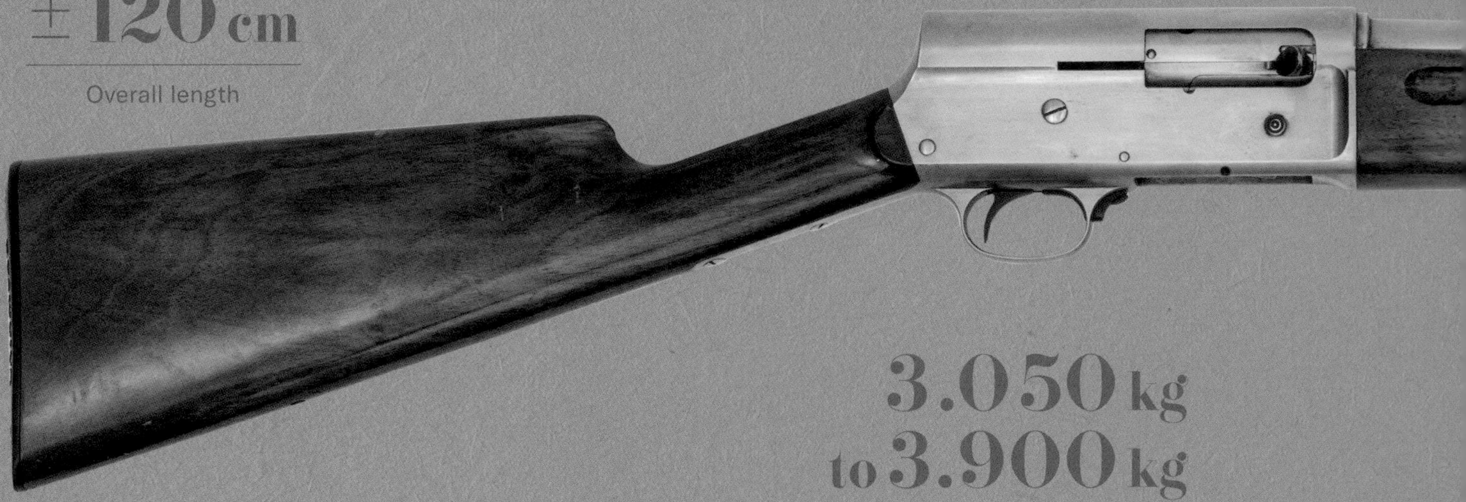

**3.050 kg
to 3.900 kg**
Weight

FN-BROWNING Semi-Automatic Rifle Model 1900, calibre .35 Rem

In 1903, with the **Auto-5 Shotgun**, John Moses Browning revolutionised the arms world by applying semi-automatic operation – with locked breech and long barrel recoil – to hunting firearms. The shotgun had a single interchangeable barrel and a single trigger, making it easier to use and to lock onto a target, when compared with shotguns that had side-by-side barrels (more common in those days). Six patents were filed for the shotgun's operation, resulting in a simple and effective firearm. When the trigger is squeezed, the breech recoils under gas pressure, pushing back the barrel, which is attached to it. The recoil compresses the recoil springs, which, when released, causes the barrel to move forward, followed by the breech, the insertion of the next cartridge and the locking of the mechanism. There is little recoil, thanks to the absorption of the gases by the

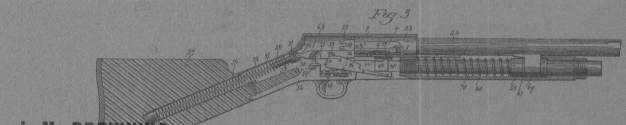

J. M. BROWNING.
AUTOMATIC FIREARM.
(Application filed Mar. 18, 1901.)

Fig. 3

65 / 70 / 75 / 80 cm

Barrel length

FN-BROWNING Shotgun Auto-5 no. 1

Fig. 4

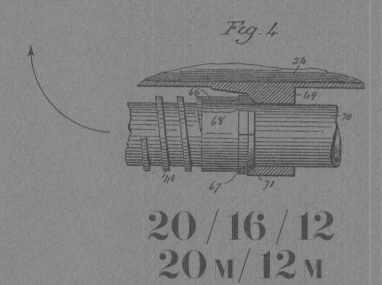

20 / 16 / 12
20 M / 12 M

Gauge

John M. Browning.
Inventor

recoil springs. This shotgun was a big success. First it was available in 12 and 16-gauges, then in 20-gauge, 20 Magnum and 12 Magnum, in a lightweight or super-lightweight version and with nine types of engravings. FN only stopped producing it in 1999, with the Final Tribute series. Five million of these shotguns were produced in a little under a century. In 1910, FN began to produce the rifled bore barrel equivalent of the Auto-5

semi-automatic shotgun. The calibre .35 Remington (9mm) semi-automatic Browning rifle was fed by a five-cartridge stripper clip and operated with a locked closure by barrel recoil. The barrel was placed in a jacket carrying the rib, as well as the front sight and the visor. The mechanism was patented twice, in 1900 and 1902.

Advertising image promoting the balance qualities of the new FN-Browning SA 22 Semi-Automatic Rifle, 1913

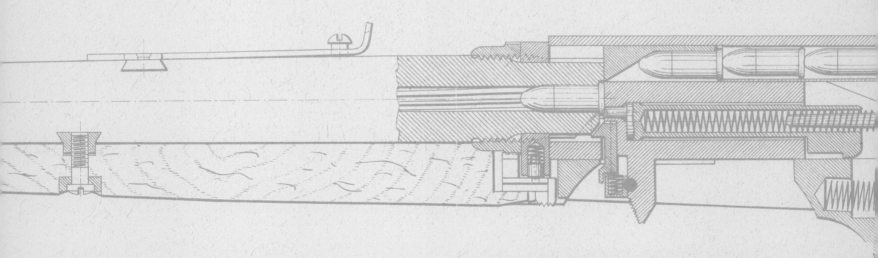

Cross-section view of the FN-Browning SA 22 Semi-Automatic Rifle, 1913
This rifle, which operated on the blowback principle, was notably fitted with a tubular magazine of eight .22 LR cartridges integrated into the stock. The weapon was a major success, with over half a million of them produced in two models up until 1972.

Advertising photograph for the FN-Browning Semi-Automatic Pistol, 'Pocket' Model 1905, 1911
This gun, a 6.35×15.5 mm Browning (.25 ACP), worked on the blowback principle. It offered three safeties, a system that could also be found on the 1910 model. The first safety comprised a moveable part, placed on the rear face of the grip, which the shooter had to push with sustained pressure in order to activate the trigger. Secondly, when the magazine was removed, an automatic safety was activated and prevented any firing of a cartridge if it remained in the chamber. Finally, an ordinary safety enabled the slide to be blocked.

↕ Details of a patent for a bicycle chainset sold to the Pope Manufacturing Company, 1894

↤ FN bicycle chainset with chain drive, 1911

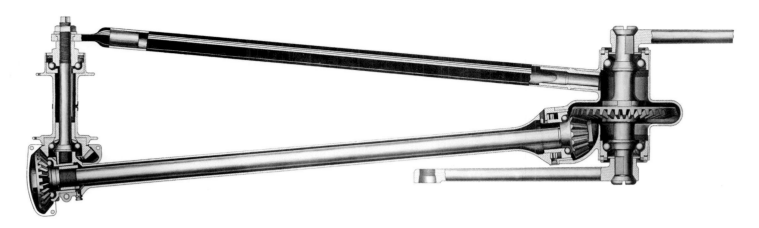

Shaft-driven system for a FN chainless model bicycle, 1912
The chainset turned a gear wheel on which a first sprocket was inserted, transferring the movement to a shaft. This shaft moved a second sprocket, turning the rear-drive wheel.

Motor power gets into gear

BICYCLES

In 1895, when FN first learned about the manufacture of bicycle frames in the United States, the bicycle looked as it does today: two wheels with spokes of equal diameter, a rear wheel driven by a chainset and a front steering wheel. Starting in 1891 onwards, bicycles were fitted with tyres that made rides more comfortable.

The bicycle was already an industrial product, but not one that everyone could afford. It would only become accessible to those who produced it later, when manufacturing costs could be brought down. The average price of a bicycle at the turn of the century was 250 Belgian francs, which was roughly equivalent to more than 300 hours of work for a worker.

FN adopted a very careful stance, typical of its approach to all mechanical manufacturing outside its usual spheres of activity. Although FN was ready to take industrial risks with firearms – as demonstrated in the alliance with Browning – it was cautious about taking on other unfamiliar products. Above all, FN lacked confidence in pushing initially beyond the limits of its engineering expertise.

For bicycles, FN restricted itself until 1898 to the manufacture of spare parts and in particular to the use of the Pope Manufacturing Company's patent. Its industrial stance was carefully planned, enabling FN to do what it did well, and thus notably building a reputation in a sector that was already highly competitive.

Within this narrow sector, this kind of cautious approach makes sense, just as much as technological audacity.

The first step was to make the small nipples that fix the wheel spokes to the rim. FN would produce tens of millions of them. Then came the spokes, which were initially very difficult to manufacture. On 22 July 1897, the Board – which had decided in April to manufacture all parts of the bicycle – looked closely at the production of bicycle frames. Three possibilities were envisaged: in-house production, the purchase of a patent from Pope Manufacturing or a patent from Pieper. After concluding that the royalties charged for each part produced would be too high, the Board chose the frame "designed by FN's technical office". This bold decision proved astute. In December 1897, during the Cycle Exhibition in Paris, and after hailing the first appearance of Belgian industry on the Paris market, *Le Figaro* highlighted the innovative nature of the frame proposed by FN (the frame had no brazing, and therefore contained no part weakened by heating).

In 1898, the company changed the gear ratio and started manufacturing completely chainless bicycles. Although the process was simple, the transmission by shaft and bevel gears required a high degree of mechanical precision. In September 1899,[59] the directors were delighted to learn that this model had received nothing but praise... and that it was already making a profit of over 100,000 Belgian francs.

Production levels were high and regular. In 1907, 10,000 bicycles with chains and 3,500 chainless bicycles were produced. The Belgian police and gendarmerie were to be regular customers of the latter model.

Bicycle assembly workshop, 1912
In its mission to diversify, Fabrique Nationale d'Armes de Guerre started producing bicycles in 1896. It made bicycles for 30 years, its star product being the famous chainless bicycle in which the drive chain was replaced by a shaft-driven system.

FN Chainless Bicycle Military Model 1914

Bracket modèle 1898.
Vélos a. Nº 1.

28   95

System
Drive shaft transmission

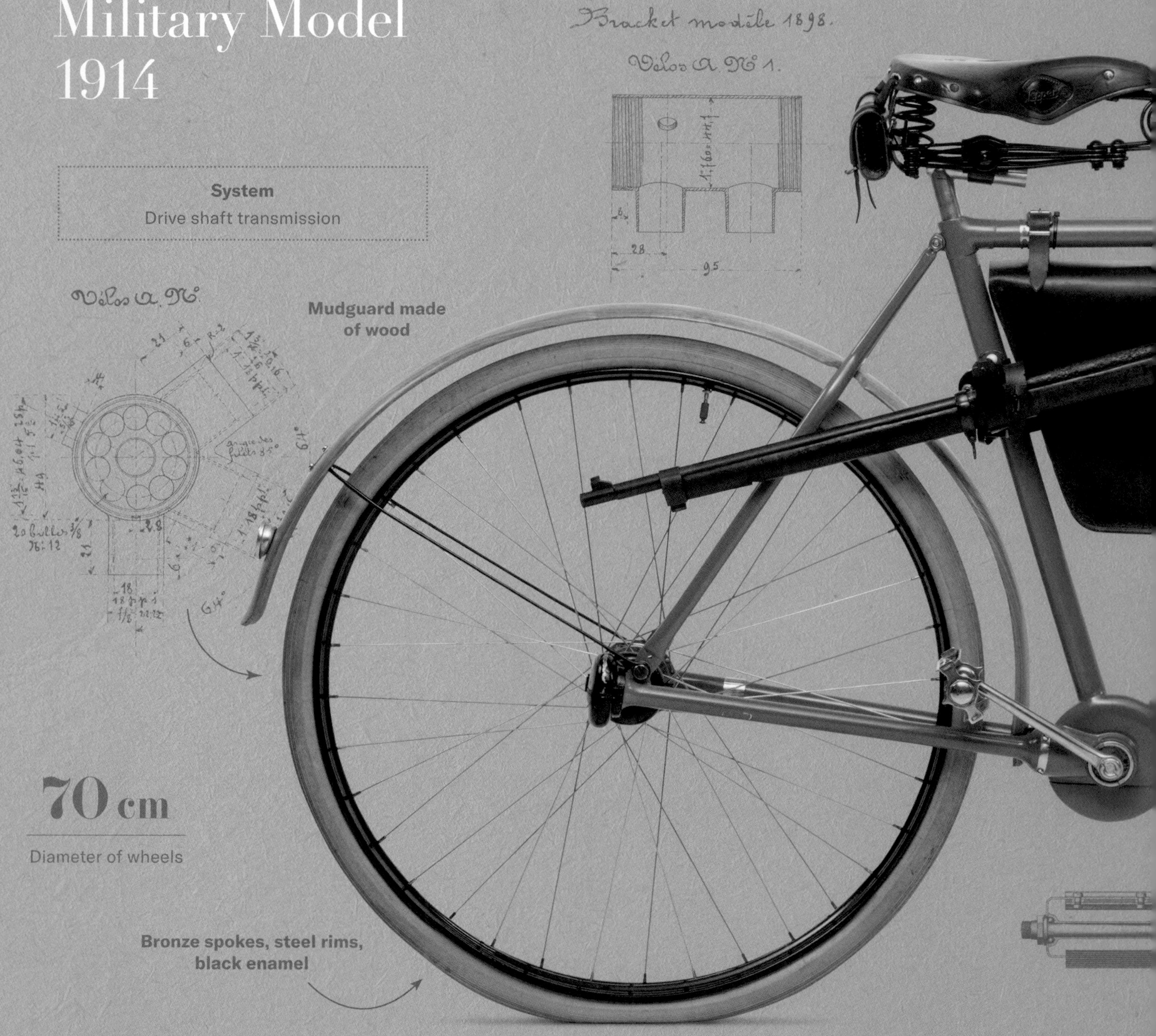

Vélos a. Nº.

Mudguard made of wood

70 cm
Diameter of wheels

Bronze spokes, steel rims, black enamel

In 1898, FN developed its own **chainless system**, or ***acatène***, bicycle, with a shaft-drive. The replacement of the chain by bevel gears was a major development for bicycles. The design of the drive system helped to distribute the forces more efficiently. The force applied to the pedals was transmitted by the cranks, the axle and the drive sprocket to a 17-tooth bevel gear, which was itself attached to one end of a ball-bearing shaft. This shaft had an intermediate sprocket cog of 26 teeth at the other end, driving the 24-tooth bevel gear on the rear hub. This robust system was also designed to address fouling due to mud and dust. In 1913, the weight of the FN chainless bicycles was reduced by using sprockets made of highly resilient chrome-nickel steel, with almost no

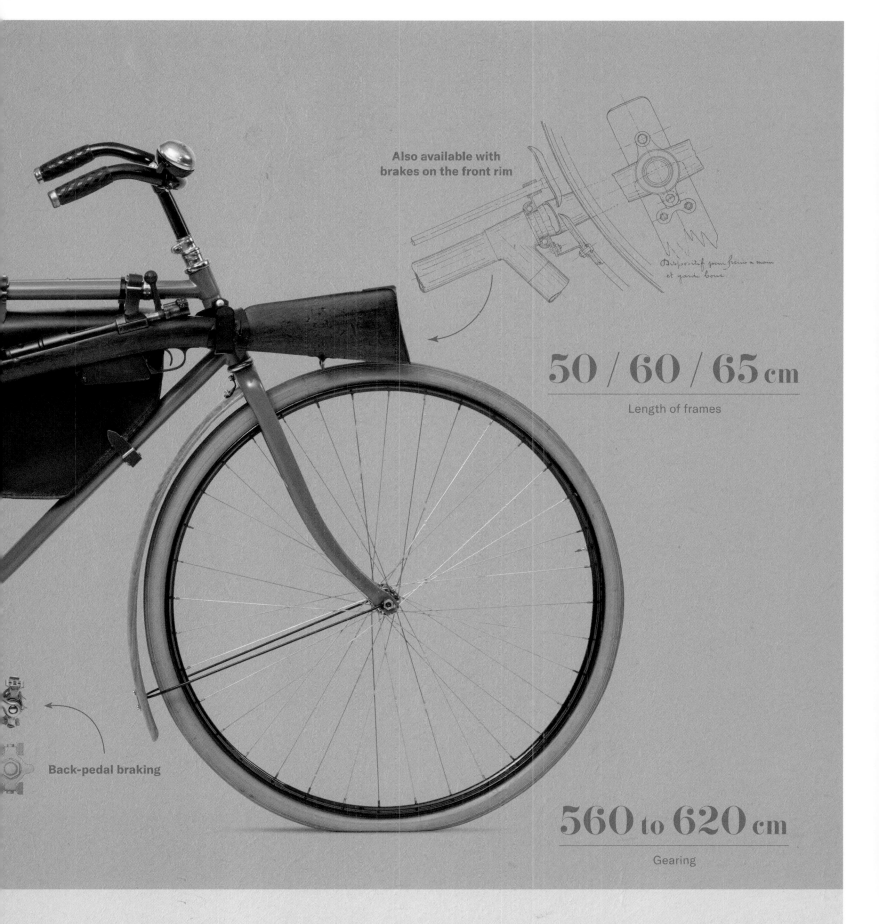

Also available with brakes on the front rim

50 / 60 / 65 cm

Length of frames

Back-pedal braking

560 to 620 cm

Gearing

wear. The advantages of this kind of steel were already well known in the automobile sector. For the frames, FN did away with the annealing – the heat treatment of the forged piece – which impaired the metal's primary qualities.

EN ROUTE TOWARDS THE AUTOMOBILE

No doubt inspired directly by FN, the *Le Petit Bleu du matin* of 19 March 1899 retraced the saga of production over the last 10 years. At the outset there were the weapons of war, which required remarkable precision:

Wishing to keep its elite personnel busy during the periods when there was no manufacturing of weapons, FN decided to undertake the construction of bicycles, first as individual parts, then as perfect complete chainless machines. The company's engineers, carried away by the automobile trend, developed a 'voiturette' [small car] that was affordable for everyone, which was to revolutionise the market.

In late 1897, the Board authorised the management to buy a motor car "as a test, and to be used for studies" and made available 15,000 Belgian francs to the management.[60] A year later, the Board urged the management to pursue this study "as actively as possible"[61] and authorised it to hire a specialist. The engineer Henri de Cosmo was chosen. In January 1899,[62] "he was actively working

FN Model 19CC car in Place de la Bourse in Brussels, 1930
The car was developed by the Italian Henri de Cosmo and produced by Fabrique Nationale. It marked the company's first phase of automobile production.

on the establishment of a first type of 'voiturette' [small car] with a three horsepower engine".

Putting this 'voiturette' [small car] on the market was a "masterstroke".[63] Its innovative design, ease of maintenance, 'admirable' suspension, geared steering, single-lever belt-driven gear change, and even its sliding seat were highly praised. The press concluded that this vehicle was equipped for long excursions.[64]

In September 1899,[65] the possibilities offered by the automobile market were set out in great detail, depending on the level of sophistication of productions. On that basis, the selling price would be lower or higher, thus generating lower or higher profits. After lengthy discussions, FN chose a middle path that seemed more in tune with its investment capacities. In the difficult financial situation that FN was going through at the time, an agreement was finally reached on a limited production of 60 or 100 vehicles, divided between FN's different workshops.

Automobile chassis assembly workshop, 1912

This cautious move into this new world was guided, as in other areas, by the primary concern of ensuring the quality of its production and bringing in the technological innovations that could shake up an already crowded market. The series of FN 'voiturette' models produced in these early years were best known for their reliability, though this was admittedly still quite limited. The FN 'voiturettes' were ranked first in many 'criterium races' organised at the time. However the key to success was not speed – only an average speed of 15 kilometres per hour was required – but rather the smallest number of breakdowns duly noted by the race commissioner.

This cautious approach had a downside too. For example, in November 1899, the Board highlighted the success of the 'voiturette' – which had come first in the Automobile Club Liégeois race – but at the same time noted that: "If we were able to supply them, we could take orders for several hundred cars."[66]

From that moment on, FN would be constantly introducing technological innovations. The aim was to adapt to customers' wishes as best it could and often even to anticipate their demands.

On 8 October 1907, FN refused an agreement with The Austin Motor Company, which was offering a contract over a period of several years to manufacture the complete chassis and spare parts, on the grounds that the success of the FN chassis allowed it "to fight under its own name'.[67] This statement was a grand response, matching the company's pre-war automotive situation. Optimistic statements were made in numerous meetings of the General Assembly and Board of Directors, from 1899 to 1914: "demand is stronger than production"; "the chassis are completely sold out"; "sales are increasing"; "the model is sold out"; "we look forward to serious profits"; "sales are assured"; "the new model is going down well"... The gamble of 1897 had largely paid off.

The Managing Director of the Fabrique Nationale d'Armes de Guerre, Mr Andri, n his FN 1200 car, 1910

FN Car
1600 Model

System
Four-cylinder engine, two blocks
of two cylinders

Between 1911 and 1912, FN produced a new car with a
smoother suspension thanks to a system of shackle
springs. It was a practical system that would be used
for over half a century. The car also had four gears and a
reverse gear, a sliding shaft drive system and a fuel tank
located at the back. Around 500 cars of this type were
sold, making the car a major success for the time.

1600 cc

Capacity

12 hp

Power

By drive shaft, 4 gears

Transmission

815 × 105
810 × 90

Tyres

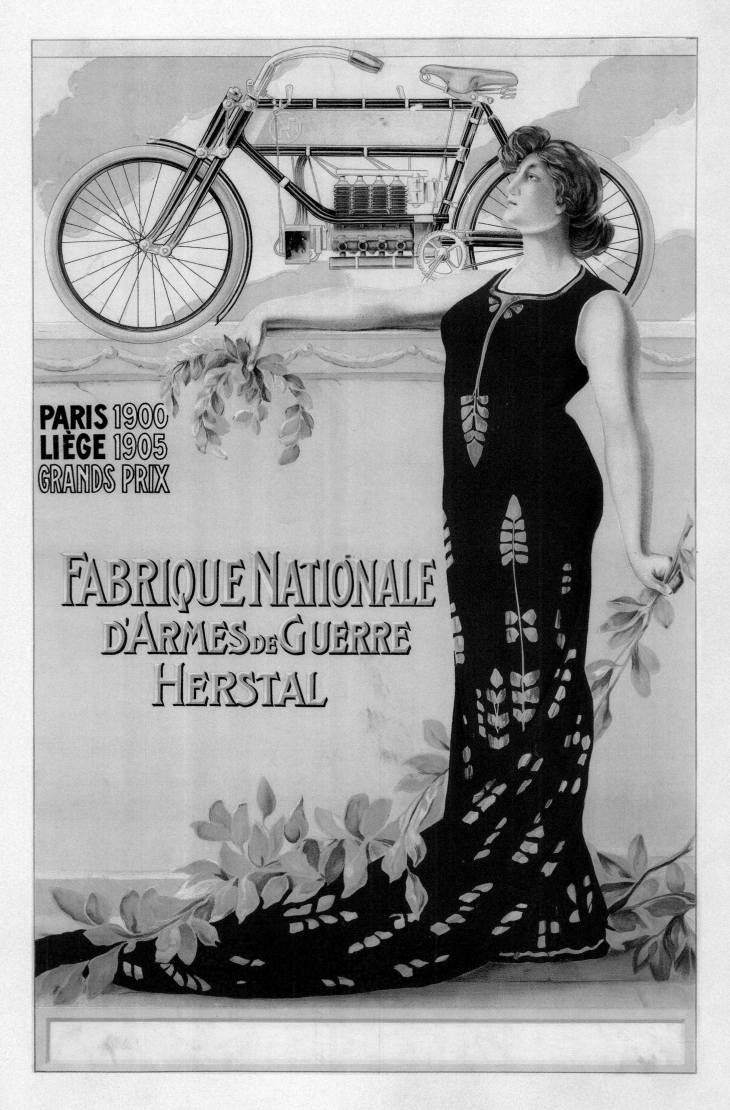

↕ **Belgian Army motorcycle couriers on four-cylinder FN motorbikes, c. 1910**
The success of FN motorcycles was not restricted to Belgium and Great Britain; it also reached the continent's fringes, with the Russian Army purchasing four-cylinder and single-cylinder models in 1913.

↤ *Fabrique Nationale d'Armes de Guerre*, lithograph printed in Liège at Auguste Bénard, around 1905
This imposing woman, decorated with victory laurels, embodied the successful new products of the Fabrique Nationale d'Armes de Guerre, and in this case the four-cylinder FN motorcycle.

MOTORCYCLES

In July 1898, after returning from the International Automobile Exhibition in Paris,[68] Fresnay, the Director, gave a talk on developments in the new motorcycle industry. He was tasked with studying the motorcycle manufacturing process and proposing quotes for production facilities and processes. In November that year, the pressure to do something had increased and the company simultaneously sped up the production of cars and launched the production of motorcycles.[69]

In 1901, the management hastened, as it put it, to create a model that seemed "well established" and put 300 of these motorcycles into production.[70] The term 'moped' would have been a better description of this first attempt. We are talking about simply adding a motor – manufactured by FN of course – to a bicycle with a reinforced frame. The transmission was through a flat leather belt on a wooden rear pulley. The idea is said to have accidentally come from one of FN's former cycle sales representatives, Joseph Houart. He was a Ghent-based mechanic who, after a few tests, had ordered 100 engines from FN to equip the same number of bicycles.[71] Improvements were made to this moped, with its capacity increased from 133 cc to 300 cc and its power from 1.25 horsepower to 2 horsepower. Six thousand of them were manufactured and sold between 1902 and 1914.

A genuine motorcycle made its debut on FN's production line in 1905. Its four-cylinder engine was quickly increased from 362 cc to 412 cc. Its transmission, which was more reliable, was shaft-driven instead of belt-driven. It had neither a clutch nor a gearbox. These would not arrive until 1909. This was the first real four-cylinder motorcycle in history to be mass produced. It remained in production until 1926.

The motorcycle quickly became very popular. In the UK, it was nicknamed 'the two-wheeled car'[72] and was exported in such large quantities that FN set up a sales outlet, FN-England.

FN continued, as was its wont, to work on improving its one- and four-cylinder models, increasing their power. The Russian Army and the cycling battalion of the Belgian Army were among the 9,000 buyers of these motorcycles between 1911 and 1914.

Motorcycle assembly workshop, 1912
In this period the Fabrique Nationale d'Armes de Guerre could produce 50 motorcycles per day.

FN 4-Cylinder Motorcycle 1913 Model

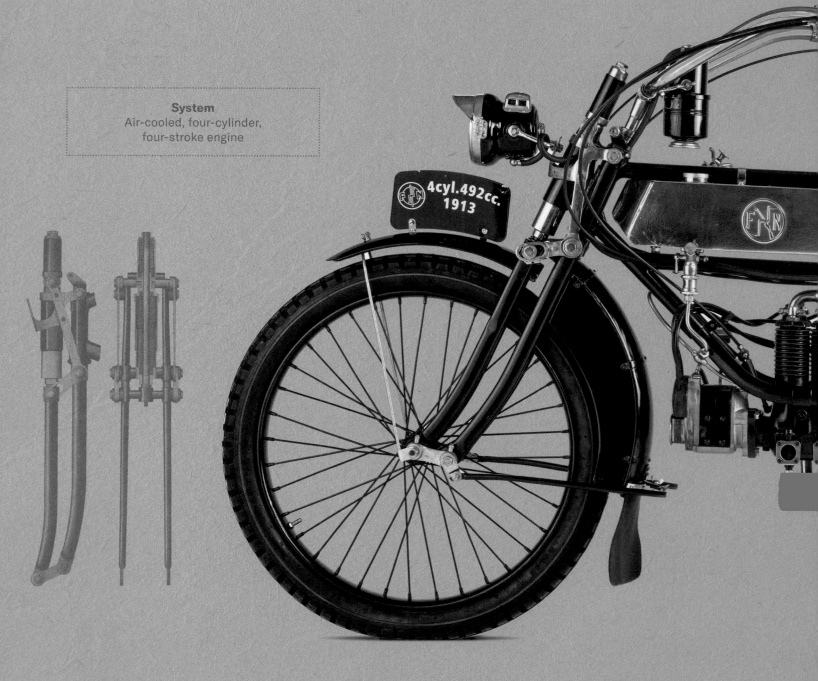

System
Air-cooled, four-cylinder, four-stroke engine

4cyl.492cc. 1913

In 1905, FN produced its own model of **four-cylinder motorcycle** with automatic intake valves, and its sophistication ensured its reputation as far away as England. Year by year, FN upgraded the motorcycle until the 1913 Model, which was certainly the one that featured the most modifications. A two-speed gear box was notably integrated as well as a new oil pump, a multi-disk clutch, fitted in the steering wheel, a longer frame and a larger capacity fuel tank with a revised shape. Lastly the motorcycle had a 1912-type handlebar: this was longer and had rippled grips, but now also featured a clutch control and decompressor controls on the left, with the handbrake and throttle control levers, including the choke, on the right.

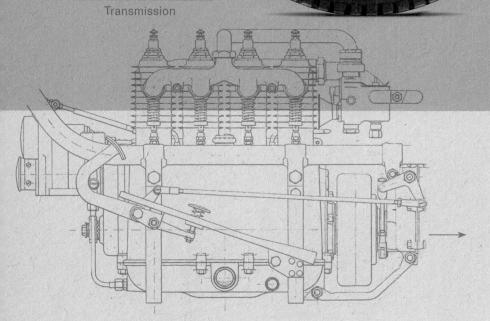

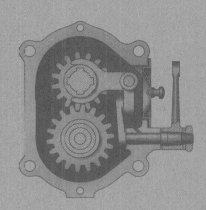

± 60 km/h
Speed

5 hp
Power

Drive shaft, two gears
Transmission

492 cc
Capacity

52.5 mm
Bore

57 mm
Stroke

MODEST BUT PROMISING PRODUCTION LEVELS

We have seen that FN took care to protect itself in those days against the hazards inherent in the war weapons market, by diversifying its mechanical production (retail firearms, bicycles, automobiles, motorcycles, etc.). It was equally keen to protect its ammunition factory by developing its civilian production, which would carry on regardless of ups and downs in the arms market. This would guarantee a minimum level of use for its machines and above all minimum levels of activity for its special-ised personnel. The contract with the Aerators Company of London, which was signed in 1897, provided for the supply of steel cartridges containing carbon dioxide to make fizzy water using siphons. This modest level of pro-duction continued until the 1950s, by which time FN had delivered several hundred million of these cartridges.

A smart combination of caution and technical progress drove the company's diversified mechanical products and was the key to its success. Its bicycles, automobiles and motorcycles all followed the same economic and technical patterns. The company's goal was to offer a practical response to the expectations of a demanding clientele, in an increasingly competitive and technologi-cally innovative world.

Firearms assembly workshop, 1921
Alongside the Browning range of firearms, Fabrique Nationale d'Armes de Guerre produced, under its own name, shotguns with side-by-side barrels (sidelocks and Anson & Deeley) in the purest European arms-making tradition.

Kitchens of the Fabrique Nationale d'Armes de Guerre, distribution of 'economic dinners' to staff, 1922

Progress and prosperity

On 11 June 1913, news broke of the sudden death, in Germany, of Victor Fris, Senator of Antwerp and Chairman of FN. A few weeks later,[73] the directors underlined his dedication, intelligence and authority in running FN's affairs. They concluded their tribute by highlighting that, under his leadership, the company had entered an era of prosperity. But was the company really doing well? Fris had been with FN since its restructuring. As a brilliant lawyer for the Loewe Group and Mauser, he had been a key figure behind that restructuring. From 1896 onwards, he attended all the meetings of the Board of Directors. In 1906, on the death of the Chairman, Charles del Marmol, he was appointed both Director and elected Chairman. Fris acted in much the same way as his predecessor. However because of circumstances, as soon as he took over the reins, things improved for FN – or at least for its shareholders. The total amount of dividends distributed rose by a factor of 4.5 in less than 10 years. Was that increase an indication of the company's good financial health? That is a moot point. On several occasions, the company had to resort to taking out loans and increasing its capital to secure new resources.[74] Nevertheless, the archives clearly show that the company was in good health within its sector. FN was very

successful in diversifying its activities. It also managed to make its mark on national and international markets, which were unfamiliar to the company in 1896. With the First World War looming, FN was now a respected player in the bicycle and automobile engineering sectors.

On 29 February 1912, in Cologne, Victor Fris chaired one of his last Board of Director meetings. In retrospect, it could be seen as a summary of his work as the boss of FN. First, the Chairman of FN was pleased to announce an increase in sales and profits and a very marked increase in orders. He then recommended increasing the engine power of FN's factory with a new gas engine, the necessary installation of a second water pump in the canal supplying FN, the purchase of more land and the establishment of a medical service for staff. He then explored the reasons for the company's success: "The worldwide reputation acquired by FN is thanks to the perfection of its manufacturing, which only a rigorous revision of all the parts has enabled us to achieve". The laboratory, which was set up in 1904, was at the core of these rigorous practices, for the analysis of raw materials as well as supervising the metallurgical operations carried out within FN.[75] He went on to suggest that this revision service should be focused on systematic research

Firearms Technical Office, 1912
The main parts of a mass-produced gun typically require a hundred or so different mechanical operations. After the prototype has been developed, multiple copies of it must be reproduced without compromising any of its qualities. They must be made at a fixed daily rate, within the desired timeframe and at an acceptable unit cost price. Mass production of a specific product therefore requires a series of preliminary studies. They aim to define the manufacturing dimensions of each of the parts as well as to determine the mechanisation process for each of the parts. Lastly, these studies pave the way to draw up and establish the tooling plans required for each operation and to define acceptable tolerances, without compromising the quality of the whole or the exact interchangeability of each of its components.

into the various areas of interest to FN. "This field is extremely vast," he added, before highlighting in order to better convince his audience, the material advantages that had already been derived from it: purchasing the most suitable raw materials (particularly types of steel) at the best price and developing techniques through the observation of physical and chemical processes (such as the hardening and annealing of brass or the influence of mercury salts on this metal). He underlined the necessary collaboration between the laboratory and all of the factory's technical departments in order to make it "not a luxury instrument, but one of considerable utility", before explaining to the Board that "a company as important as ours cannot be certain of its future if its successes depend solely on the quality of its manufacturing. The objects it produces must also have value in themselves". The Research Office, the embryo of a research and development department, would be tasked with studying and testing new products, finding new solutions, carrying out concept studies and, "with no other agenda, it will keep us in the vanguard of progress".[76]

This progress would also include some social progress, although only in a low-key way that deserves some explanation. The period – at the turn of the 19th and 20th centuries – did not have the same approach towards workers as we enjoy today. Dismissal decisions could be ruthless. Management would react immediately to strikes.[77] The hierarchical nature of things was clear and there was no way round this. Yet both the Belgian and German directors, most of whom were christian democrats, introduced a real sense of humanity up until 1914. They came up with a purely symbolic gesture, combining empathy and public relations, as soon as the new Board was in place on 22 February 1896: "An extraordinary donation of a ten-ton wagon of coal to the poor people of Herstal." The most significant contributions to a modern social policy were, starting in 1896, new pharmaceutical and medical services for the staff, all paid for by the company. Workers there were also provided insurance against workplace accidents, well before Belgian legislation covered this. In 1906, the company set up a mutual pension society for workers and employees, followed later by a savings bank.

FN therefore went far beyond the paternalism so typical of the period. In those years, the company introduced what we could call a new social movement, in which everyone's position and role in the hierarchy was clearly set out.

Central laboratory, the scales room, 1921
In 1904, the Fabrique Nationale d'Armes de Guerre was among the first Belgian companies to have an integrated laboratory for tests, controls and the reception of raw materials and products. This laboratory was inspired by the theories of Henry Le Chatelier, a professor at the Collège de France and by the example of the Paris automobile manufacturer De Dion-Bouton. Fabrique Nationale set up scientific methods to improve the quality of products manufactured, without increasing their cost price. For over 80 years, young engineers completed an internship at the central laboratory, prior to taking on responsibility within the company. In the last years of the 20th century, the laboratory's missions were decentralised.

⬍ Raymond Tournon, *La Fabrique Nationale d'Armes de Guerre*, lithograph, 1907

◂◂ Émile Dupuis, *Fabrique Nationale d'Armes de Guerre*, lithograph produced in Liège by Auguste Bénard, early 20th century
This art nouveau-style poster highlights the prizes won by Fabrique Nationale during the Paris exhibition. It also announces the forthcoming production of motorcycles.

From adverts to information

In its early years, FN sought to secure its status through 'advertisements', which are vital for any company's good reputation. Unlike FN's advertising, which would later aim to be informative, its first illustrated documents simply underscored the company's reputation. FN employed leading artists to convey that message.

At the turn of the century, Émile Berchmans made no bones about using mythology to publicise how the company was making bicycle parts.

Alongside these illustrated monuments, FN endeavoured, through numerous seemingly spontaneous press articles, to convey the many reasons for the quality of its products. It also carefully explained why, as final products, these machines were so efficient. They mentioned "delightful machines assembled with parts from the Fabrique Nationale d'Armes de Guerre"[78] or the theft of a bicycle "built with genuine parts from the Fabrique Nationale d'Armes de Guerre".[79]

A real change can be detected in those particular months. Although the company was still clearly famous for its weapons of war, it was now becoming known for its bicycle parts. It was really just marketing, which was still in its infancy.

The company would later address its audience differently. Its decision to employ Parisian artist Raymond Tournon in 1907 underlined FN's ambition to take gradual steps into what can already be seen as advertising and no longer just promotion.

References to the Renommée, which Émile Berchmans had used a decade earlier, were gone. Specific products were now the focus. In a naturalist spirit, the poster called on images to depict the four products manufactured by FN for the public. Automobiles, bicycles, motorcycles and hunting guns were all brought together in the same lively movement. The 'Armes de Guerre' [weapons of war] were still there, but far less obviously. FN's logo became a rising sun, casting the landscape in a golden light. However, the approach to the FN 'brand' was still global. The traditional lyricism of advertising still did not single out any products: this would only be introduced after the war.

In 1910, FN ran a large-scale communication campaign by taking part in the Brussels World Fair. Since 1896, the company had regularly participated in the exhibitions and fairs that were so popular at the time. In 1905, it participated in the World Fair in Liège. But it failed to stand out from other arms manufacturers or bicycle and automobile manufacturers. In truth, FN did not need to do that. Its factory in Herstal was so near that many conference attendees visiting the exhibition also took a trip to Liège. Participants in many different national and international fairs – including those for municipal employees, for technology and chemicals, for industrial property, for medicine and for commercial education, not to mention the members of 'Tourist Associations' – all joined excursions to the Cockerill factories, the Wanze Sugar Factory, and FN. Those seeking more picturesque places would visit Spa.

A few years later, FN's positioning changed. It decided to "participate the following year [1910] in the Brussels Exhibition and to erect a stand-alone pavilion, worthy of the grandeur of FN".[80] Within six months, the plans for and decoration of the pavilion were accepted and the bids approved,[81] probably retroactively, as the Exhibition's opening had already taken place on 23 April. In its coverage of the Exhibition, the newspaper *La Meuse* announced:

The Fabrique Nationale d'Armes de Guerre has decided to offer its staff an excursion to the Brussels Exhibition. Two thousand five hundred people, men and women, will take this pleasant trip. They will be divided into two

Émile Berchmans, *Les forges de Vulcain*, oil on canvas, 1910
Produced to decorate FN's pavilion at the 1910 Universal Exhibition, the work of art is now exhibited in the John Moses Browning Collection, the heart of Herstal Group's artisanal activities.

groups, which will leave by special trains on Friday 1 July and Sunday 3 July. The trip and entry to the Exhibition are completely free for all staff members.[82]

The Board of Directors was informed of the success of FN's participation in the Exhibition, both in terms of commercial positioning and social generosity:

The Director reported on the great success achieved by our Pavilion and by our products on display. [...] The sacrifices made for this participation in the Brussels Exhibition will not have been made without fruitful results. The Director informs the Board that [...] all the staff (workers and employees) have been sent, at the Company's expense, to the Brussels Exhibition. The impact of this generosity on morale was considerable.[83]

On 13 September 1910, the Board of Directors offered an almost photographic snapshot of the company's activities. The state of production and sales, as well as their evolution, were explained in detail. Over 20 products were now in production: Browning 6.35 and 7.65 mm pistols, a Browning 9 mm pistol, Browning 12 and 16-gauges shotguns, a Browning 9 mm semi-automatic rifle, Mauser rifles, an FN .22 calibre rifle, a school training rifle, gun parts for the Syndicat des pièces d'armes [Gun Parts Union], hammerless shotguns, Browning cartridges, war cartridges, Aerators cartridges, bicycles of various types, a four-cylinder motorcycle Model 1910, a two-speed motorcycle, the FN 1500 automobile chassis, the FN 1560 chassis and the FN 2400 chassis.

3 Momentum is halted

Meeting room of the Board of Administration of Fabrique Nationale d'Armes de Guerre during the German military's occupation
After the Germans took Liège in August 1914, Fabrique Nationale d'Armes de Guerre ceased its activities. Seized by the occupying forces in 1917, the company resumed normal service in November 1918 and became fully independent of the Loewe Group the following year.

The factory at a standstill

A few weeks before the world was set ablaze, everything seemed peaceful within the industrial sector. In April 1914, FN was discussing the recent installation of a new gas engine, the extension of the laboratory premises, new forging hammers and the purchase of machine tools to equip the 'automobile' workshops. There were regrets about a bank overdraft and a decision was made to limit expenditure to what was strictly necessary, but without harming the company's smooth running.

The German Army entered Liège on 16 August 1914. The factory came to a standstill. Between 1914 and 1918, FN suffered the same fate as all Belgian industrial companies. Machines were requisitioned, workers were mobilised and there was rampant inflation. These events underpinned the hard economic reality of this gloomy period. The specific circumstances of FN, which was part of a German-dominated industrial consortium, were about to get very complicated.

On 13 October 1914, the Board of Directors brought together Hagen, von Gontard and Kosegarten for the German part of the company, plus Laloux and Dallemagne for the Belgian part. The German government made clear it wanted work to resume at FN. Despite having a majority of German directors, the Board firmly stated that it was impossible to restart the factory's production. These products could "in fact all be considered in the present circumstances as articles of war". Moreover, restarting production would result in very challenging material difficulties – a shortage of money, a lack of raw materials, too few managers and workers, military occupation of the factory, closed markets, difficult conditions for transport and export, etc.

On the same day, taking into account the interests of the company and its future while wishing to keep its staff and for humanitarian reasons, the Board authorised the up-front payment of wages to the workers. One third of the salary was maintained for 'unemployed' staff and half for staff 'employed' on maintaining the factory and the packing of the requisitions.

The Board's surprising unanimity in the face of the occupying forces signalled its clear refusal to compromise plus its determination to preserve its most precious tool of production – the workers. It also revealed the respect that the directors had for one another: Hagen, Laloux and Dallemagne had been sitting together since 1899. So it was clear that the spirit of those who had built FN had managed, against all odds, to impose their rules on the enemy.

The relationship between the Belgian and German parties remained courteous over the following months, but tensions sometimes boiled over, such as when Ludwig Hagen was proposed as Chairman of the Board of Directors. Georges Laloux objected on the basis that the company traditionally had a Belgian Chairman, and pointing out how strange it would be to see a Belgian company with the title of 'national' chaired by a foreigner. Hagen accepted these reasons and stayed on as Vice-Chairman.

On 20 February 1915, the two parties clashed more sharply on a central issue. The Germans were claiming ownership of FN's successes ("it is to the Deutsche Waffen that the FN owes its prosperity, its organisation, its tools"), while the Belgian directors disagreed and pointed to the merits of the management, engineers and "good arms workers"... who were all Belgian.

Although these incidents were relatively minor in the end, it should not be forgotten that the war was a major backdrop. The Managing Director Alfred Andri, and Georges Laloux, were both imprisoned in Germany – the former for refusing to restart the factory and the latter for acts of resistance.

The work of restarting the means of production, envisaged from January 1915, began the following year. FN then began to manufacture machine tools to replace those requisitioned by the occupying forces in 1914, with many of these tools being sent to Germany. The machines from the 'Autos' ['Automobile'] Division and the 'best of the staff' were used for this purpose. The Germans undertook to ask Berlin for an assurance that these new machines would not be requisitioned and the Deutsche Waffen-und Munitionsfabriken would advance the sums necessary for this reconstruction effort.

Entrance of the factory, 1914-1918
Under German control, Fabrique Nationale d'Armes de Guerre was turned into a big repair workshop for vehicles.

As the military occupation authorities were unable to get the factory up and running again, they sequestered FN in 1917 and set up a vehicle repair workshop.

Just as the Germans were taking a tougher approach and, as the tide of war slowly turned in favour of the Allies, Belgium began to envisage a victorious post-war period. The Union financière et industrielle liégeoise (UFIL), an offshoot of the Société générale de Banque allied with other banks – including the Liège bank Nagelmackers, was set up that year under these terms:

Together with a group of friends from Liège, we have set up the Union financière et industrielle liégeoise, a limited company with a capital of 10 million francs, whose main purpose is to support industrial, commercial and mining companies and to promote their development. We hope that this organisation, in which we have taken a major interest, will be able to render valuable services after the war and provide effective help in the industrial recovery, thanks to the support it has secured in the financial and industrial circles of Liège.[84]

At the end of the conflict, German control of a Belgian arms factory was obviously no longer an option. The 5,700 German-held shares would now be administered by the Managing Director of FN, Gustave Joassart, who negotiated buying them back. In March 1919, the Deutsche Waffen-und Munitionsfabriken sold their entire shareholding to UFIL. The UFIL kept 3,900 of these shares and sold the rest to third parties. The Société générale de Banque then came onboard at Herstal.

The other war

The war on the front was in full swing, but men and women were doing their bit back home, contributing their expertise in weapons manufacturing to Belgium's Allies. This expertise was of course highly prized.

Alexandre Galopin, born in Ghent in 1879, was someone who climbed the ladder at FN step by step. He was initially hired in 1904 as the director of a brand new laboratory, which was the first research and development service to be set up by any Belgian industrial company. Subsequently, he was appointed as Managing Director.

In 1914, Galopin fled to France, where he had connections in the world of industry. Thanks to his technical knowledge and organisational knowhow, he was able to oversee a recovery in France's arms manufacturing sector. Galopin launched an innovative production process. Mechanical engineering companies were brought together into a 'group of small arms manufacturers', with each one making firearms parts on a large scale based on their available tools and capacities. The great precision of these products facilitated their assembly subsequently.

Yet Alexandre Galopin's impact on the progress of the war became really decisive in July 1915, when he founded the Manufacture d'Armes de Paris (MAP) with Gustave Joassart, who had also come from FN.[85]

The MAP, with a capital of 1.5 million Belgian francs, established its head office in Saint Denis, to the north of Paris. It assembled parts of weapons that were mass produced elsewhere and it manufactured rifles and machine guns, as well as several milling machines to be used by other manufacturers. This discreet role, behind the front lines, certainly played a decisive role. By the war's end, Galopin's team had delivered more than a quarter of the weapons made in France during the war. His expertise was also requested during the peace negotiations.

From 1915, many Belgian workers – who had been forced to flee, like many others – ended up in France, where they plugged the gaps left by French workers who had been called up to fight for their country. In 1917, a total of 22,000 Belgian workers were employed in 1,600 French companies. The Manufacture d'Armes de Saint-Étienne had up to 350 workers from the Liège area.

In the UK in 1917, according to estimates, nearly 20,000 exiled Belgian workers were employed in munitions factories.

Assembly and repair of motorbikes by German soldiers, 1914-1918

Time to take stock

On 27 December 1918, *La Meuse* told its readers:
All those who knew the Fabrique Nationale de Herstal would feel pain in their heart if they were to see it again at the moment [...]. This jewel of our mechanical industry, which its managers were legitimately proud of showing to the many visitors like a fair where the most perfect order and the most rigorous cleanliness standards prevailed, had grandiose facilities, magnificent machines, incomparable tools, and was served by an elite staff, who enabled it to compete with ease with the top factories in the world. Of all that, there remain only rooms and halls that are empty or full of old scrap metal, disordered pieces of wood, the most disparate collection of rubbish.

In various columns on the front page, the journalist painted a clear picture, initially despairing at FN's plight at the end of the war. He finally concludes that the only thing missing before normal service can be resumed, after a clear up, are the machines. FN confirmed that a month later:

Thanks to the success of the Allied forces, the country has been liberated and all the bullying measures in force and put in place by the occupier have been terminated. We have taken back ownership of our offices and workshops. The clearing up measures, the cleaning, etc. which have been actively carried out, have already made all our workshops look more normal. Things are still far from how they looked before the war. Our halls are still missing a number of machine tools that the occupier decided to seize.[86]

4 The world had changed

90 · ARS MECHANICA

Reconstruction, crises and strategies

Although 'post-war' was the expression people used at the time – such was the prevailing mood that 1914-1918 could only be the 'war to end all wars' – difficulties still piled up for industry in the period between the two wars. First there was a brief but intense period of reconstruction, which would see the business world put its production infrastructure back together again. Then came a sequence of crises and more prosperous periods, up until the collapse in 1929-1930, before a gradual return towards prosperity, which the Second World War would wreck in a matter of a few weeks.

This was a scenario that FN was also swept up in. For FN, there would be further complexities and difficulties, related to its specific activities as a weapons manufacturer in a world that yearned for peace.

Mindful of these hurdles and guided by cautious optimism, FN prepared a programme of future activities for the company, covering "precision mechanics manufacturing". This happened a few months after the armistice. This field of action was sufficiently broad to encompass weapons of war, automatic pistols, automatic rifles, FN rifles, Browning hunting guns, motorcycles, bicycles, automobiles, and cartridges, etc. The conclusion could thus be drawn that the "Company can resume without risks, and has in fact already resumed, its former activities, even if this means later increasing the production of certain new products which could be placed, for example, in the 'category' of the sewing machines or similar".[87]

Admittedly, there was a whole host of problems. The small size of Belgium's market in a world that was becoming increasingly protectionist,[88] competition that was more acute than before the war; the lack of appropriate machine tools; not to mention the new developments that had emerged since 1914 on the American continent. All these factors threatened the future. However, FN was preparing itself: "The plan to create the League of Nations[89] may well compromise the production of weapons of war, but there is a possibility that Belgium will have to start renewing some of its portable armaments and increase its supplies of cartridges." In September 1919, FN was completely reorganised for weapons of war: Belgium was planning to set up an army far larger than before the war.

Cover of the magazine FN Sports no. 14, 1931
During the inter-war period, Fabrique Nationale d'Armes de Guerre rebuilt itself and thrived again. With a new focus on precision mechanical production, the company turned its attention to competitions for engines and vehicle speed records, which strengthened its brand image.

The Belgian state made a huge financial effort to revive national prosperity. This effort was based on the hope of receiving compensation to be paid by Germany. In 1922, the courts awarded FN 59 million Belgian francs in compensation for war damage. It was paid in successive instalments by the Société nationale de Crédit à l'Industrie, formed in 1919 to facilitate the nation's reconstruction.

RECOVERING ITS INDUSTRIAL STRENGTH

In 1919, FN devoted much of its energy to restoring its range of machine tools. Some 4,300 machines had disappeared, been requisitioned or destroyed: some were returned, others were purchased abroad (notably in the United States) while others were made by the factory itself. In July 1920, 1,500 of these crucial machines were completed in Herstal. However FN also produced bicycles, motorcycles and automobiles with parts manufactured before the war, and added to this by buying new sets of machines. This would be the easiest part of this remarkable industrial endurance. Yet it was based on a premise that soon proved to be wrong. In common with many other industrialists at the time, FN's directors and management were convinced that the rebuilding of production potential alone would be enough to ensure the company's prosperity.

It is noteworthy that, throughout 1919 and 1920, the company assessed its reconstruction efforts by comparison with 1914. Hence the number of machine tools, workers and employees were compared to those prior to the war: 4,233 machines in 1920, compared to 4,888 machines used in 1914; 3,865 workers set against 3,742; and 504 employees compared to 404. Everything was happening as if 1914 had become a kind of industrial ideal, and one that should be replicated. It was noted that 653 machines "were still needed" or that the number of workers and employees was "high compared to that before the war".

The departure of the Germans, who FN were able to rely on, if needed, for management and engineering purposes, turned out to be a trickier challenge to overcome. Engineers were in short supply, as were the very good specialised workers, known as the "elite workers". Just as FN had decided to mobilise its own resources by starting to make the machine tools that it lacked, it created, in 1921, the 'FN School' to train the skilled workers that it needed for its mechanical production.

CRISIS FOLLOWS CRISIS

Now FN's levels of production could only be based on exports due to the Belgian market's small size. As such, as soon as the brief upturn of the reconstruction was over, FN had to face the ups and downs of a national and international economy in a pitiful state.

In 1921, FN was confronted with its first crisis. The company's rapid assessment of that crisis would shape its strategic choices in the years ahead. In June 1921,[90] the management noted that "due to the current crisis, it is no longer the means of production and the restoration of the factory that determine productivity, but the commercial situation". A few weeks later,[91] there was recognition that "sales are no longer limited by production capacity, but by the commercial opportunities. [...]. The successful results of the fiscal year 1920-1921 are not thanks to regular commercial activities, that is to say, those that are repeated on a daily basis in approximately equal quantities, but to a few large and profitable activities that we cannot expect to repeat continuously."

Everything that made FN strong before the war, everything that laid at the heart of its prosperity, was at a standstill in the early 1920s, and all divisions were affected.

FN recognised that the difficulties it was facing were not due to its industrial choices, but to the marketing situation for its products, which was no longer the same as before 1914. The Board of Directors made the same assessment on several occasions up until 1929. These assessments formed the backdrop for the actions taken to make the company more competitive, in an extremely difficult political and economic context until the end of the 1930s.

A key feature of the economic and commercial context facing FN – and often undermining the company's goals – was its instability. Prices, currencies and customs duties were the shifting markers of change.

Between 1914 and 1939, the retail price index[92] typically fluctuated upwards, and sometimes downwards. Overall, it grew by a factor of nine during this period. These movements resulted in wages and prices being constantly readjusted, several times a year. Between 1920 and 1939, wage adjustments were repeated, through cost-of-living bonuses, additional bonuses or just through salary increases. Sometimes though, when the price index was falling, this led to salary cuts negotiated in the employers' room or directly with the staff.

Depreciation of the Belgian franc – which had to be replaced by a new accounting currency, the Belga – was very sharp. So in 1926, the national currency was 'stabilised' at less than 15% of its 1914 value. This presented an opportunity for exporting companies. But it was a fragile one, since all European countries were tempted to adjust the value of their currencies, which would of course negate this theoretical advantage.

FN's professional school, theory course classroom, 1922
In 1921, FN's management decided to create an internal training centre for skilled toolmakers. The Ecole de la FN was tasked with creating a highly skilled labour force for different factory workshops. Monitors helped young people, who held at least a technical education certificate, to use the tools in the workshops where they would work. Theory classes also enabled students to enrich their general culture with a basic knowledge of economics, industrial design, technology and applied mechanics – over eight hours a week for 36 weeks every year. The Ecole de la FN was moved to the Hauts-Sarts site in 1980.

Import duties, coupled with the exchange rate depreciation of national currencies, were the main instrument available to protectionist countries. During the period 1919-1929, France, for instance, applied an import tax on cars that rose from 10 to 45%. An 82.5% import duty was levied on hunting guns when they entered the United States.

On 11 October 1921, at the general assembly of shareholders, the directors summarised the situation arising from the variability of the above-mentioned indicators: "The financial year [1 July 1920-30 June 1921] initially presented itself under the happiest of auspices. Yet soon afterwards, an economic crisis broke out worldwide and became increasingly serious from one day to the next. Many countries, especially the producing countries, wanted to solve this by raising economic barriers in the form of higher customs duties or import bans. [...], these difficulties, exacerbated by the instability of exchange rates, have considerably hampered trade relations." In 1922, "the economic nationalism practised since the armistice by numerous states to the detriment of rebuilding the world economy" was denounced. In 1923, a reference was made to "the outrageous customs tariff policy" and "the competition from countries with depreciated currencies".[93] In 1924, there was opposition to the "unprecedented currency crisis".[94] Up until the Great Depression of 1929-1930, customs barriers and unfavourable exchange rates were at the heart of FN's concerns.

The tools developed by the company in the inter-war period, which were a response to the hostile climate, were industrial and commercial choices based on often bold financial decisions.

FN's most important export markets before the war were now shrinking. France was becoming a protectionist citadel, exports to Germany were impossible and the United States excluded European industrial goods.

The company managed to be as ingenious as the circumstances allowed in order to win back these markets, despite a difficult economic situation lasting until the late 1930s

"A drop of genius
in a barrel of sweat"

JOHN MOSES BROWNING

New ambition for the arms sector

THE RETURN OF BROWNING

Relations between FN and the Browning family resumed rapidly, as soon as the factory restarted. On 10 June 1920, the Managing Director, Alexandre Galopin, reported "the current stay, in our factories, of Mr Browning junior [Val Browning], who brought a new trade firearm to us from America" and a new war weapon. It was in fact the Trombone .22 LR rifle and a machine gun. As the reports and tests of this rifle, both technical and commercial, proved to be positive, the Managing Director decided, in September 1920, to conclude a licence agreement with Browning on the usual terms, but whose territorial scope would be the whole world, including America. The company took advantage of the presence of Val Browning to settle the problem of the royalties paid by FN for each manufactured firearm. Expressed in depreciated Belgian francs, they had become too low; they were now to be paid in gold franc equivalent.

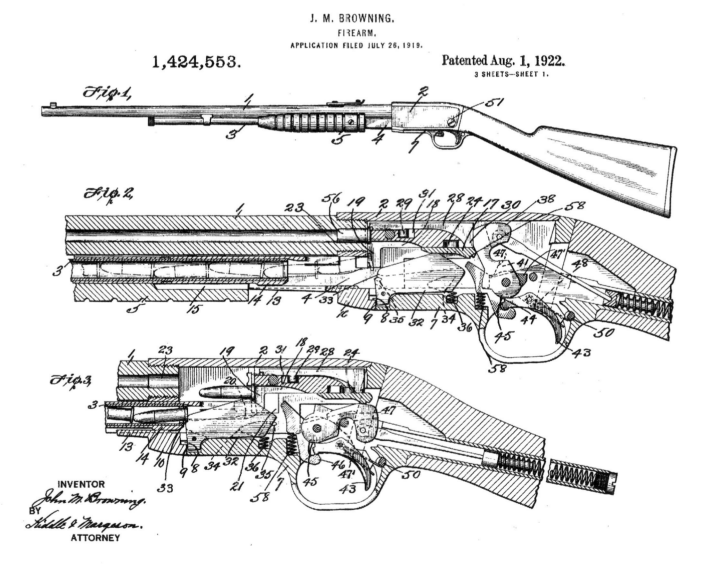

‡ Patent filed by John Moses Browning for the .22 LR Trombone Model Repeating Rifle, 1922

↔ John Moses Browning posing with his invention, the .22 LR Semi-Automatic Rifle

FN-BROWNING
Over and Under Shotgun
B25 Model

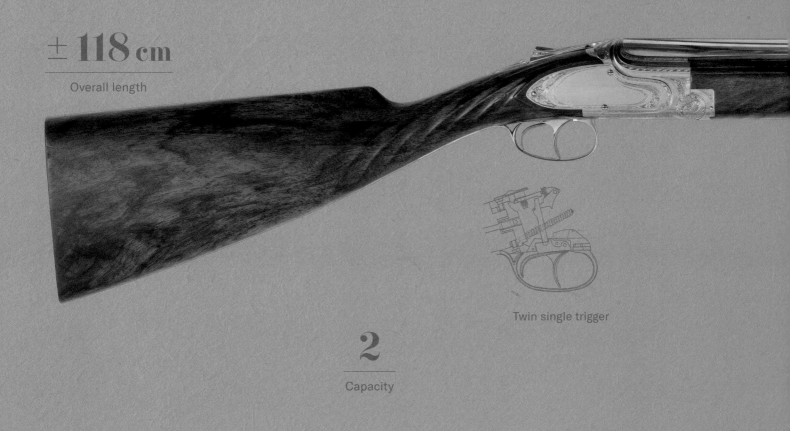

± **118 cm**

Overall length

Twin single trigger

2

Capacity

The **'Superposed' (Over and Under)** shotgun – which became known as the **B25** – was one of the last firearms developed by John Moses Browning. Three patents were filed in 1923, 1924 and 1928. While John Moses Browning did not invent the superposed barrels shotgun with box-lock action, he did improve and simplify the ejection and locking systems. Although side-by-side shotguns dominated sales, FN now mass-produced a shotgun in which the position of the two barrels (one above the other) ensured the ideal line of sight, similar to that of a shotgun with a single barrel.

The first B25 Model produced was a 'twin single trigger'. With each squeeze of the trigger, you could fire from one or the other barrel in sequence. Subsequently, Val

A. Browning introduced the single trigger, which was followed by the single, selective trigger that is still used today. In the single trigger model, the safety catch also served to switch the firing. The push button acts on the sear control mass-rod assembly and operates as a three-position selector: safety, over, under. When the gun is armed, the two percussion hammers are held by their respective sears. If the shotgun is first fired with the upper barrel, the shock generated by the recoil slightly draws back the whole sear control mass-rod assembly, pushed by its coil spring, then moves back slightly further than its initial position, thus allowing the lower right arm of the sear control rod to engage under the right sear. Squeezing the trigger releases the right hammer

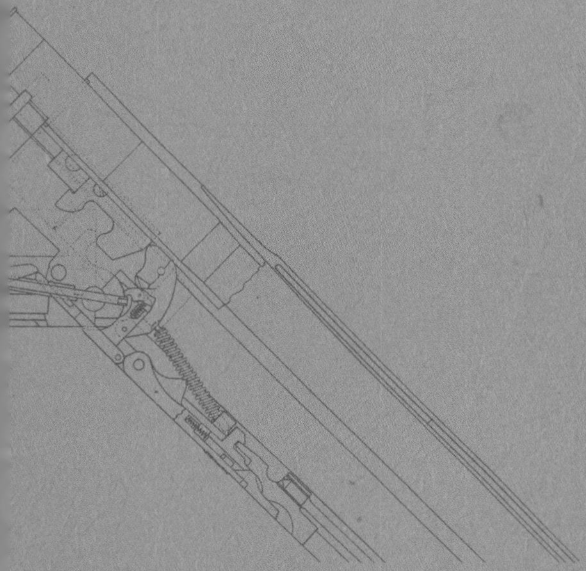

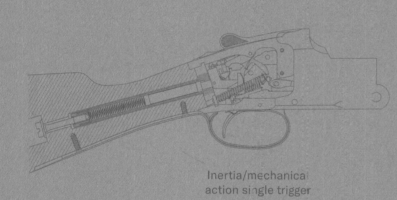

2.9kg
to**3.7**kg

Weight

B25 20-gauge Shotgun from the Water Exhibition

66 to **81** cm

Barrel length

Inertia/mechanical
action single trigger

**12 / 16 / 20 / 28 /
.410**

Gauge

and thus fires the lower barrel. Another notable feature of the B25 is a forearm fixed to the barrels and sliding along them during the dismantling. Initially designed in 12 and then 20 gauge by Val. A Browning, today it comes in 16, 28 and .410-gauges versions.

The B25 became a prestigious shotgun, like the side-by-side shotguns. So FN's arms engravers developed an engraving register based on five grades. The B25 shown here was produced for the Water Exhibition of 1939, an international event to celebrate in Liège the opening of the Albert Canal. To highlight its firearms production, FN exhibited this special edition gun there. Its engraving was completed by master engraver Félix Funken and the wood production was done by Richard Gérard. The

modernist lines, influenced by Art Deco, are a harmonious composition, with metallic parts in polished steel, an action frame fitted with imitation plates finely engraved with animals and stylised flowers, as well as a handle and handguard embellished with twisted fluting.

FN-Browning 'Superposed' model shotgun, or B25 model, hunting configuration, grade A, blued and colour-hardened, with double trigger

Browning firearms were selling well, particularly in the United States. The Brownings were regular customers of FN: 7,000 automatic rifles in 1924, and 10,000 in 1925. Sales were boosted by the persistent weakness of the Belgian franc.

John Browning travelled again to Herstal in July 1925. This time he presented a shotgun with 'superposed' barrels. This shotgun resulted in one of the rare disagreements between Browning and FN. At first, FN refused to manufacture it and suggested that it be manufactured on Browning's own account by the Anciens Établissements Pieper. Browning rejected this idea, trusting only in the workshops and employees of Herstal. This was actually rather flattering, even if it did disrupt the production schedule.

The Browning with superposed barrels went on to become one of the world's leading shotguns of the 20th century. A real icon of contemporary arms manufacturing, it is, to this day, one of the crown jewels of the Herstal Group and will soon be celebrating its 100th anniversary.

Duck hunting scene on the Korean coast, 1928
These Korean hunters, illustrating the success of Fabrique Nationale d'Armes de Guerre gun exports, are carrying the FN-Browning Auto-5 Semi-Automatic Shotgun.

John Moses Browning posing with his invention, the water-cooled .30 calibre M1917 Machine Gun.
After the First World War, John Moses Browning returned to Herstal with a new range of machine guns based on the M1917 and M1919 Models, which he developed for the US Army during the conflict. Thanks to the inventor's expertise, Fabrique Nationale d'Armes de Guerre embarked on its first wave of producing machine guns, notably in .50 calibre, which made the company a success half a century later.

John M. Browning died suddenly on Friday 26 November 1926, on the premises of FN. Staff there were very emotional, while Belgian newspapers carried sober accounts of the event with front-page headlines. American newspapers reacted more strongly, from Arizona (*Douglas Daily Dispatch*) to Alaska (*The Alaska Daily Empire*), recounting once again and, in the usual way, the inventor's career.

On 12 March 1927, the Board of Directors – meeting for the first time since the death of Browning – highlighted "conversations that had started with the heirs of the late John Browning, conversations which pointed out how in the future the Browning would no longer present themselves as 'inventors' but rather as 'general agents' of FN for the sale in the United States of weapons of various types".

Nevertheless, Val Allen, the son of John Moses, would develop the shotgun with superposed barrels invented by his father. He also created new models, such as the 'double automatic' or Twelvette, and the BAR hunting rifle, the latter in collaboration with his own son, Bruce Warren Browning, who would also contribute to the development of the T-Bolt rifle and the Nomad and Meda ist .22 calibre pistols.

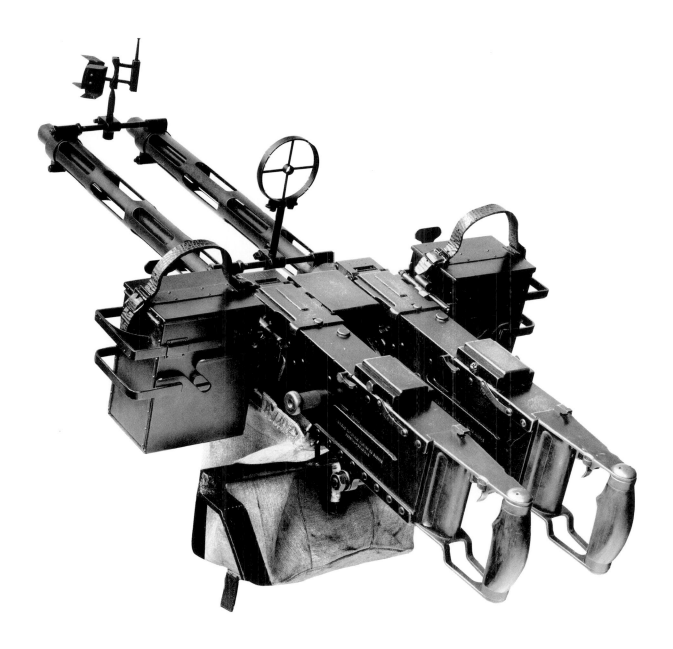

FN-Browning aircraft machine guns, 1933
This machine gun had a short barrel recoil, was air cooled and fed through articulated belts. It was a modular weapon and could be used as a turret gun or a fuselage gun. Another major asset was its adaptability to all the ammunition used at the time at a rate of fire of around 1,200 rounds per minute.

FN-Browning Baby Model Semi-Automatic Pistol, 1931
This new version of the 6.35×15.5 mm (.25 ACP) pistol was among the first firearms developed at Herstal after the death of J.M. Browning. Produced at the Fabrique Nationale d'Armes de Guerre until 1979, the gun was then produced at a subsidiary, the Manufacture d'Armes de Bayonne, until 1983.

WEAPONS OF WAR

According to Claude Gaier,[95] Gustave Joassart, who had been running FN since 1923, was convinced that weapons of war were underpinning FN's production and would continue to do so, even if the outlook did not seem very encouraging. So he encouraged FN to focus on the arms and ammunition business, and particularly to develop products that would stand out on their own merit and innovation from competing products.

Now managed by Dieudonné Saive, the design office developed new models. He thus fitted the Browning machine gun with a mechanism that could slow down the firing (1930), then with a removable barrel (1932). He modified the Browning infantry machine guns, of small and medium calibres, in order to increase their rate of fire to equip fighter planes with them. In the field of handguns, he first created a smaller model – known as the 'Baby' (1932) – and then a high-powered locked breech pistol (9 mm calibre) fitted with a magazine with two rows of cartridges (1934), the famous 'HP' [GP].

Around 1933, the ammunition factory began manufacturing armour-piercing, tracer and incendiary bullet ammunition cartridges with an average calibre of .50 (12.7 mm) and 13.2 mm).

The infantry had abandoned the biogival bullet cartridge in favour of a pointed bullet cartridge defined by FN. So the Belgian government was obliged to buy nearly 100 million of them between 1930 and 1934, with most being manufactured in the Bruges factory.

At the same time, the Belgian Army adopted the Browning machine gun with an adjustment mechanism to reduce the rate of fire. During the same period, it also procured 630 aircraft machine guns and a thousand HP pistols.

Internationally, the temptation to rely on intermediaries, as had occurred before the war, gradually disappeared. The directors themselves now became involved and they were responsible for travelling to distant and often new countries.

FN-BROWNING
Semi-Automatic Pistol
HP 35 Model

11.8 cm
Barrel length

13
Capacity

9 × 19mm Parabellum
Calibre

350 m/s
Muzzle velocity

19.7 cm
Overall length

0.900 kg
Weight

System
Semi-automatic gun, operating with a
locked breech by short recoil of the barrel

The **HP 35**, together with the Colt 1911 Model, is certainly the most legendary pistol and the one that was most used by armed forces worldwide. The project began in the 1920s, following a request from the French Army for the production of a large-capacity pistol. The first prototypes of a 15-round magazine were produced by Dieudonné Saive, based on the 1903 Model, which was already chambered in a 9 mm calibre. Results of these trials were shared with John Moses Browning, who went on to develop the design of the first model and filed a patent in 1923. He opted for a locked breech system, i.e. the firing causes the simultaneous recoil of the barrel and the slide, which subsequently move apart. The slide alone continues its course until the recoil spring reverses the movement and causes it to insert a new

round in the chamber and to resume its position. During the 1920s and in the early 1930s, Dieudonné Saive made improvements. These notably included the striking and extraction systems, a reduction of the magazine capacity to extraction system, a reduction of the magazine capacity to 13 rounds to lighten the gun plus the addition of a visible hammer so the gun could be carried in total safety with a round in the chamber, a slide stop that indicates when the magazine is empty, a fixed sight system and some elements from the Colt 1911 (with the patent having expired in 1928) such as the barrel removal pin. It took a decade or so for the model to be sold under the HP 35 name, also adapted as a 7.65 × 20 mm calibre for the French Army.

NEW MARKETS

The various treaties, which after the armistice estab-lished the conditions for a peace that was meant to last, led to the creation of new countries. They fragmented Europe especially into a multitude of states, which attracted the attention of the FN as it was assumed that these still fragile countries would need to arm themselves. This interest in new potential markets was boosted due to the fact that FN's traditional markets were closing as economic nationalism (and nationalism in general) gained strength.

The new countries that were born – or reborn – in the immediate post-war period were, from north to south: Finland, Estonia, Latvia, Lithuania, Poland, Czechoslovakia, Austria, Hungary and Yugoslavia. The Russian Empire – which officially became the USSR in 1922 – as well as the German Empire, the Austro-Hungarian Empire and the Ottoman Empire disap-peared, while France (adding Alsace), Denmark (adding Schleswig), Greece, Italy, Belgium and Romania gained new territories.

From 1919, FN became interested in most of these new states. Gustave Joassart visited the 'Kingdom of Serbs, Croats and Slovenes' – renamed the Kingdom of Yugoslavia in 1929, Poland and Romania. He was also interested in the East (Greece, Turkey) and the Far East (China, in particular). This did not prevent FN from pur-suing its relations with its older markets (Mexico, Peru, Brazil, etc.), as well as with the Netherlands and France. Consequently, FN was developing intense international activity to strengthen its trade in weapons of war.

This active involvement in searching for new markets ini-tially resulted in very small orders, or merely intentions to order: a small number of pistols for Serbia in September 1921, for example, or for Estonia in May 1922. FN was undaunted and continued "steady work with some mar-kets initially considered as rather poor".[96]

FN did not want to get involved in the geopolitical ten-sions of the time. On one of his many trips to the Baltic countries, Gustave Joassart considered the possibility of doing business with Soviet Russia.[97]

TECHNOLOGY TRANSFERS

These were essentially traditional approaches to the search for markets for selling arms. In addition, on sev-eral occasions FN tried to transfer technology and was sometimes successful. Several new countries, such as Poland or Yugoslavia, or some countries facing pressure from their neighbours, such as Romania, wanted to set up their own arsenals. They were looking at the possibil-ity, over time, of equipping their own armies.

Kragujevac arsenal

In 1923, the Kingdom of Serbs, Croats and Slovenes placed an order for 60,000 automatic pistols (9 mm short calibre), Model 10/22, plus six million cartridges. This order was increased in July of the same year to include 50,000 Mauser rifles (7.9 mm calibre) and 50 million cartridges. FN was also entrusted with install-ing a factory in Serbia, about 100 km from Belgrade. The arsenal had to be able to manufacture 200 Mauser rifles and 200,000 cartridges per day. This was FN's first engineering contract and it helped to establish the com-pany's reputation in all foreign markets. This contract was obviously deemed satisfactory by the Kingdom, as Yugoslavia ordered 40,000 Mauser rifles and 60 million cartridges in 1926.

Romanian arsenal

In September 1921, the Romanian government sought FN's expertise in organising its arsenal. The idea was to create a Romanian company that would operate the arsenal under its own management and in which FN would participate: "It is obvious that if we do not decide to take shareholdings of this kind, we could see some of the markets on which we had relied slip away from the viewpoint of 'weapons of war'." Although this statement is rather convoluted, it establishes the principle of part-nership – public-private, one might say – and makes it a rule.

Polish arsenal

On 12 January 1922, the Board of Directors was informed that "Like the Romanian government, the Polish govern-ment has requested our assistance in studies for estab-lishing an arsenal in Poland. [...] Following a recent visit to Warsaw by Mr Joassart, it seems that the Polish gov-ernment would welcome our participation, which would be restricted to a simple technical contribution and a portion of the capital, with the government itself con-tributing land, buildings, machinery (to come from the Danzig arsenals) and the rest of the capital".

Poland was recreated in 1919 and had just emerged from the conflict with Soviet Russia. Although Poland did not recover the territory of Danzig (now Gdańsk), which had been withdrawn from Germany and made a 'free city' by

Prince Leopold of Belgium (future King of the Belgians) testing the Belgian Army's new FN-Browning Model 30 automatic rifle at the Beverlo camp in Bourg-Léopold, 1931
Following an order placed by the Polish government in 1927, Fabrique Nationale d'Armes de Guerre began making automatic military guns. It initially produced firearms invented by Browning and improved them, while later inventing its own new models.

the Treaty of Versailles, it obtained economic control of it. Poland was therefore able to dismantle the powerful and formerly Prussian arsenals, but it needed technical assistance to ensure their operation.

On 18 January 1923, following a trip to Warsaw by Galopin, FN's Managing Director, and Joassart, FN's Secretary General, a two-stage partnership was outlined. Poland initially placed an order for 6,000 automatic rifles with FN for an estimated total of 15 million Belgian francs. This order was to be fulfilled in Herstal, which enabled Polish managers to be trained there. The expected profit for FN was five million Belgian francs. A joint Belgian-Polish company would then be created for the manufacture of small arms, in particular Browning automatic rifles and Mauser rifles. This company's capital could include contributions from the Polish government (land, buildings, machines and funds), money from Polish bankers, as well as... the five million Belgian francs from FN earned during the initial order. On top of this capital participation, FN

would receive 10% of the shares for its technical support. This original plan was complicated by the fact that the licence for that automatic rifle belonged to the Colt Company of Hartford, Connecticut, to whom John M. Browning had sold the licence for use in all countries during the war. An agreement was made that Colt would manufacture the automatic rifle in Hartford and pay FN (its European agent) a commission for its services. In 1924, a contract for supply and technical assistance was concluded with K. Rudzki i S-ka of Warsaw. It was not until 1927 that Poland placed an order for automatic rifles. But this order came with such strict specifications that Colt preferred not to follow through. FN lacked experience in the manufacture of a firearm as complex as the Browning automatic rifle, but agreed to take over Colt's obligations and signed the contract alone on 10 December 1927.

FN BAR
Automatic Rifle
Model D

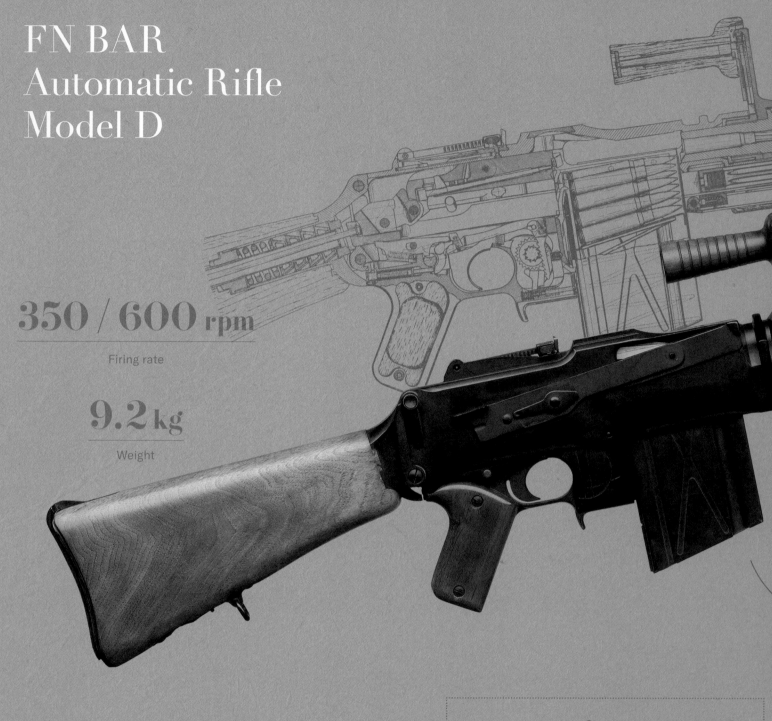

350 / 600 rpm

Firing rate

9.2 kg

Weight

System
Automatic firearm, operating with open bolt and
gas-operated system

The **BAR Automatic Rifle**, which was designed in 1916 and adopted in 1918 by the US Army, was one of the most important firearms in the US military arsenal. The BAR was relatively light and used the standard Springfield .30-06 cartridge, and was a selective-fire automatic rifle using a 20 or 40-round magazine and air cooling. This long gun was designed to marry the firepower of a machine gun with the precision and portability of an

infantry rifle. It had a gas-operated system. So when the ammunition is fired, the gas produced by the powder charge's combustion penetrates through a port drilled in the barrel inside a gas cylinder and pushes the piston head. This forces the piston head, as well as the slide and the whole closing mechanism, to move backwards. The movement leads to the extraction of the case that has been fired. When the gas has stopped pushing the

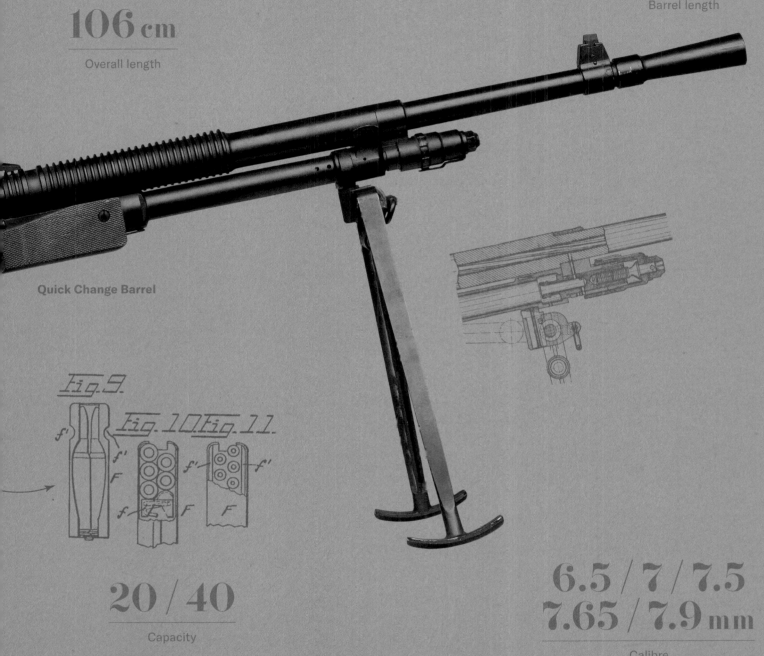

55 cm
Barrel length

106 cm
Overall length

Quick Change Barrel

Fig.9

Fig.10Fig.11

20 / 40
Capacity

**6.5 / 7 / 7.5
7.65 / 7.9 mm**
Calibre

piston, the recoil spring brings the whole mechanism forward, thus introducing a new cartridge into the chamber, locking it and firing it. The BAR Automatic Rifle was modified by FN during the period between the two world wars. Initially, the idea was to produce for Poland a new model that was very similar to the original. FN then responded to a request from the Belgian government by developing the type 30. The rifle was fitted with a gas regulator and a firing rate retarder, flanges to facilitate the barrel cooling and a pistol grip. Finally, a few years later, the Model D introduced the quick release of the barrel and the whole mechanism from the rear plus a carrying handle.

'TRADITIONAL' SALES

The examples above underline FN's ability to successfully sell its know-how on the international scene. Yet they should not overshadow the importance of 'traditional' arms orders: Brazil (1923-1924), Mexico (1926-1927), and later Persia, Ethiopia, Venezuela, the Netherlands, Estonia, Lithuania, Romania, China, and of course Belgium, would be Herstal's clients in turn, and sometimes even simultaneously. This obliged FN to be creative in its organisation of manufacturing. These orders were significant: for China, 24,220 Mauser rifles, 6,465 automatic rifles with removable barrels, 40 machine guns and more than 58 million ammunition; for Persia, 21 million ammunition; for Ethiopia, 2,000 Mauser rifles, 425 automatic rifles and nearly 16 million ammunition; for Venezuela, 16,500 Mausers and 12 million ammunition, and so on. Quantities like this should be compared to the 150,000 Mauser rifles ordered in 1889, an order that led to the creation of FN.

So it seems clear that, during the inter-war period, FN's industrial base was the manufacture of weapons of war. Selling them in a troubled economic world – protectionism, inflation, depreciation of currencies – required the company to invest most of its energy, be that in capturing markets, organising production or maintaining essential technological advances.

Mauser rifles assembly workshop, 1924
The Mauser 1924 was ordered by many countries. Hundreds of thousands were produced and they mobilised much of the company's industrial capacity.

Visit by the King of Egypt, Fuad I, to the Fabrique Nationale d'Armes de Guerre, 1927
Reflecting the power of sovereigns, the arms industry generated particular interest from heads of state. With its model factory, Fabrique Nationale d'Armes de Guerre therefore often hosted official visits, always organised with great pomp.

FN-MAP no. 3 typewriter, 1922
The Manufacture d'Armes de Paris was founded during the Great War, with the collaboration of Fabrique Nationale d'Armes de Guerre managers, and it subsequently switched to being a producer of civilian equipment. The company was part of FN from 1921 to 1936 and it produced typewriters like this, among other items.

Novel methods

In the inter-war period, FN employed a range of financial and industrial levers to establish its production in Belgium and abroad, as well as to do its best to overcome an unfavourable economic context. A combination of three economic realities became apparent at the end of the war: foreign markets were mainly closed for all of FN's products; the arms trade was expected to decline in importance; and national and international competition was very active in all areas. Besides these economic factors, there was also a feeling that FN was capable of doing everything and doing it well as far as mechanical engineering was concerned. Yet equally there was a genuine desire to maintain a considerable level of industrial activity in Herstal. These imperatives guided FN's industrial and commercial choices throughout the post-war period.

A MULTI-PURPOSE TOOL

From 1919 to 1939, FN used the acquisition of shares in other companies as a versatile tool. For example, this might involve acquiring control of a competitor, maintaining or expanding business channels, investing in a promising industry, or increasing profits. Sometimes several of these considerations were taken into account at the same time.

Most often – when a company wishes to increase its capital or when a new company is created, whether it's a new one or one that takes over the activities of an older company under better conditions – shares are acquired by subscribing to shares when the capital is increased. When FN's investments in these holdings proved disappointing, the company had no hesitation in selling its shares, even at a loss. Conversely, if a positive result resulted from an alliance – be that in terms of dividends or market share – then this alliance would be maintained or even expanded, sometimes leading to a merger and acquisition.

This multi-purpose tool was called on several times in the inter-war period: Manufacture d'Armes de Paris, Cartoucherie Russo-Belge, Établissements Bachmann, Cartoucherie Française, Cartoucherie de Soleure, Société des Machines à Sténographier Grandjean, Société Anonyme Belge de Constructions Aéronautiques (SABCA), SpA Achille Fusi de Milan, Société Belgo-Danubienne (for the Romanian market), Agence télégraphique belge, Comptoir des Mines et grands Travaux au Maroc, Compagnie Industrielle Africaine, Société des Carburants Makhonine,[98] and a few other companies being formed about which we know very little. Of course, these investments were sometimes modest and some companies quickly disappeared or remained in limbo. Yet FN's determination to open up to the European industrial market remained constant throughout this period.

Until 1928, FN pursued an active policy of acquiring holdings, before withdrawing from them, although this approach was accentuated by the crisis of 1929-1930.

LA MANUFACTURE D'ARMES DE PARIS

The involvement of FN in La Manufacture d'Armes de Paris (MAP), established with the help of Galopin and Joassart to revive the manufacture of small arms in France, is an example of this. The two companies maintained close ties. Up until 1936, FN supported MAP's activities through loans, sometimes secured by mortgages, or by taking an active stake in the company's capital. In the 1920s, MAP diversified its activities, i.e. those catching the eye of FN, and began to produce typewriters, which the company considered complementary to its own efforts in the field of calculating machines. In 1923, the production rate of typewriters (600 per month) suggested a favourable trend in production. Yet a few months later, it became obvious that sales were not keeping pace with production capacity. The future of MAP was then complicated by American dumping, and was thrown into disarray by France's nationalisation policy of 1936. After selling its shares the same year to a French industrial group, FN concluded in January 1938 that "this selling off of our French interests is rather disadvantageous...".

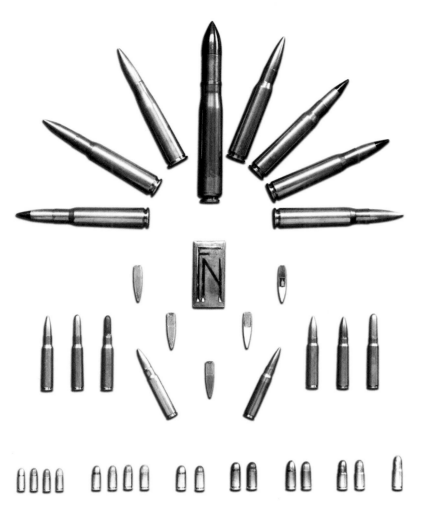

↕ **Presentation of ammunition produced by Fabrique Nationale d'Armes de Guerre during the inter-war period, 1937**
Fabrique Nationale d'Armes de Guerre developed a full range of ammunition, beyond the operating range for its own arms production. This winning strategy positioned the company among the leaders in this sector, leading to its growing monopoly abroad.

↦ **Checking of bullets in the ammunition factory, 1930**

TOWARDS A CARTRIDGE MONOPOLY

Although it had no experience in pyrotechnics, the fledgling FN began manufacturing cartridges in 1891, with considerable success. Its natural entrepreneurial urge to increase its market share led FN to pursue a proactive policy of acquiring stakes in the capital of cartridge industries immediately after the war. In 1919, it took control (albeit excluding its Moscow activities) of the Cartoucherie Russo-Belge [Russian-Belgian cartridge factory]. The aim, as underlined by the Board of Directors, was to prevent "the emergence of a more skilful and stronger competitor that could harm [FN] in dealings with the Belgian government". In the 1920s, FN also took control of the Cartoucherie française, with which it had been doing business for years. This acquisition was intended to ensure that the Herstal Company had

control of one of its most important distributors. This collaboration ceased in 1938, when France mandated that all directors be of French nationality. FN immediately sold its shares at a substantial profit.

The takeover of the Cartoucherie de Soleure, in Switzerland, demonstrates more starkly the reasons that sometimes drove FN to acquire shares in some of its competitors. In 1928,[99] FN and Cartoucherie d'Hirtenberg purchased all shares in the Cartoucherie de Soleure, with each company taking 50%. According to the head of FN, the aim was to "eliminate a competitor that was a nuisance because of its production capacity and quality". In September 1929, FN sold its shareholding.

In 1928, with a view to creating a major cartridge zone, FN responded to the Belgian government's call for the establishment of a cartridge factory in the north of the

country in order to move the production of pyrotechnics geographically away from the German border. A further aim was probably also to ensure a regional rebalancing. In exchange for this new project in Bruges, the government guaranteed major orders to FN. To succeed in this venture, FN acquired the Cartoucherie belge [Belgian cartridge factory], in which it already held a large part of the capital, as well as the Établissements Bachmann. FN acquired all the assets and liabilities of the two companies, including their installations, orders and customers, and of course their staff.[100] It benefited from a Belgian law of 1927 reducing by two thirds the fees applied to mergers of companies.[101] If this company was created by FN, the government would guarantee it an order for 112,500,000 military cartridges, to be delivered between 1930 and 1934. The necessary investment was estimated at 50 million Belgian francs, to be borrowed from the Société Générale de Banque. The virtual monopoly position provided in this way to FN would give rise to heated debate in the press. In March 1936, there was some alarm:

The managing director succinctly summarised the various attacks on our various 'Bruges' contracts in certain magazines and even in some daily newspapers. He added that this campaign, whose grounds seemed to be 'no monopoly', had in a way already harmed us because an order to transform 19,000 Mauser rifles for the Colony had been placed with the Établissements Pieper, apparently only with the aim of breaking a monopoly.

The advantage that Bruges had enjoyed would end in 1940. The approaching war meant that the impact of a competitive environment would not be felt.

DEALING WITH PROTECTIONISM

"Measures to be taken to maintain our sales in Germany. The Director points out that, before the war, FN exported a high proportion of its total exports to Germany. Since the armistice, no sales have been possible in this country, not only because of the import bans, but also because of the cost price at which our products would have been sold in Germany, given the current value of the mark." [102] [103] Only seven months after the war ended, FN considered "purchasing a small factory in Germany that would have manufactured firearms for the needs of our German customers". But after a study was carried out, the operation was deemed too costly, and the idea was temporarily shelved. However, this project remained under consideration... It's noteworthy that the "expansion" into other countries was also of great interest to FN "and would be of considerable help in solving the 'manpower' problem, a tough problem since it seems that starting from a level of 8,500 to 9,000, FN would run out of manpower in the region".[104]

In February 1929,[105] the company once again began to examine how to penetrate markets that were still closed to it. It was reported that FN had been pushed to set up small assembly workshops in the Netherlands, France, Italy and Germany in order to reduce customs charges. The company then wanted to adopt this policy systematically, suggesting that it should produce the 'bicycle part' of the motorcycles there: "This bicycle part represents approximately 65 kg out of a total weight of 105 kg. Therefore, the production of this 'bike part of a motorcycle' in the country itself would reduce by 2/3 the specific customs charges imposed in certain countries". This arrangement would be particularly advantageous in Germany, where between 5,000 and 10,000 motorcycles could be sold each year. The Board of Directors therefore authorised the rental of the Fafnir workshop for a minimum of three years[106] in Aachen, with a purchase option. FN would later create a sales office in Berlin.

When the problems of nationalism and economic patriotism were getting worse in Germany, FN had to change its name: FN Motoren GmbH became Berlin-Aachener Motorenwerke AG. Up until 1937, FN motorcycles and its 'Super Elastik' bicycle were sold and branded under this name. "The difficulties experienced by FN Motoren GmbH in Berlin and in Aachen as a result, stem not only from our foreign status, but also from our company name, which leads agents to say that, when anyone works for us, they are supporting the arms factories of a former enemy", it was stated in 1933.

In January 1937, FN sold its investments in Germany. The terms were strong enough to suggest that the directors viewed this disengagement as a welcome relief: "we have liquidated our 'motorcycle business' in Germany and we are in a way definitively finished with our factory in Aachen". [107]

EXPLORING NEW AVENUES

The purchase of shares, which, in 1928, amounted to more than 23 million Belgian francs, a sum greater than the company's capital (20 million), was sometimes motivated by concerns other than the simple desire to maintain a monopoly. In the early years following the conflict, the production of weapons of war seemed likely to decline, or even disappear, due to the weariness of Europeans and their governments. FN therefore had no choice but to pursue a bold policy of searching for new products that would call upon its expertise in 'precision mechanics', as emphasised in March 1919.

The MAP typewriters had already tempted FN. It viewed the Ellis adding typewriters as an attractive addition to this essentially civilian production line, so FN acquired the licence from its inventor.

A special division was set up within the factory, and great efforts were devoted to hiring technical and sales staff. In August 1923, FN pondered the pros and cons of closing down its adding typewriters department. The main argument for closure was that the company's success no longer depended on their production. This had been launched as a purely precautionary measure, in case one or other of FN's traditional divisions had to shut down. However, this was no longer necessary in 1923, as all the company's traditional production was benefiting from a favourable economic situation. This department was finally sold at the end of 1923 to its parent company, the Ellis Adding Typewriter Company. The overall profit of this transaction came to 200,000 Belgian francs.

Production of 12-cylinder aircraft engines for the Manufacture d'Armes de Paris, 1932

Stenography machines

FN was also interested in producing Grandjean stenography machines. In the unsettled context of the 1920s, the company decided to diversify its precision mechanics activities by focusing on promising civilian products and by acquiring a stake in any company looking for capital.

Société Anonyme Belge de Constructions Aéronautiques, SABCA

The Belgian government asked FN to participate, rather modestly and without any real enthusiasm, in the founding of SABCA in 1920: "Despite the disadvantages, the Board decided in principle to participate, for a sum to be determined and which will be in the order of 300,000 Belgian francs, in the capital of the subsidiary company that will be created by the Société nationale de Transports aériens for the maintenance and repair of army engines."[108]

On several occasions between 1929 and 1939, FN used the "shareholding" tool and its opposite – "divestment" – fairly efficiently, to achieve its multiple objectives: to gain a monopolistic position or to eliminate a competitor, without ever losing sight of its industrial mission and its commercial strategy.

THE SEARCH FOR BALANCE

FN sought to maintain a balance between its different divisions. Up until 1932, successive general assemblies[109] were informed that the company's regular operation was handled almost equally by the various divisions. In 1920: "the profits can be attributed in approximately the same proportions as before the war to each of our divisions: Armoury, Ammunition Factory, Motorcycle and Bicycle Manufacturing, and Automobile Manufacturing". In 1921: "Our divisions have continued to meet our expectations [...] and we will succeed in maintaining this satisfactory performance together." This statement was repeated, in various more or less eloquent forms, from 1922 to 1929: "If we wanted to go into some detail about each of our major divisions, we would have to tell you again that all of them have contributed to the company's profits." In 1930, it was conceded that "Our various divisions have suffered, in roughly the same measure, from the effects of the downturn in business." This statement was repeated in 1931 and 1932.

The following year, a change in tone was noticeable: "Our different divisions have been affected in different ways by the crisis." While the Arms and Ammunition Division was boosted by "many orders", the 'Sports' Division, which included all civilian 'non-gun' production, was more sensitive to economic fluctuations. In 1934, the company repeated that "our different divisions have all participated in the increase in business". The following year marked a break in this balance, which had been maintained since just after the war.

Thus it was important for FN to stress, year after year, that its mechanical manufacturing activity comprised four "divisions" of equal importance for the company's overall success. This statement was clearly correct. The diversification of activities imposed after the "reorganisation of the company" in 1896 was a stabilising factor, allowing the company to balance its results whatever the external circumstances that were imposed on it.

However, this assessment must be qualified. Although each of the divisions certainly contributed to the overall profit, they did not do so equally or with the same level of efficiency. For the inter-war period, only very partial figures on the volume of business are available. The Board of Directors occasionally gave a breakdown of activities. From these fragmentary and sometimes estimated figures, it appears that in 1920, for example, the 'arms' (civilian and military) and 'ammunition factory' divisions contributed half of the turnover of 14.1 million Belgian francs. In 1924, government business – i.e. mainly military orders – accounted for 66% of total orders. In the fiscal year 1924-1925, 'Arms' and 'Cartridges' accounted for 56% of total sales. This business split into two roughly equal halves continued until 1935, when automobile production ceased and arms sales increased as a result of widespread rearmament.

In other words, the 'main business' of FN in 1889 – the manufacture of and trade in weapons of war, modified in 1896 by the introduction of the manufacturing of civilian guns – remained largely unchanged between the two wars.

The sale of weapons of war is prone to significant swings. The state of the world, international tensions, governmental or budgetary choices, etc. can all lead to increases or decreases that are difficult to predict. Arms orders from foreign governments that may have seemed certain can be cancelled or sometimes awarded to a competitor that is politically stronger internationally. In other cases, payment for orders that have been fulfilled may be delayed for a long time.

Consequently, it is crucial for all armaments factories to have access to civilian production, in order to keep their production facilities operational and to provide work for specialised workers. The company's diversification, imposed during its reorganisation in 1896, had to be continued and expanded for totally opposite reasons. At the end of the 19th century, the aim was to restrict or prevent the manufacture of weapons of war by setting up civilian production. However, starting in 1919, civilian production was developed to effectively support a division with less regular production. This showed genuine anticipation of the uncertainty that would prevail in the markets during the inter-war period.

FN entered the 1920s and 1930s with its distinctive industrial characteristics: those of a mechanical manufacturing plant focused on precision and a dedication to a job well done.[110] This demand for excellence, stemming from the company's culture as a weapons manufacturer, was an asset for its civilian production. But it was an obstacle in competitive markets, where a lower price was often the only criterion that would secure a purchase.

Civilian arms assembly workshop, 1921

The automobile dream: the Pré-Madame factory

Alexandre Galopin was convinced that the automobile would become FN's leading product, at a time when the manufacture and sale of weapons of war were apparently on the wane. Most of the directors followed suit on this point. In 1925, Galopin reiterated to the general assembly of shareholders his "special confidence in the future of the automobile division".

Exceptional resources were deployed to upgrade and adapt the car manufacturing workshops to the new demands. Gustave Joassart, who became Managing Director in 1923, had the new Pré-Madame factory built in 1929. The new workshops were an extension of the factory, to the north-east of the original facilities. They were entirely dedicated to automobiles and motorcycles, including bodywork operations, which until then had been carried out outside Herstal. In parallel, FN purchased the workshops of the former Établissement Nagant in Liège, in order to set up the car repair and maintenance departments there.

When FN halted car production in 1935, Belgium's dream of making its own original and inventive cars also ended.

OUTSTANDING CARS

On 8 May 1919, FN was delighted with "the already numerous orders". For the cars, "this year we will manufacture 50 chassis for the 1950, 50 for the 2700 A and 30 for the 1250. By manufacturing some new parts to complete the mismatched collections, we could complete 300 units of 1250 chassis, 400 units of 2150 chassis (towards the end of 1919) and 200 units of 3800 chassis (to be released in 1920)". However, in order to produce the aluminium parts, new machines must be purchased for about 1,250,000 Belgian francs. "This investment will allow the use and sale at a good price of at least 2,000,000 (1914 value) parts in the process of being manufactured, which would be of no value if they were not made usable by the manufacture of the parts missing from the collections." The Board of Directors accepted the expense. As regards automobile manufacturing, FN first of all continued its pre-war activity, a period that was undoubtedly the golden age of the car in Belgium. In 1910, 21 car manufacturers and 327 'coachbuilders' shared a market where their inventiveness guaranteed them an enviable reputation.

◄◄ **Assembly lines for car bodies in the new buildings of the Pré-Madame, 1930**
Thanks to the new factory, the production of automobiles and motorcycles could be adapted to keep pace with the development of modern machinery.
The industrial process of continuous flow production, inspired by the Ford system, enabled product quality improvements at more affordable prices for customers.

↕ **Car upholstery workshop in the new buildings of the Pré-Madame, 1930**

The first post-war automobile show in Brussels 1920

Be proud of our industry

The consequences of the world's major upheaval are not all terrible. They have taught us some useful lessons about patriotism and how to value our factories, to love their products and to prefer them to foreign ones.
[...] Our automobile, motorcycle and cycle manufacturers, despite all the difficulties they have had to overcome to rebuild their looted and devastated factories, have managed to exhibit vehicles at the Brussels Motor Show that rival the finest of the best foreign brands. These are the sentiments felt especially when visiting the stands of the Fabrique Nationale d'Armes de Herstal which, after having re-established – thanks to huge efforts! – its enormous workshops, and after rebuilding its elite staff that had been scattered by events, is presenting to us its automobiles, motorcycles and bicycles that are just as perfect as they used to be [...].[11] *All this highlights a manufacturing perfection, a superiority which, together with the excellence of the raw materials and the high performance of the engines, comprise the main characteristics of the Fabrique Nationale de Herstal.*
La Meuse, 11 December 1920

The Pré-Madame factory during completion, 1928-1929
The factory was designed to match the period's architectural characteristics, notably including a saw-tooth roof for overhead lighting.
This superstructure with 'mushroom' type pillars could bear considerable loads and better withstand any bombardment.

The war disrupted these auspicious starts. At FN, as happened elsewhere in Belgium, automobile production ceased completely between 1914 and 1918. Other countries continued to produce these vehicles, even during the war (France, the UK, and the United States in particular). This included rolling stock for the armed forces, thus improving the technologies used. More importantly, the mass production of inexpensive vehicles for the general public, which had begun in 1908 in the United States, became the economic norm. In 1920, the Ford factories produced one million of the famous low-cost Model T. Ford's highly standardised industrial methods were quickly taken up by its competitors. Just after the war, these methods were coupled with a major push into the European market. In 1926, Chrysler, following Ford in 1922 and General Motors in 1925, set up an assembly plant in Antwerp. This led to increasing American pressure on Belgium's car manufacturers.

FN shied away from competing in an unfamiliar field, as it was inexperienced in price wars or lower quality products. The company wisely limited its operations in the automotive field to what it knew, i.e. producing luxury or semi-luxury cars, referred to at the time as 'medium-sized cars' that were technically accomplished, innovative, comfortable and satisfied customers' requirements. Nowadays we would call this a 'niche market'. In 1925, FN produced 1,500 cars and 2,000 in 1927, making it the second largest Belgian producer just behind Minerva. Belgian companies could not compete with their international competitors. In 1922, Belgium's car brands already accounted for only 15% of the total number of cars assembled in the country. In 1934, shortly before car production ended at FN, this figure fell to 3%, while total production on Belgian territory had more than doubled during the same period.

Higher customs duties on automobiles also contributed to the industry's decline. In 1935, the Belgian government, which was no longer convinced that it was possible to maintain a car manufacturing activity in Belgium, decided to encourage foreign companies. It reduced customs duties on the import of parts[112] in order to create a national assembly industry that it believed had a promising future.

8 Cylinder luxury model FN car, 1931
In 1930, Fabrique Nationale d'Armes de Guerre presented the ambassador of its automobile range, a neat combination of elegance, comfort, speed and safety.

Car manufacturing after the First World War was driven by a dual imperative: to make up for the technological setback accrued over four years due to a lack of research and production; and to take into account the shift of Belgium's economic centre of gravity towards the north of the country.[113]

In the years immediately after the war, FN was satisfied with adapting its pre-1914 models to the new expectations of its customers. The 1250 A was a development of the 1250; the 2150 was an improvement on the 1950; and the 3800 was the successor to a prototype developed in 1914, the 6900, thereby creating several genuine automobile families. As usual, FN was mainly concerned with ensuring the long-term profitability of its products. In 1930, FN set about challenging American manufacturers and asked the engineer Warnant to study

a car of American inspiration and size. It would be an eight-cylinder, 3.25-litre engine with side valves and a four-speed gearbox. The car was available as Standard or Grand Luxe. It was a powerful, comfortable and fast car. However, its outstanding road qualities were not a match for its slightly high price. Only 400 of them were produced until 1936.

In addition to these models, in 1933 FN launched the Prince Baudouin, a medium-sized car that was produced in 1,300 units until 1935. More in keeping with the times, it had an 'all-steel' body and offered four comfortable seats. It won the King's Cup at the 24-hour Francorchamps race in 1933, at an average speed of 100 km/h. A slimmer and more spacious version, the Prince Albert, was added in 1934... one year before the production of personal cars was stopped at FN.

THE INEVITABLE END OF THE AUTOMOBILE

In May 1935, when discussing the Automobiles Division, the Managing Director noted:

The very satisfactory progress of sales, since the current monthly trend was practically double that of previous months. [He] pointed out that the very clear decrease in customs duties, together with the persistent level of competitors' prices, notwithstanding the currency devaluation, led to the conclusion that it would henceforth be impossible to continue [the] activity in automobiles, as this activity could only lead to losses, given [...] the small size of the market and the impossibility of exporting [...]. He referred to the conversations that had been held with Chrysler with a view to undertaking the assembly work for the whole of Europe in Herstal on behalf of that company. Unfortunately, Chrysler rejected our proposals, believing it more advantageous to continue to carry out its own assembly in its modern Antwerp premises. In this respect, any hopes we might have had were dashed.[114] *This being the case, the only prospect for future activity in the "Sport" workshops will be the construction of special vehicles and motorcycles.*[115]

This desire to find an automobile manufacturer that would assemble its cars in Herstal would continue until 1938. It was then reported that "the agreements with Peugeot were on the right track but had not yet been signed, as the Peugeot firm seemed to have certain difficulties in making its decisions".[116]

Four-cylinder, sports coupe model, FN cabriolet car, 1300 cc, built in 1926
The 1300 was an innovative car, with a four-cylinder engine, aluminium cylinder head and overhead valves. It preceded the eight-cylinder model. However, it was more technologically advanced and was the mid-range of the FN brand, a major segment of the company's business in terms of units produced.

FN 8 Cylinder Car Luxury model

75 hp
Power

3 255 cc
Capacity

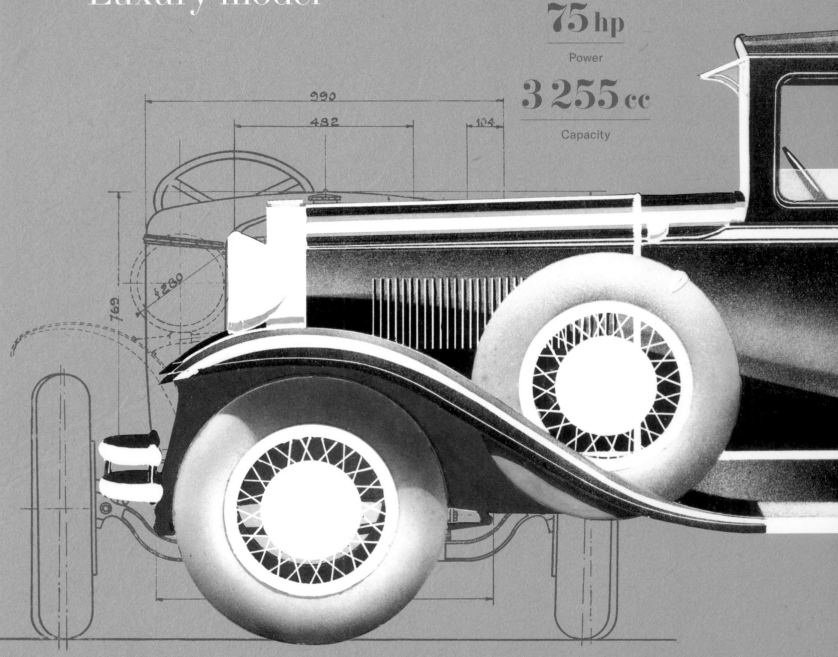

With its pure aesthetic look, the **FN 8 Cylinder** Car embodied the optimism and powerful dynamism of its period. It combined a robust look, with a tough and solidly assembled chassis. It was fast, being able to pull away in a gentle and speedy way thanks to a four-speed forward gearbox (the third one being silent) mounted on ball bearings. It was safe, featuring large-diameter brakes on the four wheels, with extendable shoes and an emergency hand brake linked to the gearbox. It was also comfortable, with five seats including an adjustable front seat, fine cloth upholstery, four doors with windows that could be wound up and down, blinds, remote control locks, windscreen wipers and many luxury accessories. The FN 8 Cylinder Car came in two models for bodywork: the standard 8 type, with a spare wheel at the back, and the luxury 8 type, with a luggage rack at the back and a spare wheel on the sides.

$\pm$ **110 km/h**

Speed

1575 kg

Weight

Drive shaft, four forward gears and one reverse gear

Transmission

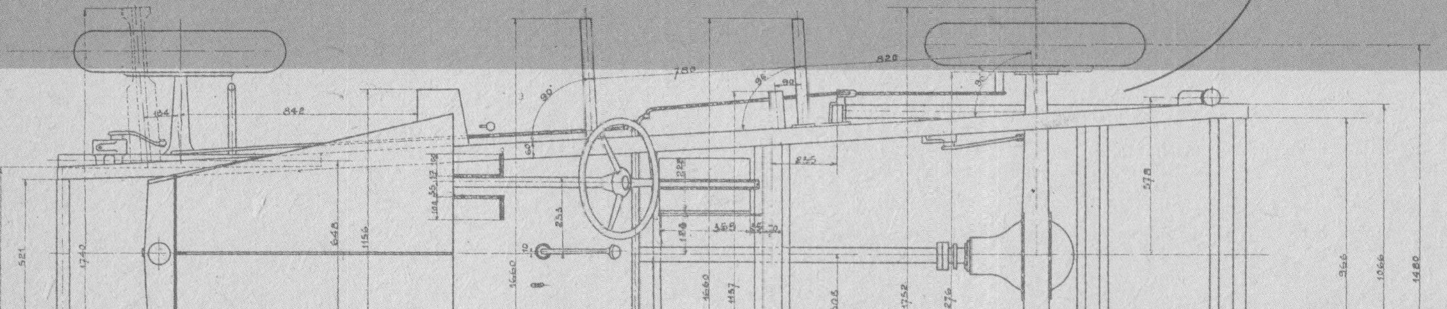

System
Single block eight-cylinder engine with lateral valves

Motorcycles: pragmatism leads to success

It should be recalled that FN had successfully started manufacturing motorcycles, starting with a first model of moped in 1901 up until the 'car on two wheels' of 1905. At the end of the First World War, FN continued to manufacture models that were available before the conflict, thus exhausting its stocks of spare parts, which were supplemented as needed: "In order to take advantage of the machined parts that are still in our workshops, we have planned to restart the production of 1,500 2-speed motorcycles, 500 four-cylinder motorcycles to be sold during 1920."[117] In March 1921, it was noted that these motorcycles were not selling well, which possibly prevented them from being sold as planned.[118] So the company decided to "rejuvenate" these machines with some "improvements".

FN once again showed that it could adopt an efficient practical approach. The two-speed motorcycle, once perfected, would become the 285 T and the four-cylinder, the 750 T. It proved successful. In May 1922, more than 800 of the transformed models were sold, 300 were on order, and the stock was practically sold out by mid-1923.

Assembly line of the FN M70 motorcycles, 1930
The M7C was a commercial success. It was mass-produced for over a decade, with numbers produced exceeding several tens of thousands. The M7C was popular and reliable and made its mark at a time when motorcycles were still considered a utilitarian vehicle. The Fabrique Nationale d'Armes de Guerre had an extensive sales network, as well as assembly workshops in Paris and Aachen. These complemented the facilities in Herstal, which were transferred in 1930 to the new Pré-Madame factory.

Advertising photograph for the FN M86 Super Sport – 600 cc motorcycle, 1936
The FN M86 Super Sport had a four-stroke single-cylinder engine block with overhead valves controlled by protected rocker arms plus a four-speed gearbox and a foot switch. It was a high-performance motorcycle for the time and could reach 130 km/h.

Advertising photograph for the FN M70 luxury motorcycle, 1931

Faced with a slowdown in sales in a market that was becoming increasingly small – eventually being reduced to only Belgium and the Netherlands – and highly competitive, from 1924 FN developed models that were more affordable. The necessary reduction in cost price was achieved by replacing the acatene (chainless) system with a chain drive. The FN 60 (346 cc) and the FN 70 (350 cc) were the first FN models featuring chain transmission. This simplification of the transmission enabled the engineers to group the engine, clutch and gearbox in a single housing, which ensured better protection of these essential components. In its barely modified 'Sport' version, the M60 achieved 17 records in 1926 in Monza, Italy, including the 24-hour record at an average speed of over 105 km/h. More than 4,700 units of the M60 were produced. As for the M70, sales of which began in 1926, more than 16,000 were produced. The steering wheel of the M70 is external and its central

external part is painted red. This led to it being nicknamed 'Moulin Rouge' [The Red Windmill]. Captain Jean Bruneteau 'conquered' the Sahara Desert in 1927 on an M70, which led to it being called the 'Sahara'.

FROM CIVILIAN TO MILITARY
The M86 motorcycle was launched in 1934. It is certainly the most well-known motorcycle of the 1930s and the most remarkable of those produced by FN. Adapted to the needs of armies that were looking for greater mobility in the field, it was mass-produced for Belgium. Argentina, Bolivia, Venezuela, Romania and Yemen also purchased small quantities. An armoured version with a side-car and driving wheel was then developed and sold to China and some other countries.

Developped in 1939, the Tricar – a three-wheeled machine – was a model between a large motorcycle with a sidecar and a small off-road vehicle. The motorcycle part was derived from the M12-1000 cc sidecar motorcycle. One of its distinctive features was that the cylinders were fed by a single carburettor with a large filter, placed very high to prevent the engine from ingesting water. It was equipped with handles at the bottom of the forks to facilitate cleaning. The motor was designed to be as quiet as possible, to allow a discreet approach. There were 14 versions of the Tricar, notably for transporting five people or a motorcycle repair shop, for supplying fuel or ammunition, or for transporting a heavy machine gun or supplies. The Belgian Army bought a little more than 300 of them, but they could not all be delivered before the outbreak of the war. A few examples were also sold to the Dutch East Indies.

A civilian version was created after the war: the Tricar T8, a true three-wheeled small truck, with the same engine as the military vehicle, capable of transporting payloads of 500 to 1,000 kg depending on the model, and which was a great commercial success.

Group of 70 chassis assembled for the FN12 Tricar, 1940

Trolleybus assembly and bodywork workshop, 1933
In addition to its automobiles and before long a replacement for them, Fabrique Nationale d'Armes de Guerre opened a heavy loads department. This produced trucks and trolleybuses, which were notably used on the streets of Liège.

Trucks and special vehicles

On 19 January 1932, in the midst of the world crisis and faced by declining sales, the Board of Directors began to look for new markets. The minutes of the day's meeting, in a paragraph entitled 'Search for work', noted that they were trying "to find work for the workshops, negotiating to all intents and purposes on various manufacturing items, in particular: Cardon Lloyd tank, Austro Daimler all-terrain truck, Gnome and Rhône aircraft engine, and trolleybus".

FN would consider any initiative, as long as it was within its sphere of activity, engineering. The various negotiations mentioned above were obviously not all successful. However, the company's 'Engines' Division then set about manufacturing special vehicles and vehicles that we would today describe as 'utility vehicles'.

FN van on C10 chassis with telescopic lifting, destined for the Belgian Ministry of Public Works, 1937

Eight-cylinder FN 2.5 T short truck with its Truck Tractor trailer, 1934

As with its other products, FN never doubted its ability to embark on new adventures, and it contributed the knowledge acquired from its other productions. Thanks to its engineers and technicians, as well as its mechanical and materials knowledge, the company was able to develop new research enabling it to create and adapt its models to the requirements of its potential customers. Its three principles were: "to know how to do (everything), as long as it is in the field of engineering", "to innovate" and "to listen to its customers".

This enthusiasm was sometimes tempered, and occasionally, as in September 1935, doubt crept in. The decision was made at that time to stop manufacturing cars and to start manufacturing special vehicles, "subject to

the establishment of a satisfactory programme of organisation and economic production, and as a sort of *trial* (i.e. subject to the expected results being confirmed in practice), the decision [was taken] to continue for two or three years [...] the manufacture of special vehicles such as tractors, trolleybuses, etc."

FN began by adapting the eight-cylinder engine of its famous car to equip a civilian truck with a 2.5 tonne payload. The eight-cylinder car was produced from 1931 to 1939 and the truck from 1932, being modified for different uses (from two to three tonnes). The eight-cylinder truck was more successful than the car and sold 900 units.

For the army, FN then designed a three tonne, six-cylinder

military truck with four-wheel drive, a real innovation. This truck's engine was later used to equip rubber-tracked tractors for the Belgian Army, made under a licence with Citroën-Kégresse.[119]

FN owed its strength and inventiveness to the constant switching between civilian and military production, between the world of cars and that of utility vehicles, and between the world of research and the commercial sector. For proof of this, one can simply mention some of the prototypes developed by FN, which often moved on to mass production.

The eight-cylinder 'Coloniale', designed in 1935, was an 'all-terrain' vehicle[120] featuring a new design, but it was too costly for the Belgian Army. From 1935, FN built ambulances on an FN C-10 chassis, using the engine of the 'Prince Albert' car (in production from late 1934 to 1936). In 1937, a truck with a forward cabin and a six-cylinder engine was produced; several hundred units were sold, sometimes as buses. The 4RM42F truck, with four cylinders and eight gears, is worth mentioning. It was the precursor of the 'Ardennes' truck that FN developed in the 1950s and had to undergo tough tests that were widely publicised on the Thier Savary stairway in Liège.

Stress testing of the FN4RM truck on
the Thier Savary stairway, in Liège, 1938

FN participates and wins

To promote its motor vehicles more widely, FN introduced a dual strategy. First, to publicly demonstrate its technical capabilities and its flair for mechanical innovation through the races and expeditions it organised or participated in. Second, to also use the new graphic arts to establish the company's image in its own time. These new requirements resulted in FN calling on professionals from the worlds of sport and graphic arts.

A DEMONSTRATION OF TECHNICAL STRENGTH

"The current development of [vehicles] of all makes is such that only a prolonged test under normal operating conditions can differentiate the various models."[121]

Since the end of the war, automobile and motorcycle races and "long-distance rallies" had mushroomed: some pitted brands against each other, others took on technical and logistical challenges for a single manufacturer. Races and rallies were designed to test and improve products as well as to promote their success to the general public, with a view to establishing the reputation of the companies. Despite their high costs, these rallies were unavoidable.

These events were very tough, regularly lengthened and their difficulties continually increased. They often came with a touch of exoticism. The 'Croisière noire', the first of its kind, organised by Citroën in 1924, made a lasting impression by skilfully combining the development of new techniques with a determination to explore unknown lands. FN, like other manufacturers, would follow this path for its motorcycles as well as for its automobiles.

FN350 motorcycle at the 12 hours of Monza race, 1926
During the competition, drivers Maarten Flinterman, Antonio Sbaiz and Guillaume Lovinfosse set a new average world speed record of 105.253 km/h.

FN M70 motorcycles between Tabankort and Bourem, in Mali, during the Sahara Rally, 1927
In the history of two-wheelers made by FN, the M70 is without doubt the model most associated with long rallies. In 1931, this motorcycle was ridden by Justine Tibesar, from Arlon and only 22 years old, to travel from Saigon to Paris, a distance of 22,000 km, and a major technical feat for the company.

LONG-DISTANCE RALLIES

The manufacturers have been preoccupied for some years with organising long-distance rallies, in order to quickly show the public the qualities of various vehicles and to interest them in the sporting nature of journeys to far-off countries.[122]

The Sahara on a motorcycle

Lieutenant Bruneteau, Mr Werens and Mr Gimié, who have just completed the first crossing of the Sahara on a motorcycle, passing through the Tanezrouft, the land of fear and thirst, and completing the magnificent Paris-Bourem-Liège rally (8,000 km), were warmly received in Liège (Quai des Ardennes).

This introduction to the film *La traversée du Sahara à moto*[123] was deliberately melodramatic, not hesitating to mention the FN 350cc. Crossing the Sahara was then already something common for special vehicles and the trans-Saharan had become a marked trail since the previous year. As for the Tanezrouft, it was no longer the desert of fear and thirst, as there was even a water and fuel depot.

Nevertheless, the feat was considerable. The three men, two French soldiers and a well-known Belgian sportsman – he had won the 1926 Liège-Bordeaux-Liège race – left Paris in May 1927 for a voyage of several thousand kilometres that would take them to Bourem in Mali, on the banks of the Niger River, near Timbuktu. Their machines were the FN 350 cc type M70 without sidecars. Consequently, their equipment in terms of spare parts and tools, oil and fuel had to be minimal.

La Wallonie followed the event's progress very closely. On 11 May 1927: "We received from Colomb-Béchard[124] a telegram signed by Lieutenant Bruneteau: 'All going well stop machines have conquered wild mountains and almost impassable roads.'" On 13 May, this message was sent from Béni Abbès [Algeria]: "All going well, stop, shift in schedules, stop, impassable roads, stop, machines holding up wonderfully." On 24 May, the newspapers widely praised the heroes and their vehicles. *La Gazette de Charleroi* wrote: "This superb determination to win and the amazing endurance and extremely low fuel consumption of the FN type M70, have overcome the desert regions where access seemed to be impossible for the motorcycle."

This exceptional race – helped by its exotic nature and colonial character – was one of a long list of events in which FN motorcycles took part. The company often won and in a number of different categories.

Le Raid au Cap

Cape Town by automobile

Le Raid au Cap [the Cape Town Rally] in 1928 aimed to
cement the company's international reputation by taking
its cars on a real journey of initiation through the African
continent. After crossing France, the drivers passed
through Algeria, Niger, Nigeria, Chad, Central African
Republic, Belgian Congo, Uganda, Kenya, Tanganyika,
and Rhodesia, before finally reaching Cape Town.
Lieutenant Robert Fabry, Lieutenant Lamarche – already
known for his long-distance rallies in Europe and North
Africa, accompanied by Hubert Carton de Wiart, min-
isterial attaché, and Roger Crouquet, journalist at *Le
Soir*, made up the team, which shared two FN 10 HP cars
of standard construction, featuring a closed car body
"entirely made in the Herstal factories", and fitted with
Englebert tyres. As in all car races, the aim was to publicise the qualities of a brand. A more subtle goal was to
open up or secure trade routes between the mainland
and colonies.

This form of advertising was echoed in the newspapers
by highly visible, multi-column inserts featuring simple graphics. The goal was simply to offer information,
communicated through short words, rather than pretty
illustrations. This information was presented in an advertising structure made up of multiple press releases and
various articles. These provided updates on the progress
of rallies, while explaining or exaggerating the difficulties and seeking to put the vehicles at the heart of the
success.

Gibraltar

From time to time, these rallies took place elsewhere,
such as the Raid Gibraltar-Bruxelles [Gibraltar-Brussels
Rally] organised alone by FN in 1931. Its goal was to go
faster than trains. Two drivers, Georges and Collignon,
and two passengers travelled to Brussels from the Rock
(2,342 km) in 39 hours and 2 minutes, which beat the
faster trains by eight and a half hours: "Faster than
Grands Express trains" became the baseline for FN
adverts for a while. The objective was to test a production car, an eight-cylinder car with no special fittings.
Its bonnet had been sealed at the start. Because the
public was no longer interested in the exploits of specially equipped vehicles, simply preferring to learn about
a product that they could buy.

RACING: AN AGE OF PROFESSIONALS

The races, which were designed to pit the marques against each other and to rank them according to their results, received extensive coverage in daily newspapers' sports pages. This meant that manufacturers were obliged to take part in races, if they wanted to stay in the public eye.

Sometimes the races stretched for many miles. At Liège-Madrid-Liège in 1930, the partnership of Weerens and Mathot drove 3,300 km in 65 hours and 24 minutes, in an FN 11 horsepower. The Alpine Cup involved a route packed with mountain passes, with production cars: Munich-Innsbruck-St. Moritz-Turin-Nice-Geneva-Berne, 2,400 km of winding climbs and descents. FN won the gold medal in 1931, with its drivers E. Collignon, T. Georges and Ch. Charlier. Nicknamed 'The Road Marathon', the Liège-Rome-Liège race was also very demanding. Of

the 22 crews involved, only six returned to Liège, including the FN drivers at the wheel of an eight-cylinder car, which came first.

The track races were no less difficult, such as the Grand Prix de Spa-Francorchamps where FN won the King's Cup on several occasions.[125] This prize was awarded to the best manufacturer, the one having collected the most points with its vehicles entered in the race and which was not necessarily the winner of the race.

Motorcycle racing became truly professional in the early 1930s. Like its competitor Saroléa (Herstal) and like most manufacturers, FN created a department responsible for developing technical and staff strategies to win the most prestigious races. The main aim was to achieve technical advances that could then be adapted to commercial products.

FN's works team comprised engineers and technicians

The three Prince Baudouin four-cylinder FN cars that won at the King's Cup in the Spa 24-hour race on the Spa-Francorchamps circuit, 1933

in charge of developing powerful and robust machines.[126] The drivers were just as important as the technicians and FN tried to hire the best.

In 1930, the company went looking for what were then called 'works drivers'. It invited several of these sports-men to race on the Montlhéry racetrack near Paris on the same 350 cc motorcycle, with the fastest one being recruited. Jules Tacheny was the winner, signing his contract with FN a few days later. Becoming a works driver was a big achievement for people who were previously only amateurs and accustomed to local races. Their regular salary provided them with a more comfortable life. Above all, they benefited from top-notch equipment and technical assistance to pursue their passion .. and to help their company win.

Edmond Claessens, Pol (Léopold) Demeuter, Noir (Erik Haps), René Milhoux, Edison and Antoine Collette were among these legendary motorcycle riders. In 1931, FN booked the Montlhéry circuit for two days. In very difficult weather conditions – it was late October – FN had its drivers Jules Tacheny and René Milhoux race on a 350 cc machine that the engineers had streamlined and lightened. The plan assigned to them was simple: to keep going as long as possible at the highest speed. On that day, FN smashed 41 world records.

Pol Demeuter and Noir died in accidents, while riding at the Germany Grand Prix in Chemnitz, in 1934.

The following year, distressed by the death of two of its riders, and hit by the financial crisis, FN started designing less-expensive and more marketable motorcycles. In the eyes of the company, competition became less important.

The star driver René Milhoux, wearing aerodynamic tails, on the streamlined FN 500, preparing for the world speed record at Bonheiden, 17 April 1934
René Milhoux died in 2003. He notched up over 300 victories on Belgian motorcycles, making him one of the 20th century's greatest motorcycle riders.

Milo Martinet, cover of the brochure for the FN 11 CV car, 1930

Graphic design

Alongside this purely informative advertising – and just as it had been forced by its competitors to participate in the many races and rallies, or even to organise them itself – FN was obliged to feature prominently in the extraordinary graphic movement that would shape the advertising industry between 1920 and 1940. The company succeeded, thanks to the talent of graphic designers employed after the armistice. They replaced the artists whose work had previously ensured the marque's visibility. Milo Martinet, Albert Chavepeyer and Auguste Mambour were among those who successfully brought FN into the European fold of a new aestheticism.

Posters no longer sought to explain or demonstrate. A simplified, formatted and stylised drawing, as was customary at the time, could no longer depict a model of a motorcycle or car, whose movement and speed were better indicated by a few lines. However, the letters F and N were made very visible, in bold, sometimes with a relief effect. In other examples, the company's logo (criss-crossing pedals and rifle) took up to two thirds of the image.

The company's bold statement using only the letters 'FN' was very effective graphically. However, one could legitimately ask why the words 'weapons of war' had disappeared almost completely, except in strictly informative advertisements or in advertising inserts. Most likely the company wanted to stop talking about this. In the pre-war period, civilian production relied on FN's reputation as an arms-maker. Yet between 1920 and 1940, the 'FN' brand acquired a new autonomy, which then supported the production of weapons with its own reputation.

Marcello Nizzoli, advertising poster for FN motorcycles, lithograph, early 1930s

Rearmament

In the 1930s, international tension was at its height. In Europe, the Spanish Civil War, Nazi Germany, Mussolini's Italy and the French Popular Front gave rise to concerns in the Belgian government, which was pursuing a fragile policy of neutrality. The government had to preserve the coalitions that supported it, but also to prepare itself actively and discreetly for a war that was feared and was felt to be inevitable. Internationally, the Sino-Japanese war, which began in 1937, and other conflicts, amplified this sense of insecurity.

The manufacture of weapons of war soared and some commentators noted the "FN's return to its core business". Yet this statement was inaccurate. FN began

manufacturing weapons of war at the end of the first conflict, as soon as the German shareholders had withdrawn and the obligation to refrain from competing with factories on the other side of the Rhine was lifted.[127] "Government business" – whether Belgian or foreign – did indeed represent a significant part of the company's activity between the two wars.

In the mid-1930s, governments all over the world were rearming. FN – which had been able to innovate and adapt to the new conflicts and develop high-quality products – benefited from this international tension, which undermined the hopes for peace of 1919. The company could offer solutions adapted to all governments

eager to equip their armies, as the range of its output was so extensive.

FN's aircraft machine guns were recognised worldwide as reliable weapons: their smooth operation and rapid rate of fire put them well above what the competition could offer. Starting in 1936, FN manufactured, under licence from the Swedish company Bofors Aktiebolaget, a 40 mm calibre automatic anti-aircraft gun with manual control. The Belgian Army received 150 of these guns.[128]

The factory could also produce ammunition in 20 mm, 13.2 mm and .50 calibres, with cartridges for the last two calibres that could be loaded with an explosive bullet.

Meeting of Fabrique Nationale d'Armes de Guerre sales representatives for weapons, 1930
From left to right and from bottom to top: Mr Le Personne (London); Mr Duclos, Cartoucherie Française [French Ammunition Factory] (Paris); Mr John Browning, Mr Gustave Joassart, FN Managing Director; Mr Perrin, Cartoucherie Française (Paris); Mr Édouard Schroeder (Liège); Mr Pommerenke, Deputy Director FN; Mr Lecocq, FN Secretary General; Mr Henri Stassart, FN; Mr A. Fusi (Milan); Mr A. Kind (Hunstig); Mr Herscht (Maastricht); Mr Hubert Stassart, FN; Mr Sjoedall (Göteborg); Mr Caillier, Cartoucherie Française (Paris); Mr Genschow (Berlin); Mr Novotny (Prague); Mr Wattlet, FN; Mr Zultner (Vienna); Mr Dufrasne, FN; Mr Hengelhaupt, Firme Genschow (Vienna); Mr Bancelin, Cartoucherie Française (Paris); Mr Tamagno (Milan); Mr Demey, FN; Mr Laloux, FN; Mr Srebrny, Firme Chasseurs de Varsovie (Varsovie [Warsaw]); Mr Herman, FN; Mr Max Schroecer (Liège).

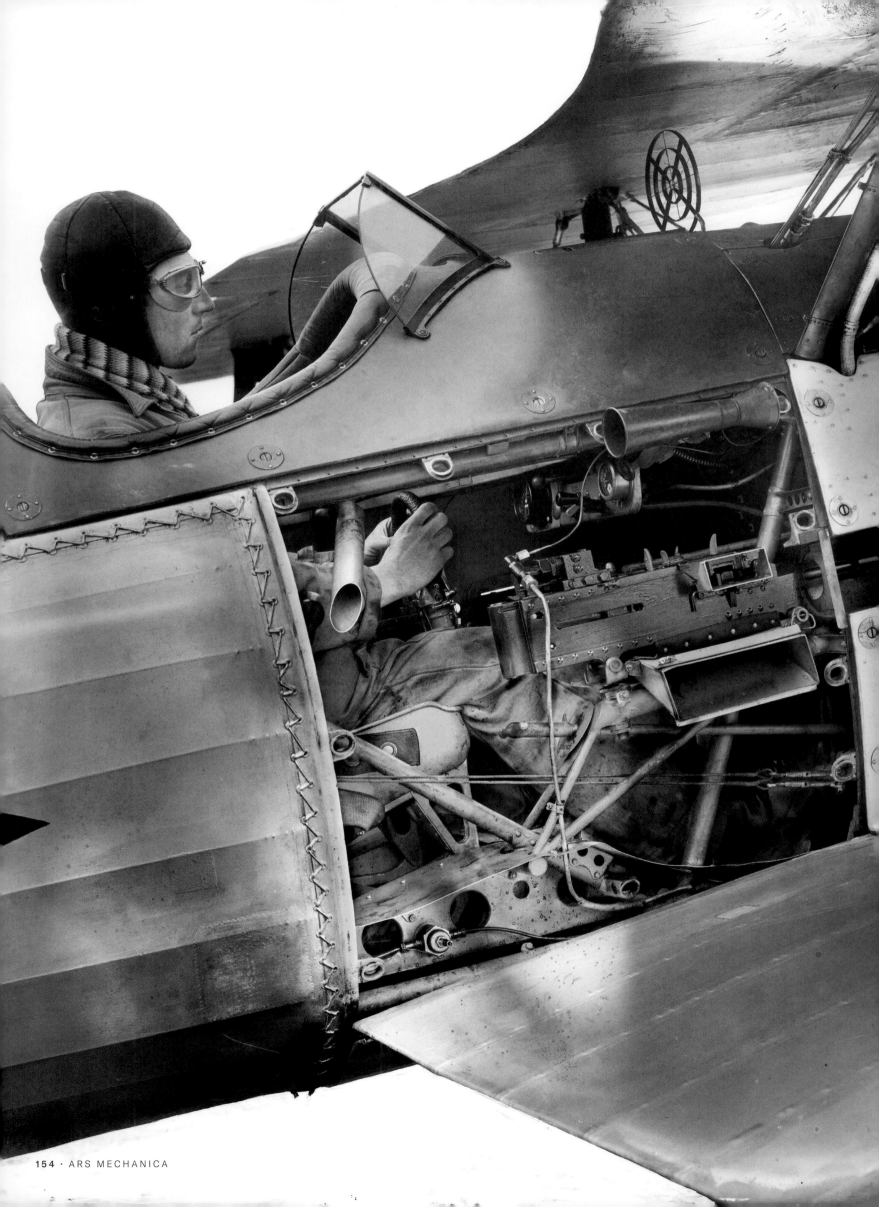

‡ **M86 military FN motorcycle, Argentine type, with a side car and a FN automatic machine rifle, 1937**

◄● **Fairey Fox plane fitted with a FN-Browning Axial Machine Gun, 1934**
Adaptation of FN-Browning machine guns for aircraft from the 1930s led to a specialisation that has today made the Herstal Group a leader in Integrated Weapon Systems.

China – at war with Japan and simultaneously engaged in an internal war – was FN's most important customer in those years. China purchased 164,500 Mauser rifles, 5,000 HP pistols, nearly 6,500 automatic rifles, 357 machine guns, 87 million infantry rounds, 13 million pistol rounds, 1,375,000 rounds of 13.2 mm and .50 calibre ammunition.

In the 1930s, and without counting deliveries to China, FN sold more than 2,000 automatic rifles to the Belgian Army; 3,000 small-calibre machine guns (Belgium, the Netherlands, Greece, Yugoslavia, Finland and Portugal); 4,200 Browning aircraft machine guns (France, Lithuania and Romania); 431 heavy machine guns (Romania, Greece and the Netherlands); 56,500 high-powered pistols (Belgium, Lithuania, Estonia, Peru, etc.); 40,000 Model 1922 pistols (the Netherlands and Yugoslavia); 14,000 Mauser rifles (France, Greece, Lithuania, Venezuela, Peru, Paraguay and Yemen); 204 million infantry rounds of various calibres (Belgium, Belgian Congo, France, Yugoslavia, Greece, Romania, Lithuania, Persia, Yemen, Paraguay, Peru, Venezuela, Uruguay, Bolivia, etc.); 18 million pistol rounds (Lithuania, Peru and Bolivia); three million medium-calibre rounds of 12.7 mm and 13.2 mm (the

Netherlands, Belgium, Romania and Norway) and several hundred thousand cartridge cases of the calibres 40 mm DTCA, 47 mm anti-tank, 75 mm OC, 75 mm DTCA mod. 1927 and 1936, 105 mm L43, 120 mm, 150 mm L43 and L17, 155 mm mod. 1924 and 88 mm.

Although this inventory is incomplete, it gives an idea of the company's great strength on the eve of the war. This strength was global and based on dozens of client countries. It also relied on the high technical quality, perfect quality and reliability of FN's war weapons production. This worldwide reputation was built by exceptional engineers and workers in less than two decades. As a reminder, it was only in 1919 that FN regained its freedom to manufacture and sell weapons of war. In just 20 years, FN had established itself on the international scene.

54ᵉ ANNÉE. — N° 132.

JOURNAL QUOTIDIEN

4 PAGES — 40 Cᵐᵉˢ
Grand-Duché : 45 centimes. - France : 75 centimes.

ABONNEMENTS

Ville ou { pour la province on } Trois mois 29.20
Province { s'abonne à la Poste }

Congo Belge { Un an 170.00
{ Six mois 85.00

Étranger : Prix suivant conventions

Agglomération bruxelloise :
On peut s'abonner par mois : 10 fr.

LE SOIR

RÉDACTION : 21, PLACE DE LOUVAIN, 21
VENTE ET ABONNEM.: PLACE DE LOUVAIN, 23
ANNONCES : AGENCE ROSSEL, R. ROYALE. 122

□ SALLE DE DÉPÊCHES : 124, RUE ROYALE -:- BRUXELLES □

BUREAUX ET FILS SPÉCIAUX : -----
PARIS : 10. RUE DE LA BOURSE. 10
LONDRES : 50. FLEET STREET. E. C. 4

SAMEDI 11 MAI 1940. Edit. ★★

5 ÉDITIONS { ★★★★ à 14 h. 00
{ ★★★ à 17 h. 00
{ ★★ à 18 h. 30
{ ● Edit. de Nuit
{ ● Dernières Edit. de Nuit

Quelle que soit l'heure à laquelle
vous achetiez « LE SOIR », demandez
au vendeur la dernière édition parue.

TÉLÉPHONES { 17.74.80 ou 17.77.50
{ (20 lignes)

Compte Chèques postaux N° 5875

Les manuscrits non insérés ne sont pas rendus

FILS SPÉCIAUX AVEC PARIS ET LONDRES

TELEPHONE ZURICH. GENEVE, BELGRADE, AMSTERDAM, ROME, ETC... — T.S.F. CONGO

LE NOUVEAU CRIME HITLÉRIEN

L'Allemagne attaque la Belgique, la Hollande et le Luxembourg

La brutale agression n'a été précédée d'aucun ultimatum, d'aucune démarche diplomatique

La Belgique fait appel à la France et à l'Angleterre

Les troupes franco-britanniques sont en marche

Vive le Roi !
Vive la Belgique !

Le nouveau crime est consommé.

Renouvelant son odieuse agression de 1914, l'Allemagne attaque la Belgique indépendante et loyale, sous prétexte qu'une invasion alliée était imminente à travers notre territoire.

Nos deux voisins, les Pays-Bas et le Grand-Duché, partagent notre sort.

Avec le souci d'observer jusqu'au scrupule, notre neutralité proclamée, le gouvernement belge a maintenu le pays dans une ligne politique loyale et franche.

Peu importe à l'Allemagne qui ne s'embarrasse pas de telles considérations.

Elle entend utiliser, une fois de plus, la Belgique comme glacis militaire pour ses attaques contre l'Angleterre et la France.

Nous répondons, comme en 1914 : « On ne passe pas ! »

C'est notre vie même, c'est notre honneur qui sont en jeu.

Nous résisterons, jusqu'au bout.

Notre appel aux Alliés a été entendu. Leurs troupes accourent à notre aide.

Tous les Belges se serrent autour du Roi pour la défense de la Patrie.

Vive le Roi !
Vive la Belgique !

LE ROI LEOPOLD.

Le crime allemand

« L'agression qui vient d'être accomplie est peut-être plus odieuse encore que celle de 1914 »
« La Belgique est résolue à se défendre. »

Déclaration de M. Spaak à l'Ambassadeur d'Allemagne

M. Spaak, ministre des Affaires étrangères et du Commerce extérieur, a reçu, vendredi matin, à 8 h. 30, la visite de M. von Bülow-Schwante, ambassadeur d'Allemagne.

M. Spaak a immédiatement donné à l'ambassadeur lecture de la déclaration suivante :

Monsieur l'ambassadeur,

L'armée allemande vient d'assaillir notre pays. C'est la deuxième fois en vingt-cinq ans que l'Allemagne commet contre la Belgique neutre et loyale une criminelle agression.

Celle qui vient d'être accomplie est peut-être plus odieuse encore que celle de 1914.

Aucun ultimatum, aucune note, aucune protestation n'ont jamais été présentés au gouvernement belge.

C'est par l'attaque même que celui-ci a appris que l'Allemagne violait les engagements pris par elle le 13 octobre 1937 et, spontanément, renouvelés au début de la guerre.

L'agression de l'Allemagne, dépourvue de toute justification, heurtera violemment la conscience universelle. Le Reich en portera la responsabilité devant l'Histoire.

La Belgique est résolue à se défendre.

Sa cause, qui se confond avec celle du Droit, ne peut être vaincue.

L'ambassadeur d'Allemagne a alors remis à M. Spaak une note ainsi conçue :

Le gouvernement du Reich m'a chargé de vous déclarer ce qui suit :

Afin de devancer l'invasion préparée de la Belgique, de la Hollande et du Luxembourg, par l'Angleterre et la France, dirigée nettement contre l'Allemagne, le gouvernement du Reich se voit obligé d'assurer la neutralité des trois pays précités, par les armes.

A cet effet, le gouvernement du Reich mettra en ligne une force armée de la plus grande importance, de sorte que toute résistance sera inutile.

Le gouvernement du Reich garantit le territoire européen et colonial de la Belgique, ainsi que sa dynastie, pour autant qu'aucune résistance ne soit opposée.

Dans le cas contraire, la Belgique risquerait la destruction de son pays et la perte de son indépendance.

Il est donc dans l'intérêt même de la Belgique d'adresser un appel à sa population, en vue de faire cesser toute résistance et de donner les instructions nécessaires aux autorités, pour que celles-ci prennent contact avec le commandement militaire allemand.

M. Spaak a immédiatement fait observer à l'ambassadeur que la déclaration dont il venait de lui donner connaissance constituait déjà, par avance, une réponse négative à ses demandes formulées.

M. Spaak a ensuite fait savoir à l'ambassadeur que Bruxelles est une ville ouverte, qu'aucune troupe ne s'y trouve, et qu'aucune troupe ne la traversera.

L'AGRESSION

Des aérodromes ont été bombardés

Alertés déjà jeudi soir, les ministres ont siégé en permanence toute la nuit pour suivre le développement des événements.

Vendredi à 6 h. 15, M. Pierlot, Premier ministre, annonce aux journalistes que la Belgique est attaquée et qu'elle fait appel à ses garantes, la France et l'Angleterre.

Immédiatement après, MM. Pierlot, Spaak et le général Denis se sont rendus chez le Roi.

Puis un Conseil de cabinet eut lieu sous la présidence de M. Pierlot. A l'issue de la réunion, les ministres annoncent que le pays est attaqué sur toute sa frontière de l'Est ; que la mobilisation générale et l'état de siège sont décrétés ; que les Chambres se réuniront vendredi ; qu'un appel a été adressé à nos garants.

* * *

A 7 h. 15, M. Cudahy, ambassadeur des États-Unis, est reçu par le ministre des Affaires étrangères.

A 7 h. 45, tous les ministres ont rega-

gné leurs départements respectifs pour vaquer aux devoirs de leur charge.

* * *

Sortant du département des Affaires étrangères, M. Pierlot nous dit en substance :

« Les Allemands nous ont attaqués sans avertissement. Ils ne nous ont envoyé aucun ultimatum ni aucune démarche diplomatique. »

M. Delfosse, ministre des Communications, ajoute, un peu après ces précisions :

« L'agression allemande a débuté à 4 h. 30. Les aérodromes ont été bombardés. Des parachutistes sont descendus. L'armée franco-britannique a fait appel aux Alliés qui ont répondu. La mobilisation générale a été décrétée. L'état de siège est proclamé. »

Attention aux parachutistes et aux saboteurs

Le ministère de la Défense nationale communique un avis à la population :

Il est possible que l'ennemi fasse atterrir à l'intérieur du pays des parachutistes isolés ou en groupe, chargés de saboter les moyens de communication et les installations vitales pour l'armée et le pays.

Il est du devoir de chacun d'aider les autorités à découvrir et à capturer ces parachutistes sur notre sol.

Toute personne a pour devoir :

1. — D'informer immédiatement les autorités militaires, la gendarmerie ou la police locale de toute descente de parachutistes;

2. — De signaler aux autorités

Les attaques aériennes

Des avions allemands ont pénétré cette nuit au-dessus du territoire belge et ont bombardé l'aérodrome d'Evere.

Des obus sont tombés en plusieurs endroits de l'agglomération.

Des incendies ont éclaté en divers endroits des casernes d'Etterbeek et dans des maisons des environs.

Une bombe est tombée également place Madou, occasionnant un incendie et des dégâts dans les environs.

Depuis 4 h. 30, la région anversoise est survolée de nombreux avions. La D. T. C. A. tire vigoureusement.

* * *

militaires, la gendarmerie ou la police locale les personnes suspectes qui circulent aux abords des ouvrages d'art ou des installations importantes.

Toute hésitation ou tout retard à renseigner les autorités peut avoir des conséquences graves pour l'armée ou pour la population.

Attention aux parachutistes et aux saboteurs. Prévenez les autorités.

Signaux d'alerte

L'alerte est donnée par les sirènes au moyen d'un signal modulé d'une durée de deux minutes. La fin d'alerte est donnée par un signal continu de deux minutes.

Beaucoup de passants croient pouvoir quitter leur refuge dès que le signal modulé cesse. Il faut attendre le signal continu de fin d'alerte.

Bruxelles, ville ouverte

Au cours de l'entretien qu'il a eu avec M. von Bülow-Schwante, ambassadeur d'Allemagne, M. Spaak, ministre des Affaires étrangères, a fait une déclaration au nom du gouvernement disant que Bruxelles est une ville ouverte et qu'il n'y avait pas de troupe et que dans ce sens il a protesté énergiquement contre le bombardement, dont la capitale avait été l'objet cette nuit.

Le Gouvernement décrète la mobilisation générale et l'état de siège

Ainsi que nous le disons plus haut, la mobilisation générale et l'état de siège ont été décrétés.

Cette proclamation comporte le transfert aux autorités militaires des pouvoirs de police dévolus aux autorités civiles et notamment aux gouverneurs de provinces.

* * *

vince et aux bourgmestres. Le Roi parlera incessamment au micro.

Toute la zone neutre est gardée militairement. Quiconque veut passer doit montrer une autorisation.

Les légations sont sous la garde de piquets.

5 Hostilities resume

Leave or stay?

It took 20 years to go from the victory of 1918 – and the hopes of universal peace that it raised – to an open conflict, a moment when history really stuttered. For FN, these two decades saw the withdrawal of its German shareholders, the gradual acquisition of its industrial and commercial freedom, plus its expansion, particularly in the field of weapons of war. Belgium's geographical location, between two major players in the coming war – the Nazi Reich and France – made its involvement in the conflict inevitable and highly predictable.

The geographical location of FN, little more than a stone's throw from Germany, was a cause for concern. With production speeding up, the idea of retreating to other regions of the country was considered. FN had slightly less than 6,000 workers in January 1939, a little more than 7,000 in September, nearly 8,000 in November, and 9,000 in December...

In May 1939, before the outbreak of war, the Board of Directors reported on formal and informal negotiations with France. The official talks covered preparing for a withdrawal of FN to France once the war had broken out and for the duration of the war. The unofficial

negotiations concerned an immediate withdrawal of the company, without waiting for the eventuality of a conflict, for an unlimited period. Although FN was prepared to withdraw if a conflict broke out, it was opposed to any immediate move to France.

A note addressed on 1 May 1939 to Edouard Daladier, President of the French Council, indicated that it would be advisable to build in the provinces, as soon as possible, a production unit for the Browning 7.65 mm machine gun, capable of producing 500 units per month.[129] This factory should also have the necessary space to accommodate the Herstal factory in case of an invasion. Consequently, it seemed a matter of urgency to order the machine tools necessary for FN, knowing that the factory had already reserved 20,000 working hours per month for this purpose. There were plans to use 150 railway carriages to transport these machine tools, and 450 others for the withdrawal of FN to France. Although the rapid invasion of the Liège region prevented the French from following through on their plans, Belgian workers were already recorded in the French arms factories as early as 19 May 1940.

Edition of the newspaper *Le Soir*, dated 11 May 1940

Belgians exiled in England, 1943
This photograph shows Gustave Joassart, seated on the right, and René Laloux, standing on the far right-hand side. Several Fabrique Nationale d'Armes de Guerre managers moved to England to contribute to the Allies' war effort. They were incorporated into the Ministry of Supply, Armament Design Department, Small Arms and Machine Guns section, Cheshunt, near Enfield.

Other measures or additional decisions were envisaged by FN: the transfer of equipment to "refuge stores" in Bruges and Anzegem, near Kortrijk, or even in the vicinity of Paris. On 12 May 1940, the German Army entered Liège. Belgium surrendered on 28 May and France on 18 June. No sooner had the move begun than it became pointless. Only the ammunition factory in Bruges – led by General Quintin – managed to evacuate its machines and equipment to an arsenal in Toulouse, thanks to an exceptional logistical effort (four entire trains were chartered). Yet it was all in vain, since the armistice was signed the day after production could have begun in south-western France.

Until the end of the 'phoney war' on 10 May 1940, FN continued its activities by managing, as responsible manufacturers, a highly unusual situation. The Board of Directors met regularly; weighed up the moral interest of the company in making it known that it was prepared to limit its margins; anticipated the likely positive results of the financial year; noted that the law on extraordinary profits would be applied for the first time to FN at the end of the financial year; and decided that a dividend of 60 Belgian francs would be paid to the shareholders when peace was restored.[130]

A very odd typewritten document totally contradicts this apparent indifference. Entitled *Secret Annex to the Minutes of the Board Meeting of 19 March 1940*, it was meant to be "filed separately" and a handwritten note specified: "To be attached lightly without looking at it". It would appear to be FN's implementation of the decree-law of 2 February 1940 on the administration, in wartime, of commercial companies:

A. For circulation to the Directors and Members of the Management Staff in the event of an invasion.

The Board notes that at least three Directors (Messrs. Galopin, Laloux, Nagelmackers) consider that [...] must remain in the Country in the event of invasion. As such, these directors will continue to exercise, as best they can, their general mandate to defend and safeguard the company's assets [...].

On the other hand, as the general interest of the Country and of the Company requires that in case of invasion the Belgian expatriate Industry be represented abroad as widely as possible and that it should be as active as possible, the Board unanimously decides that [...] the Managing Director, Mr G. Joassart, as well as the Assistant Director, Mr R. Laloux, and as many of their technical staff as possible should, before the invasion, go abroad to work freely within the framework of the statutory corporate activity, it being understood:

- *that their possible "business" must be in the form of a "special company" of the local nationality and to which they will contribute the necessary financial means [...] from FN's own financial means [...];*
- *that they shall report on their management to the Board of Directors as soon as possible;*
- *that they must respect the statutory provisions which [...] require two signatures.*

Item B concerns the financial means of the company and in particular the bank assets: only 10 to 12 million Belgian francs are kept in Liège and two million in Bruges, partly in banks and partly in private safes. Twenty-five million are kept in Brussels to pay the suppliers, the rest being placed in foreign banks.

Finally, item C covers the payment of sums owed by external debtors, which must never be made at the registered office with provision of a receipt (there was justifiable wariness of any occupant).

Hence, everything was foreseen to ensure the survival of FN in the event of an invasion of Belgium. However, the German Army's campaign lasted only 18 days, which meant that this detailed plan simply could not be carried out.

A handwritten *addendum* dated 4 September 1940, signed by all the directors, declared that the "Board unanimously decides to consider as null and void, as of this day, the decisions previously taken in application of the decree-law of 2.2.40".

FN under foreign control

On 4 September 1940, FN came under the control of the Kommissarische Verwaltung. The managers of FN were ordered to restart the factory. The Chairman of the Board of Directors, Alexandre Galopin, the Vice-Chairman and the Director refused. Weapons and ammunition were therefore seized, and a decision was made to place the company under the supervision of the DWM – the Deutsche Waffen- und Munitionsfabriken – which had run FN since the reorganisation of the company in 1896 up until the end of the First World War. This "leap into the past" was remarkable. Franz Scharpinet, director of the DWM, held all the powers at that time: those of the management, the Board of Directors and the general assembly of shareholders. It was with him that Gustave Joassart negotiated the possibility of keeping a section within the company made up of a few employees responsible for looking after FN's interests within the DWM group. He also succeeded in maintaining two independent offices outside the company, but in Liège. One was in charge of the research for hunting guns and motor vehicles, for the day when peace would return. The other handled the administrative liquidation of the files opened before the war.

Gustave Joassart continued the struggle in Great Britain, where he arrived in the summer of 1942, and became Under Secretary of State in the Pierlot government in exile (1940-1944). René Laloux and Dieudonné Saive, as well as other FN managers and engineers, joined the UK Ministry of Munitions.

The ordinary general assembly of 26 October 1944 wanted to forget the war: it looked, even if this was already in the past, at the review of the 1939-1940 financial year.

As you will no doubt recall, this was a year of intense activity. Our cooperation in military production for Belgium and other countries that later became our allies was most important. The results were therefore positive and, in normal times, you would have been invited to vote on the payment of a dividend on the shares. Unfortunately, however, what would have seemed normal when our country had not yet experienced the horrors of the invasion and the destruction of the war cannot be envisaged at the present time.

Then came 10 May 1940: the war spread to our country and the total invasion of its territory. Our duty was clear: it consisted in refusing any work that could benefit the enemy armies. We are justifiably proud to say that we fulfilled this duty without hesitation and without reservation. This was the implementation of a unanimous decision taken long ago by the Council. [...]

Our attitude in refusing to work, asserted from the very first days, was firmly maintained in spite of the pressures and threats to which we have been constantly subjected [...]. Soon the requisitions were extended indiscriminately to all of our production, while a takeover of our factories was taking place that would ultimately result in our company being placed under 'Kommissarische Verwaltung'.

FN in a German military-industrial complex

Unlike in 1914, when the Germans only asked FN to produce for their war effort (in vain, as it turned out), in 1940 they wanted to include the company in an economic context that they believed would return after the conflict. Having no doubt about their victory and its continuity for

decades to come, they included FN in the German military-industrial complex; Herstal had to concentrate on partial tasks, for obvious security reasons, such as the manufacture of parts for land and air vehicles. Between 1940 and 1944, more than 300,000 9 mm HP pistols were produced under the name 'Pistole 640b' – the 'b' indicating its Belgian manufacture – as well as 360,000 10/22 pistols in 7.65 and 9 mm short calibres. Weapons leaving the factory were marked with the Nazi eagle seal. The occupied factory also produced parts for the Mauser 98K rifle and the Walther P38, as well as a dozen of the Mark 108 gun parts for arming the first jet plane, the Messerschmitt 262 (1944). The ammunition factory produced millions of parts for Sweden, which remained neutral. This was a form of barter to ensure Germany a supply of iron ore.

The objective was to use the skills of FN within a group with complementary productions. In addition to Herstal, this included the Karlsruhe factory in Germany and the Poznań arsenal in occupied Poland, under the authority of the DWM.

A difficult recovery

In the summer of 1942, Franz Scharpinet was replaced by Dr Hol , an independent official of DWM. He reneged on all the agreements concluded in 1940, resulting in closure of the Liège offices and the management being placed under house arrest.

Thus, the restart of the factory was arduous, and could only be done with the collaboration of managers sympathetic to Nazi Germany, under strictly German management. The Service du Travail Obligatoire [Compulsory Work Service][131] provided a large part of the 12,000 workers then employed by FN. The factory had never had so many workers. The machine-tools were operated beyond their capacity and the workers, who had no particular qualifications or enthusiasm, were unable or unwilling to ensure the quality of production. There was some sabotage, albeit on a limited scale. An examination of the parts – abandoned by the Germans after they were defeated – showed their poor quality: excessive hardening made the parts brittle.

6 Start of the post-war period

A time for tributes

The Germans began to evacuate Liège and the factory in July 1944. They tried to take the most valuable machines and tools with them, but the Allies' rapid advance put a halt to the looting. American troops crossed into Belgium on 1 September 1944. That same day, the Germans handed over their control of FN and left the city. On 7 September, the Belgian management team regained control of the factory.

However, the worst was yet to come. In October 1944, while people calmly talked about the reconstruction of the factory, little did they know that, one month later, and until January 1945, the factory would be the target of retaliatory fire from German V1 and V2 'flying bombs'. Six people were killed, people were wounded and the resulting damage was very extensive.[132]

The Board of Directors met on 7 October 1944 at the home of Georges Laloux, a clear sign that the Herstal headquarters were in no fit state to receive its directors. The meeting began with a farewell eulogy for Alexandre Galopin, Chairman of the Board of Directors since 1932. He had been murdered in February 1944 by Belgian Nazi sympathisers. The meeting emphasised that Galopin had joined the FN in 1904 and he had risen through the ranks thanks to his technical and scientific skills as well as his social skills. They promised to keep his memory alive and so they did.

After this gloomy introduction, many compliments were exchanged. The purpose was to show that both those who had stayed in Belgium during the occupation and those who had left had all worked for FN to the best of their ability. The former's dedication made it possible to defend the company's social interests. Gustave Joassart, who had returned from Great Britain, René Laloux, Jean Vogels, Édouard Dufrasne and Dieudonné Saive, through their work at the Ministry of Supply, Small Arms Design Department, "certainly increased the standing of the Company with the United Nations Military Authorities".[133]

The factory in ruins, 1945
Between November 1944 and January 1945, German V1 and V2 flying bombs ripped out the heart of Herstal's factory, destroying the general management's buildings on the site.

Marching ahead despite the difficulties

[The] *resumption of activity was inevitably slow, because it depended on reorganisation of the outside world and Belgium's integration into this new world. Long delays appeared to be inevitable for this reorganisation of the world economy.*[134]

In October 1944, a broad outline of what FN would be like in a 'new world' was sketched out. In 1919, reliance had been placed on carefully reconstructing the means of production and on the unchanging nature of economic models. Yet a new world was now emerging. Rather than lamenting the disarray of foreign markets, as in the 1920s, there was talk of building 'customs unions' with neighbouring countries. These would be the embryo of what the ECSC (European Coal and Steel Community)[135] achieved seven years later. Instead of expecting substantial "war reparations", they anticipated "a fair compensation but characterised by a spirit of solidarity". Cash flow problems were expected for "a long period". However the wishes of the Labour Council, which called for a general increase in salaries, including, it was pointed out, the salary of the Managing Director, were readily accepted. Over and above the generous principles of the time, FN immediately adopted a pragmatic approach to redeveloping its activities.

The Board of Directors acknowledged the damage suffered. It noted that some workshops would have to be completely demolished and then rebuilt once more. There were questions as to whether they should be rebuilt identically or almost so, or differently, based on a general modernisation plan like this: homogeneous blocks, wider aisles, buildings with a ground floor, but partially with an upper floor for offices, cloakroom, canteen, etc. More light, more attractive appearance, this plan could be implemented in stages.

The Board, believing that FN has a duty to itself, and that marching "ahead" as it has often done is even a condition of its future success, expressed a strong preference for the principle of an overall plan for renovating the factories over a period of many years. The decision itself and the scope of this decision will depend on the solution to the problem of war damage, the long-term economic outlook, and the amount of money to be spent on renovating the stock of machine tools, etc.[136]

The Board was strongly calling for modernisation of the factory. It also forcefully highlighted the values of solidarity and expressing confidence in the future. Yet these declarations could not and did not seek to hide the fact that FN was facing major difficulties from 1945 onwards. Early in 1945, although 60% of the administrative and technical staff had returned to FN, the same could not be said for certain key positions. The number of engineers, in particular, was reduced by two-thirds compared to 1940. There were only 240 skilled workers, down from 700 before the war. However the Board noted that "there was no work for more". There were only 2,100 workers in the factory, all skills combined.

The Galopin Doctrine

Alexandre Galopin, Managing Director of FN and later Governor of the Société Générale de Belgique, developed an economic-political doctrine of 'give and take' from the start of the German occupation. This doctrine was based on the recognition that Belgium could not feed itself without imports and that these imports could only come from Germany or from occupied countries. Belgium therefore had to maintain its industrial production at a reasonable level so that its exports to Germany could provide it with sufficient resources to pay for its vital imports. Maintaining activity also helped to avoid Belgian workers being deported to Germany – as had happened during the First World War – and ensured that Belgium's industrial infrastructure would be preserved for the post-war period. This economic doctrine made no moral claims, but it was in line with the enemy's objectives. Compared to 1914-1918, looting was no longer on the agenda and Germany preferred to exploit the occupied countries' industries according to its needs.[137]

The factory in ruins, 1944
After the bombardment of 24 December 1944, it became clear that a huge amount of reconstruction work would be needed.

The reasons for the "exodus" of managers were familiar to FN which, in 1944, declared: "the difficulties encountered in the 'recovery' of engineers and other senior staff who had found, during the war, a job in other factories which, recognising their services, gave them a much better deal than they had at FN".[138]

The Board noted that 1,738 machine tools had been removed, 909 had been recovered or were recoverable in Belgium and 829 were irretrievably lost, including 822 in Germany.[139]

An advertisement published repeatedly in local newspapers revealed some frustration at the lack of raw materials and tools in the factory at the time:

We are looking for sales or rental offers for lathes, milling machines of various types, shaping vices, flat surface and circular grinding machines, with a view to equipping tool workshops. Fabrique Nationale d'Armes de Guerre, Herstal.[140]

In the post-war years, FN was continually searching for financial support. Until January 1950 – when it was granted substantial compensation for war damages[141] – the sometimes very difficult search for the capital so essential to its development took up a great deal of the management's energy. The problem was not so much the expenses associated with rebuilding the factory, but rather the absolute necessity to ensure sufficient working capital to start production while awaiting payments that were sometimes overdue. All avenues were explored and abandoned, ranging from the "vague and uncertain" possibility of an American loan to the issue of bonds at a time that was deemed inappropriate. In the end, it was the Société Nationale de Crédit à l'Industrie that allowed the company to "stay afloat", through treasury advances on anticipated "war damages". In successive instalments, 320 million Belgian francs were lent to FN, a sum that was increased by loans from other banking institutions (50 million from the Swiss Banking Corporation and 50 million from the Caisse Générale d'Épargne et de Retraite).

The three big problems (Sequestration Office – War Damage Compensation – Settlement of accounts with the Belgian Government) are all three practically at a standstill. It's a real deadlock.[142]

With these words, on 6 June 1945, the Managing Director introduced a long chapter dealing with the difficulties that the administrations raised on the complex path of reconstruction. Sometimes he did so in an amusing yet exasperated manner, such as when he evoked the Office des Séquestres, responsible for administering the property left behind by DWM at the time of the Liberation: "We provided volumes of declarations [...] We frequently insisted that the Office 'remove from our homes' the enemy's property that was weighing us down. But nothing stirs the Office anymore. For months now, it has not reacted, does not even acknowledge receipt, and does not give us any acknowledgement of our reports." Sometimes, he was more serious: "The problem of war damages is of great concern to us, since it is essential for us to return to a healthy financial position and it governs all our reconstruction possibilities. Independently of the evaluations and expert assessments (object by object, machine by machine, machine remaining in Herstal but damaged or extremely worn, machine disappeared or taken to Germany, tools disappeared, buildings transformed, etc.) we have increased our efforts to hasten the enactment of legislation on the subject." The Belgian State still owed 39 million Belgian francs to the factory for cancelled pre-war orders and for the damages caused by forced exodus. Securing payment was no small matter, as the files had only just been opened by the administration.

Bright prospects

The most surprising thing about this period of renewed peace was certainly the calmness tinged with optimism driving the country's economic leaders. We must rebuild, forge new links between nations and guard against future confrontations: it's a huge task, but we're working hard on it.

In 1945, the company noted that its monthly losses amounted to five or six million Belgian francs, although the medium-term industrial prospects looked good. Government orders were expected from Belgium, France, China and Great Britain, and the company was already planning to become one of the links in the United Nations arms industry via the latter country. For civilian firearms, motorcycles and trucks, there was also moderate optimism.

Production of 5,000 jerry cans for the US Army, 1945
The factory initially resumed operations with simple products, pending the overhaul of all the machine tools, many of which had been scattered or destroyed by the occupying forces.

It was hoped that the State would grant war damages, which FN estimated at the time to be around 250 million Belgian francs. To this could be added 26 million Belgian francs in compensation for damage caused by the forced flight of staff in 1940.

The launch of the Bruges ammunition factory, the subject of much attention and investment in the 1930s, was encouraged in 1928 by the Belgian government. In exchange for setting up the factory, FN was guaranteed a preferential right to government orders.

The strategic aspect of this choice – to move production away from the German border – did not have the hoped-for effect due to the enemy's rapid advance. The factory became the target of Allied air raids and was largely destroyed. Once the war was over, FN postponed its reconstruction, believing that the Bruges plant had lost all military relevance. The Minister of National Defence very quickly agreed to this decision.

This decision not to rebuild was symbolic of the new and pragmatic spirit at the time. It contrasted with the years following the First World War, when there was certainty that the world must be rebuilt identically in order to get back on track.

THE URGENT NEED TO PRODUCE: ORDNANCE INDUSTRIAL SERVICE DIVISION

The Ordnance Industrial Service Division was a US Army corps specifically responsible for providing logistical support to combat forces. This division provided weapons and ammunition to active units and was responsible for their maintenance.

Through its intermediary, FN was awarded a contract in 1945 to repair and recondition 50,000 machine guns and one million rifles. This order was later almost doubled. Thanks to these orders, the company was able to maintain a moderate level of activity in the first months after the Liberation, a period marked by the difficulty of obtaining raw materials and the scarcity of labour. In addition to these orders, 5,000 jerry cans were supplied. This modest production, requiring no specific know-how, underlined a spirit of resilience.

Cleaning, repair and reconditioning of a set of Browning machine guns of the US Army, 1945

In the midst of the work

The 6 June 1945 reconstruction plan

Under a general outline including some enhancements and improvements (widening of alleys, etc.), the Board approved:

- the repair of the Administration buildings and the building on Rue Voie de Liège (2.5 million Belgian francs)
- the reconstruction of the building (completely destroyed) located in front of the main offices (8 million Belgian francs)
- the transformation into a workshop (materials to be supplied from Bruges) of the completely destroyed hangars of the Pré-Madame (2.2 million Belgian francs)
- the reconstruction of a test bench at the Pré-Madame (the existing one at the old factory having been destroyed by the occupant) which necessitates the completion of the unfinished part of the Pré-Madame (2 million Belgian francs)
- the various current and daily building works which are carried out by our general service department at the rate of approximately 4 million Belgian francs per month.

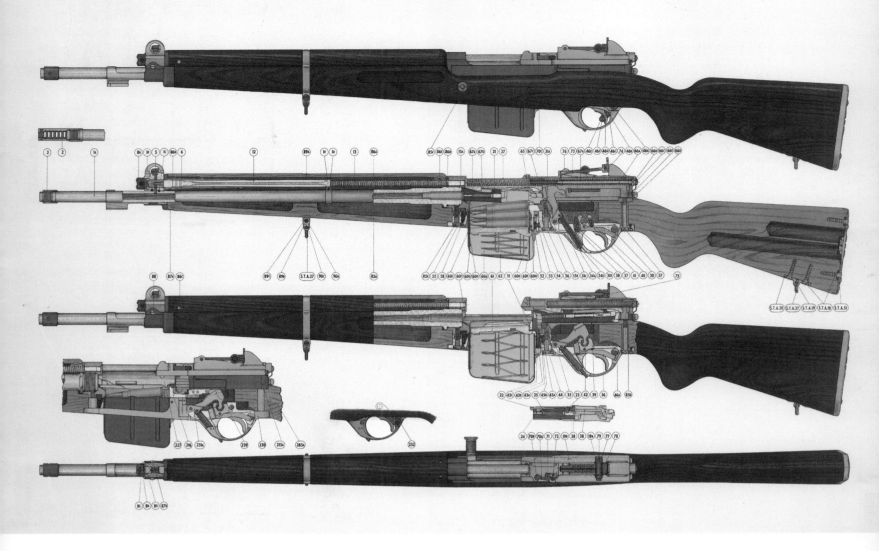

BROWNING, A PRIORITY

The first mission undertaken by Joassart, Laloux and Vogels in 1944, when peace had just returned, was to spend three months in the United States, visiting factories and renewing contacts – "commercially very useful", as they put it – with the J.M and M.S Browning Company.[143] Val Browning – the son of John Moses – visited Herstal several times and the number of sales between Belgium and the United States increased until it was higher than it was before the war.

A GREAT EFFORT, QUITE OBVIOUSLY

On 6 June 1945, the Board of Directors considered the future of the company, as it was taking shape amid the reconstruction work and occasional orders from Allied armies. The directors needed to estimate the date when the company would be able to return to its "core business". The forecasts made that day would require "a great effort, quite obviously" but also "a rather favourable combination of circumstances".

For the armoury, an estimated 80,000 hours would be needed to rebuild the specialised tools required to manufacture the weapons that were to be put back on the market. This would represent about eight months of work, at the pace the FN was operating in the summer of 1945. Rebuilding the weapons workshops would

enable the successive production of two types of pistols, the semi-automatic 12-gauge shotgun and semi-automatic rifle; then, around April 1946, the 16-gauge semi-automatic shotgun, the Mauser, the Baby pistol, the Trombone rifle, and finally, at the end of 1946-1947, the superposed and the new Anson Hammerless. The commercial prospects looked promising and the company's agents were "very keen and eager to work: there will be demand".

For vehicle production at the Pré-Madame plant, efforts were focused on the rapid marketing – around June 1946 – of a commercial Tricar with a payload of 1,000 kilos. It was derived from the military Tricar that the FN had been producing since the late 1930s. A motorcycle would then be produced, followed by a five-tonne truck. The directors knew that the sale of these products would be difficult, as the company's network of sales agents was "fragmented". Moreover, the range of products was too limited to supply and seriously interest customers.

With the return of peace, the trade in weapons of war became more difficult. In Great Britain, the negotiations that had begun were stalling. The likely disappearance of the Belgian Ministry of Armaments further slowed the talks. In China, negotiations were continuing, without much hope: the Chinese were waiting for a decision

by the international community that would allow them to obtain supplies of Mauser rifles requisitioned from Germany – the "spoils of war" – free of charge, rather than ordering them from Herstal. Furthermore, no forecasts could be made for wartime ammunition, as only Browning ammunition had been put back into production for the time being.

FRESH FIELDS

In 1946 and 1947, FN began manufacturing agricultural equipment and created a department responsible for these products in the Pré-Madame factory. This was a time of reconstruction, and above all of rebuilding and expanding the means of agricultural production, which were vital to the country's recovery.

A milking machine epitomised how the company operated. After assessing the agricultural sector's needs for machines in order to increase productivity – remember, this was 1947 – FN set about designing a milking machine. It would take into account all the technical and hygienic needs of this kind of work, such as ease of use, operational safety and easy cleaning. FN came up with a stainless steel receptacle, in two complementary shells where the joint had been totally removed by polishing and ideally designed to leave no residue.

This machine was so perfect that it won the most prestigious design award, the Signe d'Or, in 1957. It has since entered the collections of several museums, including the Design Museum in Ghent.[144]

Also discussed was the possible manufacture of Wilmet tools, the assembly of an English tractor with a motorcycle engine, and the shaping of refrigerator shelves by using body presses.

To take stock of what FN was like just after the post-war period and until 1950, one can only draw an impressionistic picture. This picture was made up of small brushstrokes of hope and disappointment, applied against a reasonably optimistic background. Nothing was certain. The economic context was changing and the geopolitical situation was developing around the Cold War. As ever, the company's objective was to maintain its industrial base until the markets it served were more stable.

As happened after the First World War, FN's survival strategy involved subcontracting or manufacturing simpler products – as these required only a small part of the company's know-how. The main difficulty facing the company was undoubtedly the slowness of the administrative and judicial system. Overcoming these difficulties would take five years.[145]

Gloster Meteor aircraft, 1951
In 1948, Belgium and the Netherlands purchased British Gloster Meteor jet aircraft. This technological leap forward was achieved thanks to Fabrique Nationale d'Armes de Guerre, which was tasked with producing the Derwent V turbojet engines, under licence from Rolls-Royce Aero Engines.

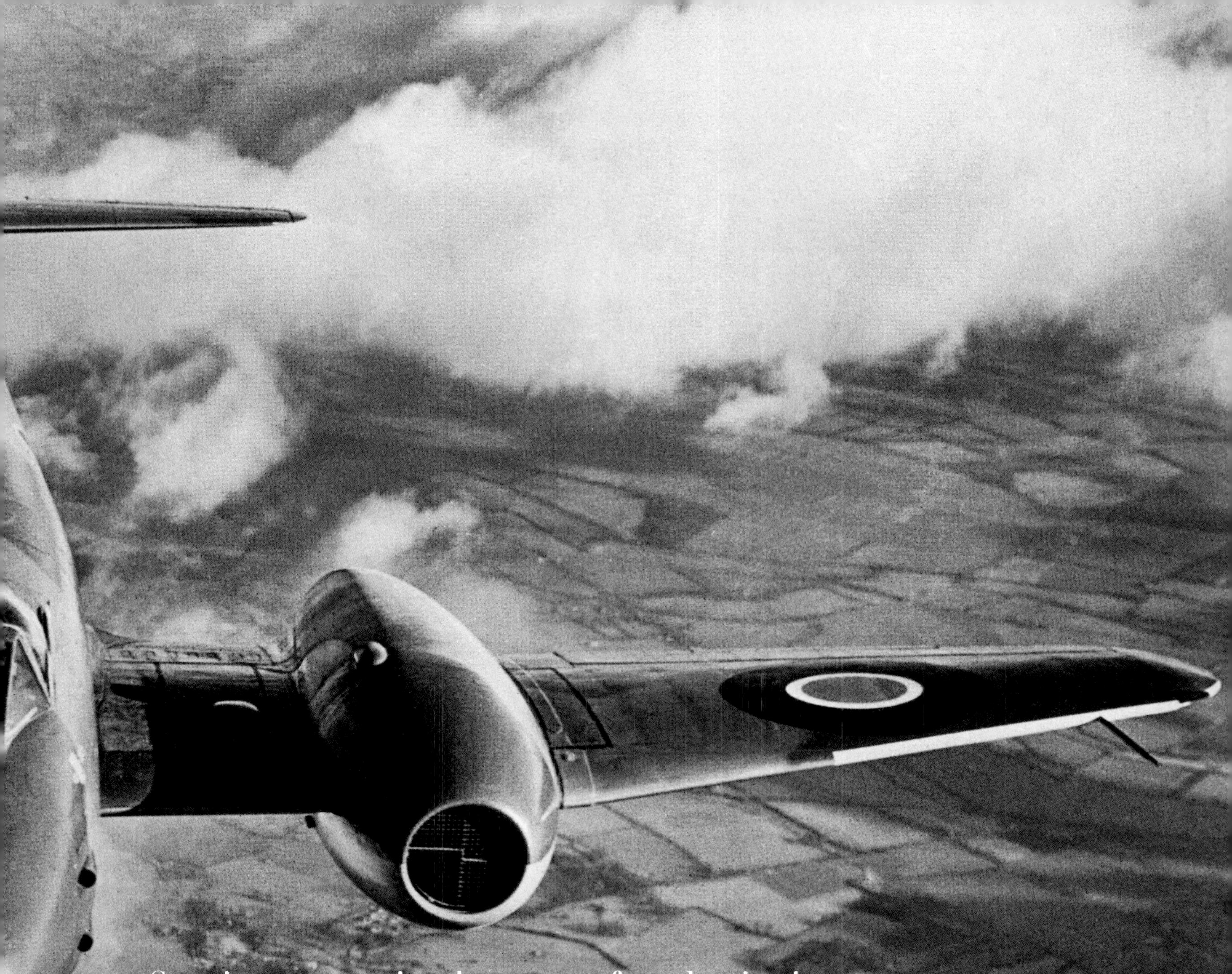

Starting over again: the power of mechanisation

From 1945 onwards, FN made it clear that its future "would inevitably lie within the field of mechanised products that are its strength".[146]

Mechanisation was indeed the path to a return of optimism at Herstal. The front page headline of *La Meuse* on 12 November 1949 ran: "FN is going to build jet engines" and continued: "The Rolls Royce Company has announced, with the British government's approval, that it has granted licences to the Belgian government for the manufacture of 'Derwent' jet engines and for these engines to be constructed by FN in Belgium. Belgian engineers have done traineeships at Rolls Royce factories in Derby [...]. At first, the engines will be constructed using individual parts supplied by the Derby factories, but over time they will be fully constructed by FN."

On 10 June 1948, manufacture of these engines was mentioned to the Board for the first time, though in very guarded terms. In the chapter on 'Ongoing business', after reference was made to FN's hopes in relation to motorcycles, trucks, firearms and cartridges, it was simply noted that "the Belgian Air Force's purchasing services were insistent that we start the manufacture, based on 200 per year, from a total of 1,000 Derwent Rolls-Royce jet plane engines".

In September that year, the Managing Director noted the proposal that had been made to FN by Belgian and Dutch Air Force authorities to manufacture 760 Rolls-Royce Derwent V engines over a period from 1950 to 1952 and for a sum of almost one billion Belgian francs. It was immediately made clear that it would be impossible to offer the Belgian government prices that were as low as those of the British manufacturer. That was because the production levels were lower, the investment was considerable and the training period was fairly long.

On the same day, signalling a new optimistic outlook, a decision was made to call an extraordinary General Assembly to extend FN's corporate term by 30 years and to triple the company's capital (from 105 million to 315 million Belgian francs) without issuing any new shares.

In 1954, Pré-Madame facilities, which were initially used for producing cars and motorcycles, had successfully made the shift to aeronautical technology and celebrated production of the 1,000th Derwent V turbojet engine.

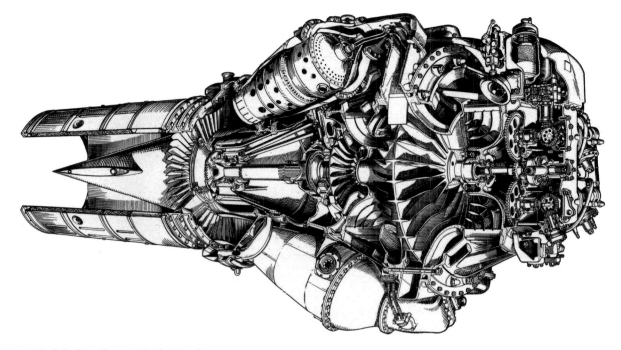

Diagram of the Rolls-Royce Derwent V turbojet engine
The Derwent V was a turbojet engine with a single-stage centrifugal compressor, nine combustion chambers and an axial flow turbine. It was 2.11 m long, weighed 567 kg and was designed to achieve 15.56 KN of thrust, with speeds between 200 and 960 km/h.

The engine business was raised again in more detail on 3 November 1948. The Managing Director, Gustave Joassart, "contrary to the custom, which was to discuss virtually everything about the company at each session", requested permission to focus on just a few points in the rest of the meeting so that it could explore in-depth the manufacture of plane engines.

An order for 860 engines was then placed with FN, with this order coming under the Brussels Treaty[147] and in the framework of the orders being distributed between the three Benelux countries.[148] FN welcomed this proposal while emphasising its reassuring aspects. There would be continuity in these orders, facilitating the necessary investment without the anxiety that they would serve no useful purpose later. There was less of a technical risk, given the value of the licensor and given the technical input that it would contribute to FN. Moreover, Rolls-Royce would provide all the documentation making it possible to quantify all elements of the issue. Lastly, a cancellation clause was drawn up, enabling FN to get out of its commitments with relative ease.

In order to embark on this new adventure, FN asked for two things. One was that a 'premium' of around 10% to 15% would be added to the price of directly purchasing the product from Rolls-Royce. The second was that the Belgian government would make an advance payment of 125 million Belgian francs to cover the initial set-up costs.

On 20 December 1948, FN noted that the business was feasible given satisfactory conditions and prices. Three phases then ensued: the first phase was to assemble Rolls-Royce parts in FN's workshops, the next phase was to assemble Rolls-Royce and FN parts, while the final phase was to assemble the 100% FN parts.

The price of engines manufactured during this third phase was 18.25% higher than the price charged by Rolls-Royce in Derby. In 1949, Belgium made an advance payment of 138 million Belgian francs to buy raw materials and the Netherlands paid out 80 million Belgian francs, though without any specific allocation. The Belgian State also contributed its help, via the Société Nationale de Crédit à l'Industrie (SNCI) [a state-owned corporate lending specialist at the time], based on introducing new industries into Belgium. It also undertook to quickly find a solution underpinning progress at last on the company's war damages file.

This 'engine business' was undoubtedly the most promising for FN. However, in the late 1940s, it was not the only one keeping the company busy. In 1948, FN once more became a company with global stature in terms of firearms and equipping armies. The total value of orders placed with the company rose to 685 million Belgian francs, including 589 million Belgian francs which were 'government orders' split between the biggest customers that year: Egypt, Venezuela, the Netherlands, Argentina and Belgium, followed by China, Peru, and Syria among others.

Furthermore, negotiations for new orders or related orders were at a very advanced stage with the Belgian Ministry of Defence, the Netherlands and the Dutch overseas territories,[149] Argentina, Ecuador, Paraguay, Venezuela, Guatemala, Egypt and Turkey. Other negotiations took place in the following months with Iran, Saudi Arabia, China (mainland), and Brazil, etc.

FN was already planning to reorganise the Pré-Madame factory to build plane engines there, a site where the company was still producing motorcycles and trucks. In 1948, sales of motorcycles were described as "fairly good", but trucks were only being sold to administrations (National Defence and SNCFB).[150]

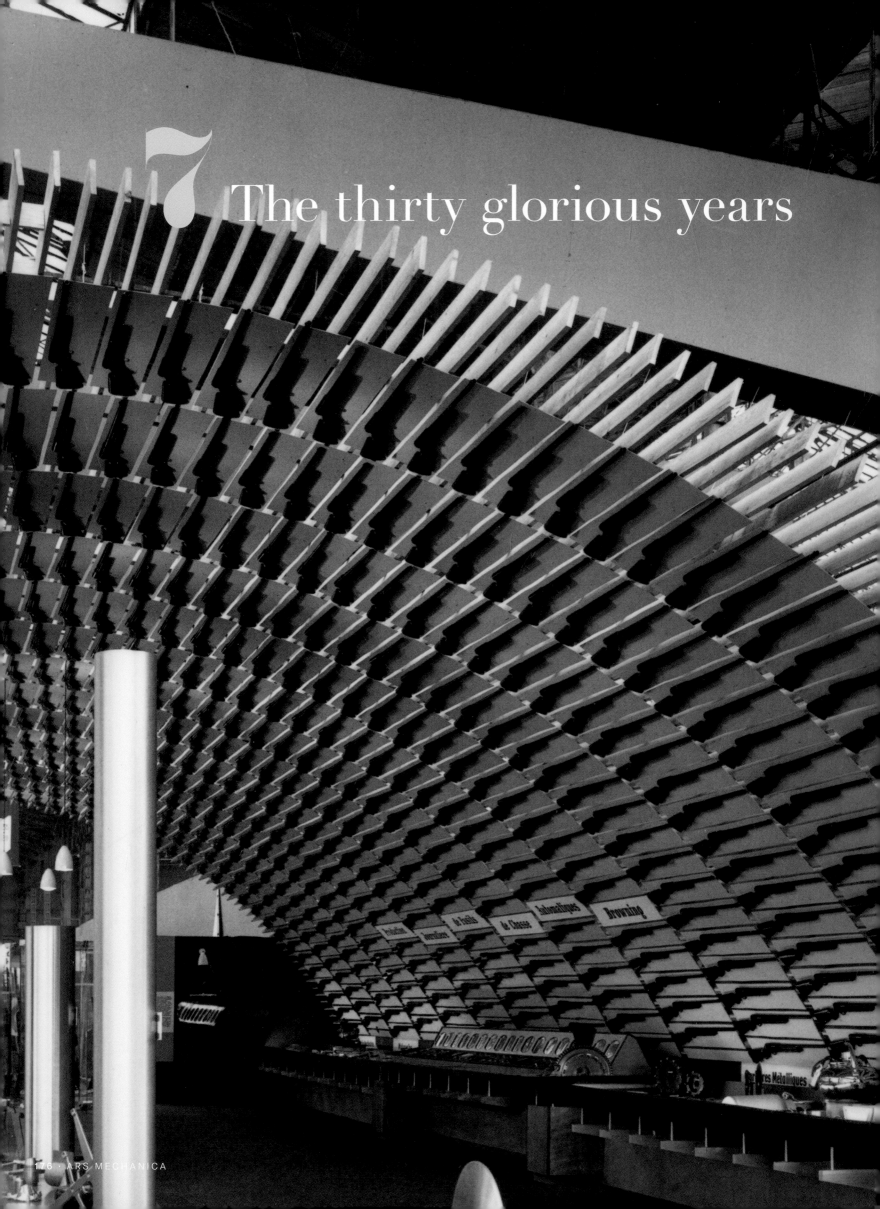

A fresh start

1950 was an exceptional year for FN, for various reasons. It truly marked the end of the huge post-war difficulties. On 20 April, the Managing Director was in a position to announce that "turnover has now reached that of the good years of 1937/38 and 1938/39". This enthusiasm was slightly tempered by the fact 'firearms' activity hit a ceiling that year...after continual progress since the war. This news was backed up during the General Assembly of October 1950: "Only a few traces of the events of 1940-1945 live on in our company. The general picture continues to be encouraging and allows us to face the immediate future with confidence".

The war damages saga finally ended on 9 January 1950 when 350 million Belgian francs were paid out to FN. The amount was deemed insufficient, but it was viewed positively because it ended the uncertainty; it was paid in cash. Meanwhile a loan facility for more than 500 million Belgian francs was opened at the SNCI and on very favourable terms. However, FN would 'only' borrow 220 million Belgian francs on these terms. This was because the loans had to be paid back within eight years, which the company considered too short a time for any further use of this facility. This single loan was an advantageous replacement for all the debts that FN had with various banks. The burden on FN had considerably lightened.

The new Managing Director, René Laloux, quickly replaced Gustave Joassart, who resigned. Laloux was delighted with the success of his recent trip to the US: "Although he had been told about it beforehand, he was struck by the tremendous reputation enjoyed by our hunting rifles over there". The General Assembly of 1951 learned with satisfaction that "hunting guns are a very important component of our production in which interest is especially dynamic because a large part of it is exported to the 'dollar zone'. The popularity and reputation of the 'Belgian Browning' and the 'FN Mauser Actions' are considerable in the US".

However, a few clouds marred this idyllic picture of progress. The General Assembly of 1950 voted for distribution of a dividend "to compensate, to a certain extent, for the lack of remuneration from capital for so many years"[151] It was then added that it would no doubt be hard to maintain future dividends at this level.

FN stand at the Foire internationale Mines, Métallurgie, Mécanique et Électricité de Liège *[International Mines, Metallurgy, Mechanics and Electricity Fair]*, 1959

Extension work at the Pré-Madame factory, 1952
Fabrique Nationale d'Armes de Guerre tailored its industrial equipment to the new demands of the booming aeronautics sector.

FN is transformed

The war and the post-war period were in the past. FN could now face the future with a justified sense of optimism. It was buoyed by the success of manufacturing Rolls-Royce plane engines, and finally had at least some assurance about its financial future. Moreover, the company could count on renewed support from Belgian and foreign governments and was reassured by the strong sales of Browning firearms in the US.

The Société Générale had always wanted to boost the presence of industry figures from within its Board of Directors. So it brought in two engineers of its own choosing in 1948. In 1952, "picking up on an old tradition", the Board of Directors appointed Paul Gillet, Governor of the Société Générale de Belgique, as its Chairman.

Georges Laloux, 86 years old, representing the old guard in FN, was discharged of his duties and warmly thanked. Backed up by Société Générale, FN was thus ready for battle and would go on to achieve success after success. In those days, FN had over 12,000 workers and employees. This large workforce made it more difficult to share information within the company and harder to strengthen the cohesion between the men and women who worked there.

There had been a long series of employment-related changes since the war. Nevertheless, efforts were made to maintain a sort of benevolent paternalism This was seen as an instrument to resolve differences between bosses and workers, outside or in parallel with the

Visit to the Fabrique Nationale d'Armes de Guerre by King Baudouin of Belgium 1953
With the impact of the war years fast receding, Fabrique Nationale d'Armes de Guerre once again became a major partner capable of handling the new challenges brought by the Cold War.

negotiating structures set up for the regulations.[152] The first edition of *La Revue FN*, a monthly publication for FN staff, came out in July 1953. The editorial column defined its aims, chief among them to "make everyone aware that they belonged to a big family". The publication was also designed to ensure that information flowed between workers from different services or workshops and ultimately to provide information about the company to people working there.

The editorial column was drafted in three languages: French, Dutch and... Italian. Since the end of the war, labour had been in short supply. The Belgian government appealed for Italian immigrants,[153] particularly in the Liège region, where the coal, steel and engineering

sectors benefited from this labour force, in order to drive the national economy's recovery.

By 1976, there had been 228 editions of the publication filled with technical information and messages to staff, underlining the employers' desire to maintain the cohesion of what a minister had one day called "one big family". This internal flow of communication has been maintained, with few interruptions and under different names, up until the present day: *Journal FN* (111 editions until 1989) *Infos GIAT Industries* (from 1993 to 1994); *La Lettre de Herstal* (1995 and 1996); *Herstal 2000* and *Made in Belgium*, which relaunched a regular in-house publication from 1998 to this day.

A risk-taking period

We have seen how successful the 'plane engine business' was, even though it was a highly risky gamble due to the technical complexity of the manufactured products. FN made more than one thousand of them, to the satisfaction of their buyers (the governments of Belgium and the Netherlands). The company took a major technological leap forward, thanks to this self-set challenge. This would prove to be key in the company's development.

In 1954, as this production drew to an end, the company obtained a Rolls-Royce licence to manufacture Avon model axial compressor turbo-jet engines. These complex engines were built for the Belgian and Dutch air forces and equipped the two countries' Hawker-Siddeley Hunter fighter planes. FN was also entrusted with the repair of engines bought by the Belgian State and with some of those that the Netherlands had bought. FN's reputation in advanced technologies was now assured.

Assembly of the J79 engine on an F104 Starfighter at the General Electric Aircraft Engines factory, 1961
In the early 1960s, Fabrique Nationale d'Armes de Guerre was appointed to carry out nearly half of the European programme to produce General Electric J79 engines, in particular the 'hot and moving' parts. These engines were used in the new F104G Starfighter, selected by several NATO countries to align themselves with the US. The US had gained considerable experience in producing fighter aircraft since the Korean War.

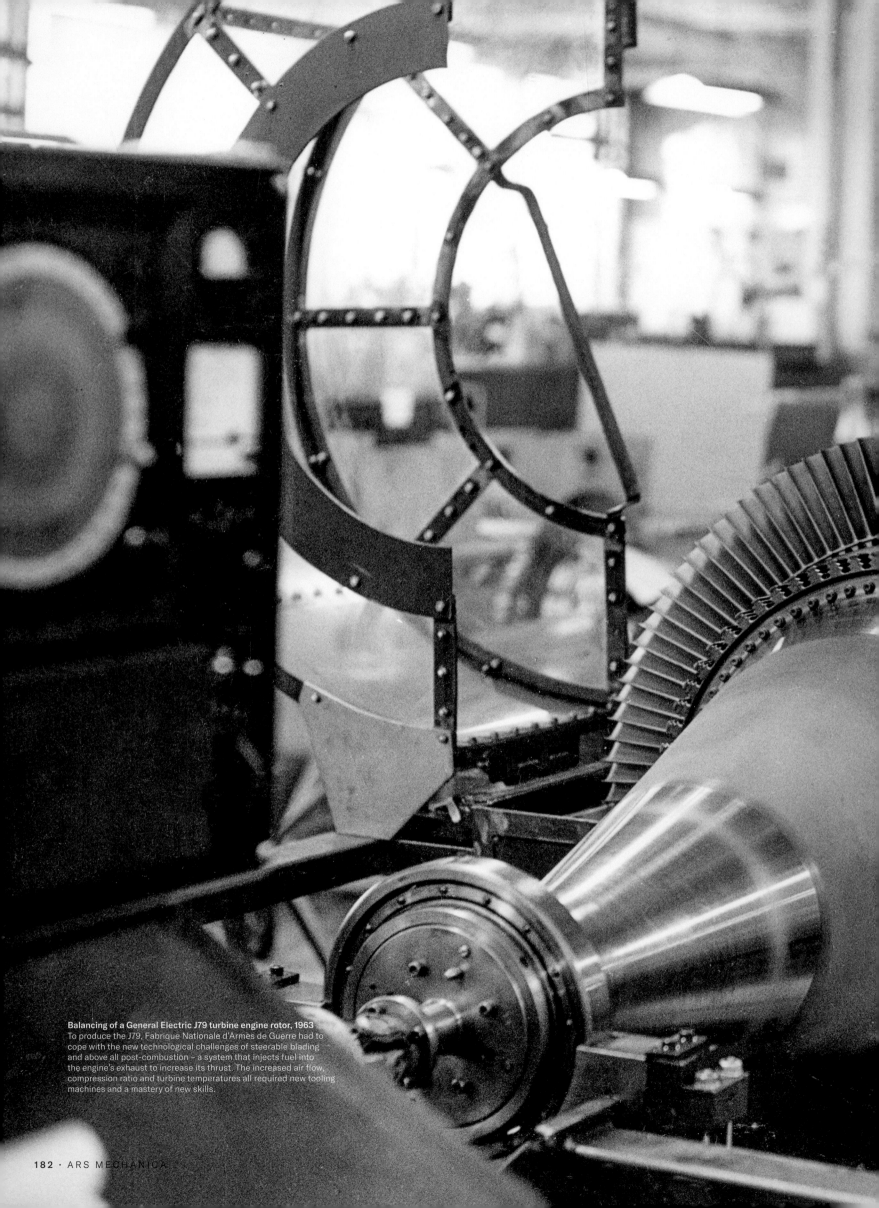

Balancing of a General Electric J79 turbine engine rotor, 1963
To produce the J79, Fabrique Nationale d'Armes de Guerre had to
cope with the new technological challenges of steerable blading
and above all post-combustion – a system that injects fuel into
the engine's exhaust to increase its thrust. The increased air flow,
compression ratio and turbine temperatures all required new tooling
machines and a mastery of new skills.

Assembly of the Atar 9c engine, 1971
In 1968, the first contact between Fabrique Nationale d'Armes de Guerre and the French engine manufacturer Snecma was established to provide Belgium with Mirage 5 aircraft from Dassault. FN was tasked with subcontractor production of the Snecma Atar 9c engine, which led to the company improving its techniques in the electrochemical machining of turbine blades.

When FN was created in 1889, its directors had wisely anticipated the initial contract's signature for the supply of 150,000 Mauser rifles. The company had even been created before the agreement with the Belgian government was made and before the model of repeating rifle had been chosen.

A new FN adventure was also launched by foresight like this and by the same 'proactive attitude' as well as confidence of the same kind. This adventure revolved around manufacturing a large proportion of the engines (General Electric J79 axial compressor turbo-jet engine) for the F104G Starfighter fighter/bombers. Germany, Belgium, the Netherlands and Italy decided to equip their air forces with this plane. Following a protocol signed by the four countries on 5 March 1960, the aircraft's production was shared among plants located in each of the countries.

FN's reputation in aeronautics was now firmly established, enabling it to win an order for a major part of the engine, with the rest going to BMW and FIAT. This was only the beginning of transnational cooperation, which did not commit the countries towards their suppliers. Yet FN, usually cautious, boldly took a risk. Within months, licensing agreements and technical assistance contracts were signed, workshops were equipped with machines, parts and accessories were bought and the production process began in June 1960. This was all based on just a letter from the Belgian government announcing its intention to order a certain number of engines from the company. A provisional contract was only signed in August 1961. For months, FN had done the manufacturing, under its own responsibility, and paid for all the expenses from its own funds. The first engine had long since been delivered, when the final agreement was reached between the States and their suppliers in October 1962. This line of business was a success for FN and underlined a genuine industrial appetite. In surface area, the 'engines' would be the biggest part of the factory for some time.

Dassault Mirage 5 aircraft, 1970

From 1950 to 1964, Fabrique Nationale d'Armes de Guerre produced one billion three hundred million infantry cartridges. At least half of them were NATO standard 7.62 mm calibre.

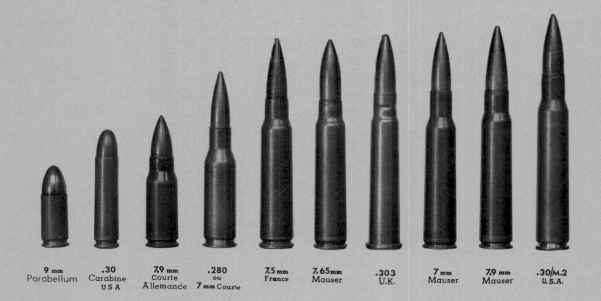

9 mm
Parabellum

.30
Carabine
USA

7,9 mm
Courte
Allemande

.280
ou
7 mm Courte

7,5 mm
France

7,65 mm
Mauser

.303
U.K.

7 mm
Mauser

7,9 mm
Mauser

.30/M.2
U.S.A.

Before the standardisation of ammunition in NATO, Fabrique Nationale d'Armes de Guerre met the needs of each country by supplying a variety of military cartridges of different calibres.

FN, a global supplier of ammunition

The North Atlantic Treaty Organization (NATO) is a collective security political and military organisation created in 1949 by the North Atlantic Treaty, signed in Washington on 4 April 1949.[154]

Before long, the signatory countries had agreed on the need to renew and standardise their armaments. In 1950, FN was invited to take part in an ammunition standard study thanks to the UK, building on their strong, friendly and professional network developed during the war. The aim of the study was to replace the US Army's .30-06 calibre ammunition, the British Army's .303 calibre ammunition, the French Army's 7.5 mm calibre ammunition and the Belgian Army's 7.65 mm calibre ammunition with a single type of ammunition and to equip all the Alliance armies with it. With so many countries involved, it was no surprise that views diverged on what this ammunition should be regarding its power or size. Each country fiercely defended solutions that matched the characteristics of its own ammunition.

FN skilfully chose a middle path. It proposed a 7.62 mm lead core bullet mounted on a 51 mm American cartridge case. This solution outperformed others in ballistic tests held in 1953.

In January 1954, the US, UK, Canada, France and Belgium agreed, calling on the Ottawa Agreement, to recommend the adoption of this cartridge by all NATO countries. Without delay, manufacturing supervisory mechanisms were put in place to ensure that the standards issued by NATO were strictly respected wherever that ammunition

was produced. FN naturally had a clear advantage over the large number of companies lining up to manufacture this product, and it was the first to be certified.

The General Assembly of 1954 announced that "FN has actively taken part in the development of the 7.62 mm ammunition which was adopted by NATO last December (1953). It is delighted to have been able to contribute, during international discussions, to reconciling points of view which, in 1951 – one may well remember – were far apart. This new ammunition is now being regularly manufactured in our ammunition factory".

In 1955, the General Assembly added: "The reports that we have received concerning the precision and efficiency of the new NATO 7.62 mm cartridge, in whose development our factory has made a considerable contribution, are very favourable."[155] It went further still in 1959: "The use of the 7.62 mm cartridge is spreading worldwide at a speed that its first backers had undoubtedly not predicted and this small but powerful piece of ammunition is now on track to replace all those available since the start of the century."[156]

FN received orders from 44 countries for these 7.62 mm calibre cartridges. Production sometimes reached levels as high as 650,000 pieces of ammunition per day. In 1960, production levels were never lower than 10 million per month. The success of this ammunition can be attributed to FN's sensible risk-taking, as was the case in 1891.

NATO 7.62 × 51 mm ammunition SS77 Model

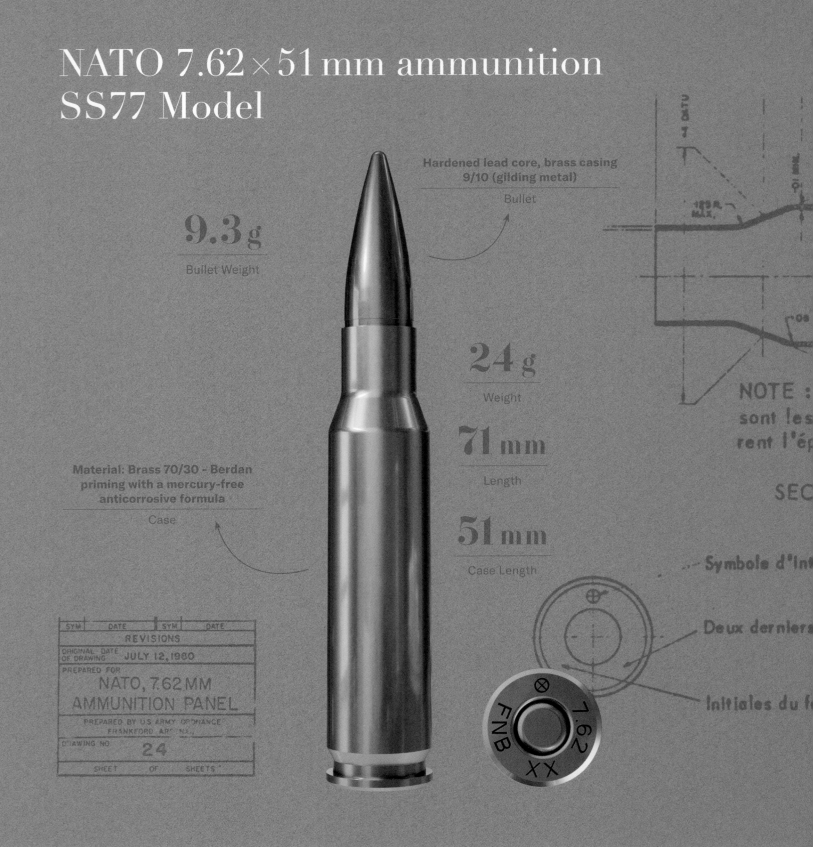

9.3 g
Bullet Weight

Material: Brass 70/30 – Berdan priming with a mercury-free anticorrosive formula
Case

Hardened lead core, brass casing 9/10 (gilding metal)
Bullet

24 g
Weight

71 mm
Length

51 mm
Case Length

SYM	DATE	SYM	DATE
	REVISIONS		

ORIGINAL DATE OF DRAWING JULY 12, 1960
PREPARED FOR
NATO, 7.62 MM
AMMUNITION PANEL
PREPARED BY U.S. ARMY ORDNANCE FRANKFORD ARSENAL
DRAWING NO. 24
SHEET OF SHEETS

NOTE :
sont les
rent l'ép

SEC

Symbole d'Int

Deux dernier

Initiales du f

The **NATO 7.62 × 51 mm calibre ammunition** was developed after the Second World War, taking into account new trends. As traditional infantry ammunition had shown itself to be overly powerful for average combat distances, it became apparent that the kinetic energy of the projectile could be slightly reduced. Moreover, the increasingly widespread use of automatic firearms and the demands of modern combat, from the perspective of the soldier and logistics, meant there was a need for reductions in equipment weight and bulkiness. The 7.62 mm ammunition proposed by FN was adopted by NATO, as it met these conditions by reducing kinetic energy by 12%, reducing the total length of the cartridge by 13% and reducing the ammunition's weight by 9%. Furthermore, the SS77 rebated rim cartridge case has excellent ballistic characteristics, with the power to pierce objects 1,200 metres away.

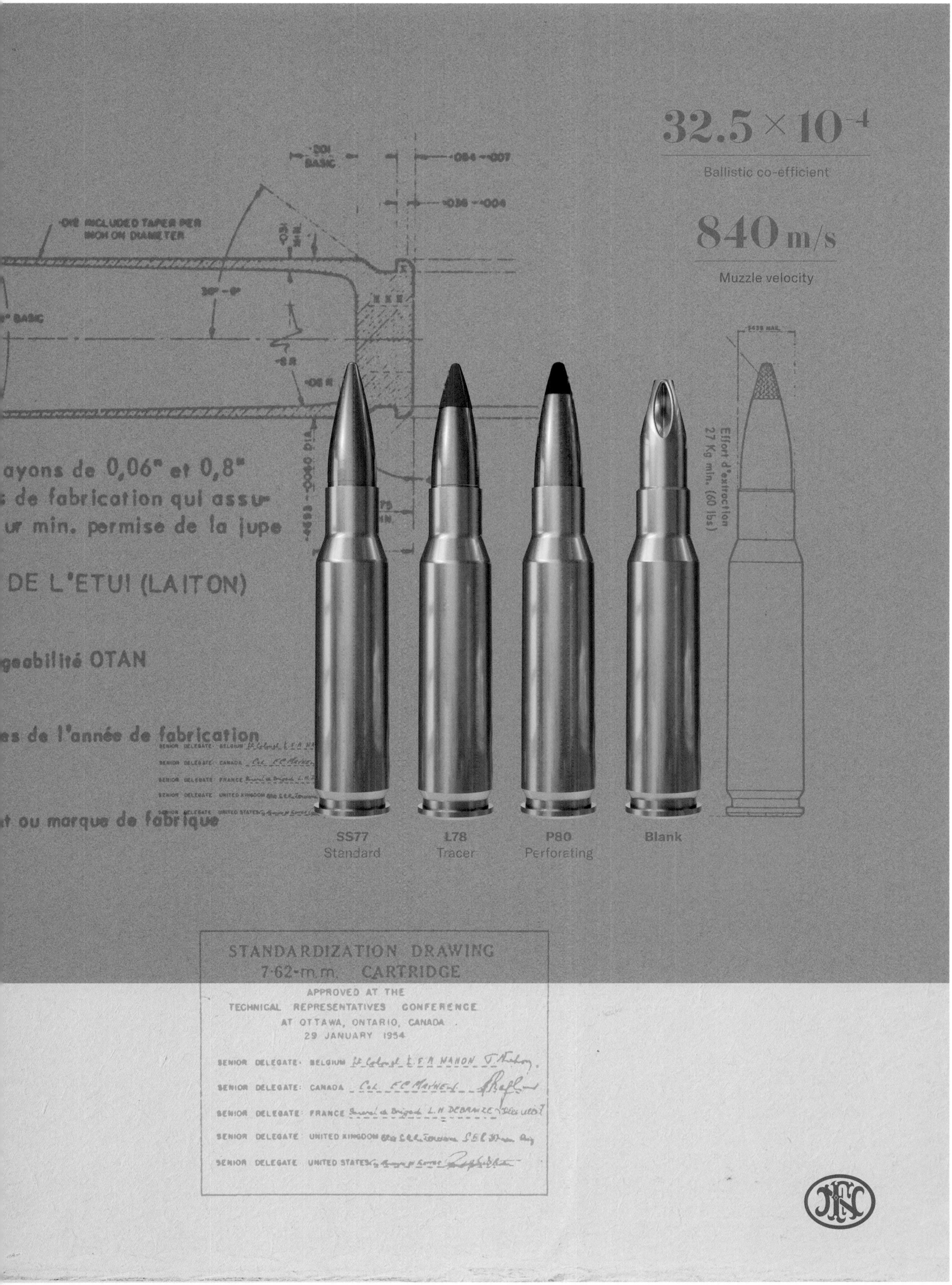

32.5×10^{-4}

Ballistic co-efficient

$840\,\text{m/s}$

Muzzle velocity

-OHE INCLUDED TAPER PER
INCH ON DIAMETER

ayons de 0,06" et 0,8"
s de fabrication qui assu-
ur min. permise de la jupe

DE L'ETUI (LAITON)

geabilité OTAN

es de l'année de fabrication

t ou marque de fabrique

Effort d'extraction
27 Kg min. (60 lbs)

SS77
Standard

L78
Tracer

P80
Perforating

Blank

"I was very glad to find that the weapon was in harmony with certain important practical and tactical conceptions, to which my own lengthy experience has led me."

WINSTON CHURCHILL

Winston Churchill, photograph by Yousuf Karsh, 1941
Winston Churchill was praised for having played a major role in the Allies' victory over the Axis powers during the Second World War, but his military and political career stretched back years before 1939. Since the late 19th century, Churchill had constantly been conditioned to think about war. His knowledge of tactical concepts, theatres of operation and weapons gave him the authority to express his views on how to optimise the British Army's military equipment. He would thus play a key role in defending, before the House of Commons, the British Army's adoption of the FN FAL Light Automatic Rifle.

The FAL, *The right arm of the free world*

After the Second World War, the concept of a self-loading assault rifle took hold. The German Sturmgewehr 44 had made a strong impression in the final years of the war. Its powerful 7.92 × 33 mm *kurz* calibre cartridge offered a good compromise between the calibres of machine guns and rifles. The USSR seized on this concept of an intermediate cartridge to produce the 7.62 × 39 mm AK-47. Simple, efficient and easy to mass produce, the AK-47 operates by gas-operated handling with a shot selector. The members of NATO then agreed on the need to identify a universal calibre that could replace the wide variety of munitions and firearms used during the war. Once such a calibre had been standardised, a universal infantry rifle could be adopted. There were two choices back then: the British model and the American model. The US favoured a standard .30-06 (7.62 × 63 mm) calibre cartridge used for the M1 Garand, their main semi-automatic rifle during the Second World War. The British focused their efforts on a new and really innovative intermediate cartridge: the .280. Two distinct rifles matched these two proposals: the future American M14 and the British EM-2. In Belgium, FN Herstal had been working for many years on developing a new semi-automatic infantry rifle. Dieudonné Joseph Saive, an assistant to John Moses Browning, developed and patented in 1936 a semi-automatic gas-operated rifle, producing a first version of this in 1937. In 1941, he took refuge in England and took his plans for his 1937 model with him and continued to work there. A new prototype was produced under the name EXP-1. This prototype led to the SAFN 49 (semi-automatic FN 1949). Even before the SAFN could be put into production, Dieudonné Saive and Ernest Vervier developed the FN Carabine prototype in 1947. This would become the Fusil Automatique Léger (FAL) or Light Automatic Rifle. The SAFN was not therefore the ancestor of the Light Automatic Rifle but rather a cousin, with which it shared a common ancestor, the EXP-1.

The FAL is an air-cooled, magazine-fed, gas-operated rifle; its first versions were a selective-fire rifle. The FAL was robust and reliable. It offered greater range and accuracy than its Soviet and Chinese rivals. The rifle was originally chambered in 7.92 × 33mm *kurz*, but it developed through use of the British intermediate .280 calibre cartridge. This led to more interest in the rifle, which also made a strong impression during the tests it underwent in Fort Benning, Georgia. In 1953, during tests under extreme conditions in Alaska, it was judged to be better than the T44 (M14); both guns were chambered in 7.62 × 51 mm. The Infantry Board recommended a limited purchase of the FN rifle and recommended halting development of the T44.

In 1954, 7.62 × 51 mm ammunition was selected as NATO's standard ammunition. This choice spelt the end for the British EM-2, chambered in .280. FN then definitively adapted to the 7.62 × 51 mm. That same year, Canada ordered 2,000 Light Automatic Rifles from FN for testing with its troops and became the first country to adopt it officially, under the name C1. Over 90 nations would go on to adopt the rifle. However the US would choose another gun, the M14. The M14 was offered to other NATO countries but unsuccessfully. The 'free world', with the exception of the US, turned to the Light Automatic Rifle.

Dieudonné Saive

On 19 February 1954, the newspaper *Le Soir* dedicated a whole page to "The person who we don't speak about: Dieudonné Saive, inventor of the Belgian automatic rifle." This article retraced the career of the engineer from 1932, when FN tasked him with picking up studies of the semi-automatic rifle. These studies had been abandoned in 1914.

It was a complex problem. There was a need for a robust gun that was easy to construct, easy to dismantle, to maintain and to repair. The projectile had to have sufficient speed and weight. Accuracy was required, both for shooting in bursts and shot by shot. Finally, the cost price had to be acceptable. Saive worked on it for four years. War broke out again. Saive went back to Herstal. The factory was in German hands. The Belgians only obtained permission, which was seen as derisory by the Wehrmacht, to study 'new hunting gun' projects![157] *In his little office, Saive drew up plans for an excellent sparrow hunting rifle and, when he thought that everything was ready, disappeared!... He reappeared in England in August 1941. He brought with him, putting his life in danger, plans for the 'Atlantic rifle' finished at Herstal, under the enemy's nose. Cleverly, he had cut up the 'blueprints' into pieces, scrunched up into a ball.*

FN FAL
Light Automatic Rifle
Model 1

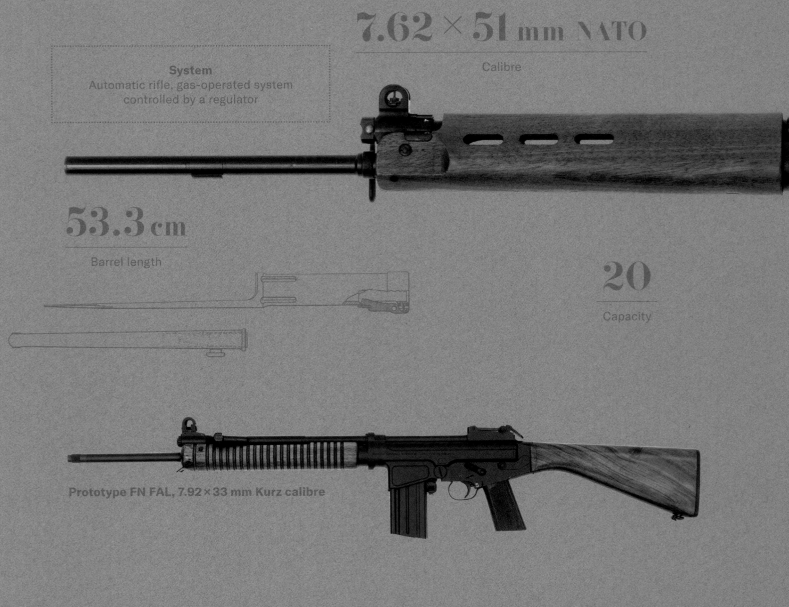

7.62 × 51 mm NATO
Calibre

System
Automatic rifle, gas-operated system
controlled by a regulator

53.3 cm
Barrel length

20
Capacity

Prototype FN FAL, 7.92 × 33 mm Kurz calibre

After the Second World War, the modernisation of the infantry based on the experience of previous combat led to new demands for armaments. The idea was to replace the widespread use of repeating and bolt-action guns with a semi-automatic rifle, while increasing the magazine capacity. The rifles also had to be able to fire in automatic mode, to occasionally replace machine guns, yet remain lightweight and adapted to combat within a range of 600 metres. The choice of the standard 7.62 × 51 mm calibre ammunition would limit these last two parameters in favour of greater stopping power, equivalent to that of a machine gun.

In the 1950s, FN responded to these expectations with the **Fusil Automatique Léger (FAL)** (Light Automatic Rifle, in English). It was a rifle that could be used in semi-automatic and automatic modes. It was robust, lightweight, offered good stability, and the recoil was not excessive. The rifle was gas-operated, and this was adjustable thanks to a regulator. When the ammunition was fired, the gas moved into the gas cylinder through the barrel vent and pushed a piston, which drove the breech block backwards. The slide then unlocked the breech block, before being returned to the forward position by the recoil spring and locked. By placing the gas cylinder

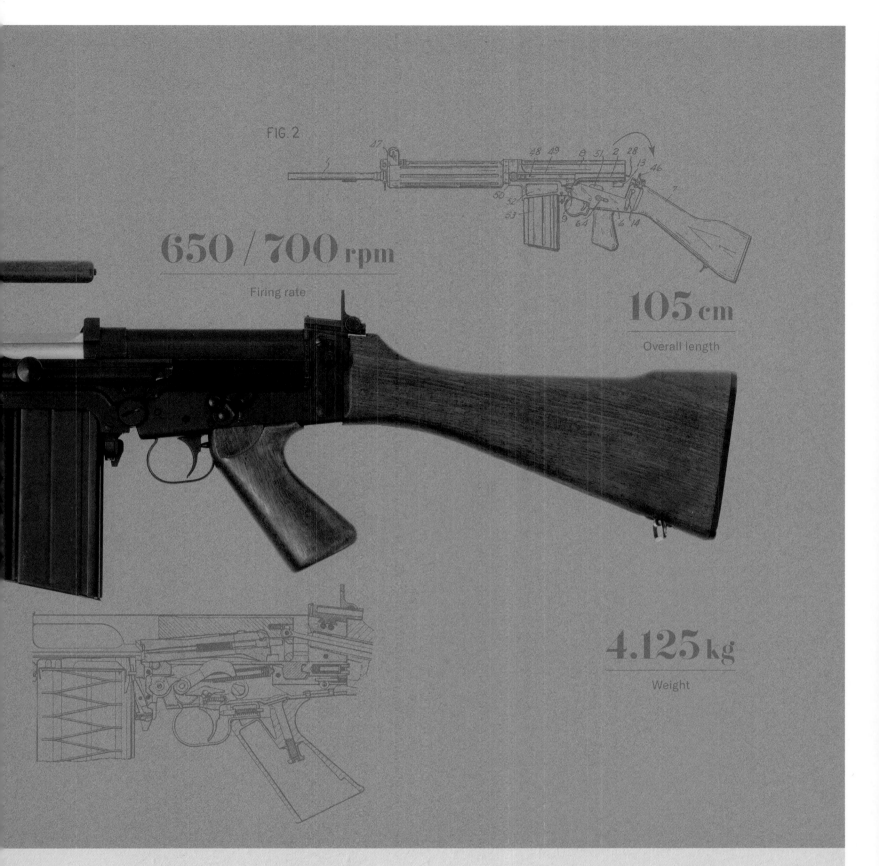

FIG. 2

650 / 700 rpm

Firing rate

105 cm

Overall length

4.125 kg

Weight

above the barrel and, as learned from an in-depth study of the general line, the recoil has a limited tendency to raise the rifle. This is a key factor in ensuring stability and accuracy. The positive locking of the breech block, before squeezing the trigger, also improves accuracy, as the mass pushed forward is reduced. When the magazine is empty, the breech is held back by a bolt catch. Initial arming of the rifle is done via a cocking handle located on the left, but immobile during the firing, ensuring the shooter's safety. The rifle was available with a carrying handle and had a tilting guard, for easy maintenance. The first models of the FAL had a wooden frame before

plastic was introduced in the 1960s. Numerous versions were produced to meet the needs of clients, particularly following requests from Commonwealth countries. Finally, FN produced a version with a 1.6 kg heavy barrel and a bipod, the FALO, to allow users to maintain firing for an extended period.

THE UK ADOPTS THE FAL

In 1954, the War Office recommended to the UK's Prime Minister Winston Churchill that the FAL should be adopted for the British Army. He personally tested the FAL and defended it in the House of Commons, though not without difficulty, and this led to its adoption in the UK.

As a token of thanks to Winston Churchill, FN offered him a personalised FAL. The Weapons Division suggested, for the occasion, to make the rifle out of bronze rather than parkerizing it as was customary. Talented engravers, who were more used to working on hunting rifles than war rifles, were called in. On the right-hand side of the receiver, the inscription recalling the reason for the gift was engraved and inlaid with gold: "This rifle was made by the workers of FN and was offered to the Right Honourable Sir Winston Churchill, K. G., O.M., C.H., P.C., M.P., Prime Minister of the UK, out of respect and admiration. Liège (Belgium), 25 March 1955." This inscription was completed with a sentence that Churchill had spoken in the House of Commons during the parliamentary debate: "I was very glad to find that the weapon was in harmony with certain important practical and tactical conceptions, to which my own long experience had led me." The rifle was offered to Churchill by the Managing Director of FN, René Laloux.

MORE SUCCESS IN EUROPE

FN would soon enjoy further success in Europe. The Bundeswehr (federal defence) was the Federal Republic of Germany's national army and was created in 1955 – under the auspices of NATO. Re-equipping this army came with a heavy symbolic burden, it goes without saying. Germany had been the enemy a very short while ago, the occupying power, and had previously been FN's owner between 1896 and 1918. When the company's senior management were faced with the situation of supplying Germany with many Light Automatic Rifles, their thoughts doubtlessly turned to the twists and turns of history.

The contract for the manufacture of 100,000 guns for infantrymen was signed in Koblenz on 13 November 1956. At the same time, 20 million cartridges were ordered to equip them "pending the restart of German industry".[158] There was a certain urgency surrounding this production process. The first conscripts were called up on 1 April 1957.

FN invested 20 million Belgian francs in buying supplementary machine tools. It also decided to work in two teams, to keep pace with the manufacturing target of 8,000 firearms per month.

⇕ The Coldstream Guards on parade carrying the FN FAL, known in the UK as the L1A1 SLR, c. 1960

↦ Troops airlifted from the Federal Republic of Germany, carrying the FN FAL
In 1956, the newly created army of the Federal Republic of Germany was equipped with the FN FAL. This was highly symbolic. For the first time in its history, the country of Mauser, Haenel and Walther was mass purchasing a gun that had not been produced in Germany.

The MAG

Portability and firepower are key distinguishing features of the different types of firearms. A handgun s easy to carry, but has limited firepower. However, a mach ne gun – a firearm fed by a belt or a magazine – is capable of powerful sustained fire but is heavy and bulky, as is its ammunition. This gun is therefore difficult to move around on the battlefield and several soldiers are needed to operate it. A good machine gun is one that does the best job of combining ease of handling and firepower.

Historically, FN MAG is basically a gun that successfully found the best balance between mobility and firepower. During the Second World War, it became clear that troops had to be equipped with a machine gur that was both light enough to provide covering fire for a platoon but also capable of delivering sustained suppressive fire. In 1957, during the General Assembly of shareholders, there was talk of this new gun for the first time: "Following the study on equipment designed to fire the NATO standard cartridge, our technical services have recently developed a belt-fed gas-operated machine gun. The great interest that this gun generated in Sweden and in the UK where we presented it bodes well for its future."[159]

The firearm operates by using the combustion gases from the powder in each cartridge at a certain point in the barrel. A gas regulator enables the rate of fire to be adjusted to approximately 600 or 1,000 rounds per minute. Based on the principles of the German Army's MG34, this air-cooled weapon has an interchangeable barrel. It is usually delivered with two barrels, so that one can be fired while the other cools.

Orders were made for the gun, shortly after presentation of the prototype. A decision was made on where to mass produce it from 1959. As the Board noted, "this plant has provided plenty of work for our design offices and tool shops".

The FN MAG equipped armies from over 80 countries and provided reliable service on all kinds of terrain. The peak of its success came in 1976, when it was adopted by the US and called the M240.

Thanks to numerous updates, variants and optional equipment, the FN MAG kept pace with tactical and technological evolutions, while retaining the simplicity and reliability that made it an efficient firearm. Sophisticated mounting systems (bipods, tripods, anti-aircraft modules, etc.) allowed for better control of the MAG, including mounting it on land, naval and air vehicles. Since 1978, FN has specialised in mounting these machine guns on land vehicles, aircraft and naval vessels in increasingly sophisticated configurations. Its longstanding success can be ascribed to this ability to adapt to the actual combat situation.

As soon as they were marketed, the FAL and the MAG became standards of military weaponry.

Ernest Vervier

In the mid-1950s, Ernest Vervier took over from Dieudonné Saive to lead the research and development department. Besides designing the MAG, he launched the early research into the 5.56 mm calibre range of weapons.

Robust and reliable, the FN MAG quickly made its mark as the best General Purpose Machine Gun (GPMG). Its optimal combination of mobility and firepower made it a real battle-winning weapon.

FN MAG
General Purpose Machine Gun
Model 58

54.5 cm
Barrel length

840 m/s
Muzzle velocity

7.62 × 51 mm NATO
Calibre

Belt feed, open metal links
Feed mechanism

The adoption of the 7.62 × 51 mm calibre cartridge by NATO and the FAL by many of NATO's members resulted in FN designing, at the end of the 1950s, a new gas-operated machine gun, categorised as lightweight at the time. The **MAG machine gun** is a locked, belt-fed, open-bolt, gas-operated and air-cooled automatic firearm. The principle of the mechanism is that the gas, which passes through the vent in the regulator – which allows the firing rate to be adjusted – and from there into the gas cylinder, launches the piston and slide assembly backwards. The

locking system with tilting lever mechanism developed by John Moses Browning for the BAR was adopted. But the mechanism was reversed, with the bolt's support point in the receiver now below the breech, to allow the belt to be released by opening the cover. Through the cam, the slide pulls up and back on the rear end of the bolt and forces it to move back. This extraction system makes the operation smoother and more suitable for firing in adverse conditions. The gun is fed by a belt consisting of open links, with the cartridge pushed forward and fed

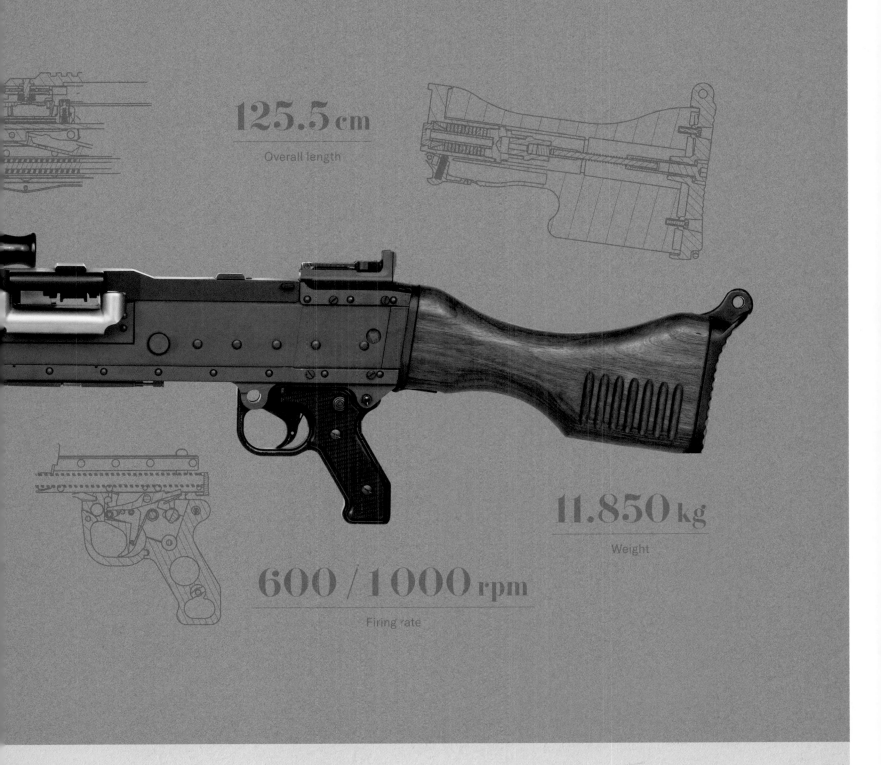

125.5 cm

Overall length

11.850 kg

Weight

600 / 1000 rpm

Firing rate

directly into the chamber. Ejection is from the bottom of the gun. The barrel can be speedily disassembled, locked by a nut with differential threads to avoid any play, and comes with a carrying handle. Safety is ensured by the firing pin, which is attached to the slide, preventing firing before the mechanism is completely locked. Finally, the shooting is done with open bolt, which prevents the 'cook off' effect.

This gun's robustness, reliability, versatility, firing volume and its increased range have made it the world benchmark for support machine guns. It is still being sold nearly 70 years later, thanks to its evolutions.

Firearms assembly workshop, c. 1960
This assembly line produced the Browning Double Automatic shotgun, known as the Twelvette. Here we see two models of receiver: standard and light alloy.
This 12-gauge double barrel shotgun featured a locked breech system operating with short recoil of the barrel.

Expansion from 1950 to 1970

Between 1950 and 1970, while there was moderate inflation (consumer prices 'only' doubled in the period), the overall dividend[160] paid out by FN increased threefold. But the wage bill (salaries, allowances and social security contributions) rose spectacularly, by a factor of 4.5, from 691 million Belgian francs to over three billion Belgian francs.

FN's remarkable expansion between 1950 and 1970 was mainly attributable to the success of the plane engines, the 7.62 mm ammunition, the FAL and the MAG. However, the company's management were convinced that none of its other products, which had become traditional, should be abandoned. FN knew from experience that the military weapons business is unstable, that orders can be cancelled or delayed or that difficulties with payments can arise.

FN AND BROWNING: A SUCCESSFUL COUPLE

Loyal American customers

The success of FN-made hunting guns was largely because of American customers' loyalty, which continued after the war. This was highlighted several times by the Board of Directors:

According to a US customs note for the period from January to August 1950, Belgium – i.e. in fact FN – accounted for 83% of the imports of guns of all kinds to the US (pistols, hunting rifles and rifles).[161]

Confirming the trend of Browning sales in the US, firearms import statistics for the first four months of the year [1962] showed considerable progress in imports from Belgium, which stood in contrast to the stagnation in imports from other countries in the world.[162]

To give a clearer idea of the US market's significance for FN, we can highlight how a continuous stream of exports to the Browning company was set up from the end of the Second World War. In 1963 for example, FN delivered to Browning its millionth semi-automatic Auto-5 shotgun to be produced since the war's end.[163]

In February 1950, Val Browning and FN negotiated the granting of a global licence for a new automatic hunting shotgun, the Twelvette, together with an order for 25,000 of these shotguns for the US, amounting to nearly 52 million Belgian francs.

These excellent commercial relations, benefiting both parties, were clearly based on great trust. When FN was obliged to announce to Val Browning that the orders linked to European rearmament would considerably

slow down deliveries of the new semi-automatic shotgun, Mr Browning "took it well" but asked for "an increase in the supplies of five-shot automatic shotguns and over and under shotguns in exchange".[164] On 17 September of the same year, the Board happily noted a new order from Browning for 50,000 automatic shotguns and 6,000 Browning B25 shotguns, which was added to the previous delayed order. Incidentally, Browning even accepted a price increase. To meet these orders, the production rate had to be increased and reached a daily level of 192 semi-automatic shotguns and 23 shotguns.

Although more anecdotal, something else epitomised the links between the Brownings and Herstal. At the end of a Board meeting on 10 March 1952, the directors, "accompanied by Messieurs Val A. Browning[165] and John V. Browning, son and grandson of the late John M. Browning, walked to the plaque which had been fixed in the corridor of the ground floor of the administrative building to recall the death 25 years ago of the great inventor and partner of FN".[166]

The sales of FN products to the Brownings increased considerably in the mid-1950s. The turnover generated by the US orders rose from 89 million Belgian francs in 1955 to 122 million Belgian francs two years later and 275 million Belgian francs in 1958. Orders became increasingly significant and FN regularly hiked its prices.[167]

Val and Bruce Browning, the inventors

Nonetheless, a few clouds were gathering. In September 1960, FN was informed of the Browning Company's intention to buy firearms from other suppliers and objected to that.[163] In December that year, a decision to buy Finnish guns was firming up. The plan was to order a repeating rifle from the firm Sako. However, John V. Browning insisted on bringing FN into this project "in order to keep the FN stamp on guns that would at least be assembled at Herstal".

Around the same time, Val Browning's creativity seemed almost limitless: "Mr Val Allen Browning, who arrived from the US on 18 February [1957], is currently working on developing a pump-action rifle and a gas-operated automatic rifle based on the two-shot automatic rifle."[169] These two guns were completed in 1958. In 1962, Val's son Bruce ('the young Bruce', as he was called at FN) brought along plans for a .22 calibre pistol that he had invented. The Board immediately agreed that this was "a very popular gun in the US and that its absence from FN's range was cruelly felt in Belgian imports into this country".[170]

BROWNING BAR
Semi-Automatic Rifle
MK1 Model

System
Semi-automatic rifle, gas-operated and locked
by a seven-lugs rotating bolt

4 + 1
Capacity

± 112 cm
Overall length

**Detachable magazine
fixed on a rotatable stand**

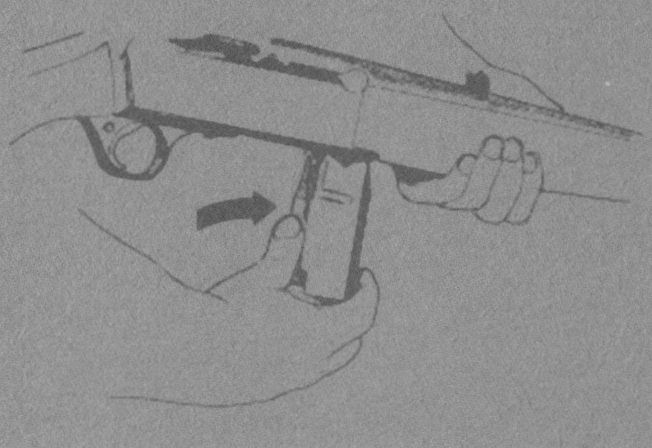

The **Browning Automatic Rifle**, which would be more widely known as the BAR, was also the work of Bruce Browning and FN engineers. It was one of the revelations of hunting firearms production in the 1960s. Specialists described the BAR as "faultless". Its ability not to jam and its firepower made it an efficient and reliable gun. It accomplished various goals that were previously thought to be incompatible. It didn't seem possible to create a rifle that was automatic, powerful and accurate all at the same time. For years, it had no competitors, as this type of semi-automatic rifle is so difficult to develop.

The company's ability to continually develop new models and types enabled it to retain a leading position in the market for semi-automatic hunting rifles.
The BAR operates by using gas from the fired cartridge. These gases, passing through the gas port, push back a cam by means of a piston. This cam unlocks the head of a rotating breech with seven lugs and pushes it back, resulting in the ejection, the rearming of the hammer and the chambering of a new cartridge when the breech is closed. This mechanism is complex, but is highly reliable and works with unparalleled speed in the shooting

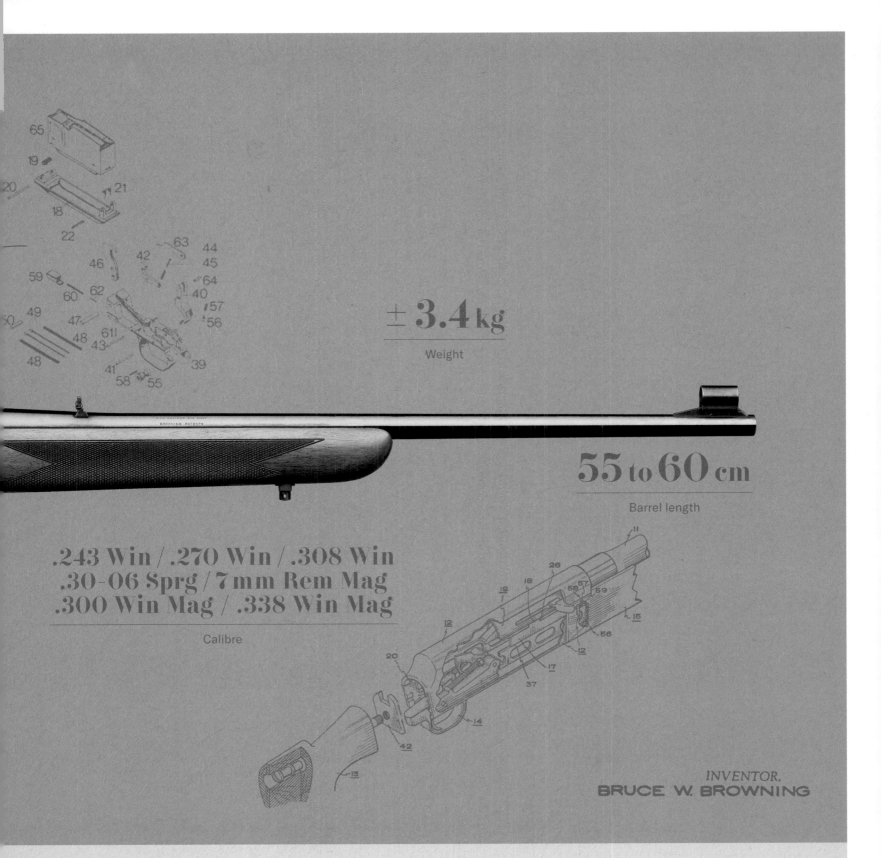

± 3.4 kg

Weight

55 to 60 cm

Barrel length

.243 Win / .270 Win / .308 Win
.30-06 Sprg / 7 mm Rem Mag
.300 Win Mag / .338 Win Mag

Calibre

INVENTOR.
BRUCE W. BROWNING

sequence. Even today, the BAR's cold-hammered barrel and the perfect rigidity of the barrel-breech assembly make it one of the most valued and best-selling semi-automatic rifles for big game hunting. Browning offers multiple versions of the BAR, including luxury models, a version for left-handed shooters and composite stock models. The range was further expanded with the Short Track and Long Track versions.

More than a million BAR rifles have been sold since 1966, with hardly any mechanical modifications. After more than 50 years, it continues to be a benchmark in its category. Assembled in Portugal, it has become FN Viana's flagship product.

BROWNING

Group of Japanese arms manufacturers visiting the Fabrique Nationale d'Armes de Guerre, 1971
Miroku's history commenced in 1893, when Karaji Miroku established himself as a gunsmith in Kochi on the island of Shikoku. In the early 20th century, Miroku first took an interest in Western methods of producing hunting guns and the production of harpoons. In 1946, a new Miroku company was set up, but it could only restart the production of hunting guns after the ban was lifted in 1951. Miroku began searching for new models to produce with a view to maximising the profitability of the factory, which had 300 workers in 1960. The company then produced shotguns – such as the over and under gun – as a subcontractor or under its own brand name from 1963.

Changing relations between FN and Browning

In 1961, the 6,283 shares of the Browning Company's capital were distributed among the direct descendants of the inventor. Back then, for various reasons, they wanted to put an end to the family nature of the company. Several options were examined, including a merger with Bell & Howell,[171] a company that was highly valued on the New York Stock Exchange. René Laloux was asked to give his opinion and strongly advised against a merger, despite it being financially attractive for those holding Browning shares. Other solutions were envisaged. In particular, issuing Browning shares in the marketplace after FN had committed to maintaining its sales conditions for the US firm. The possibility of FN buying into Browning's capital was also studied.

It was of course FN's commitment to maintain its sales conditions that posed a problem. FN knew too well how volatile the arms market was, so would only commit itself for the medium term. After intense negotiations and some friction that would leave scars (there was even talk of 'less trusting'[172] relations between the two companies), a ten-year 'Supply Agreement' was accepted by FN, albeit after fairly tough discussions. The clauses that FN found unfavourable were abandoned. It was clear

that Browning was not in a position to alienate its practically only supplier. FN also bought into the capital of the Browning Arms Company, with over 23 million Belgian francs, which accounted for around 3% of this company's capital.

So the nature of relations between Browning and FN had changed. From being amicable and profitable, it had become commercial... and profitable.

After a "1962 season that was overall very favourable" in the US, a new order from Browning was announced in March 1963: for 160,786 firearms worth nearly 500 million Belgian francs. The Chairman and Managing Director, René Laloux, immediately wrote to John Val Browning to plead with him to reduce its order by 10,000 guns in order to ensure that the FN workshops had "more regular activity"; Browning agreed to this. This goodwill was appreciated by FN, which authorised the sale, outside the US and under the Browning name, of new sports articles which the US company had recently launched. However, this was done on the condition that the brand be accompanied by the words "with a distinctive addition".

The value of the Browning orders to FN continued to rise. In 1965, 168,751 guns were thus sold plus an order of 50,000 Model 65 .22 calibre rifles, for which delivery

Visit of the Crown Prince of Japan Akihito to the *Hôtel de Ville* in Liège, 1953
The Belgian authorities offered their distinguished guest an FN-Browning Auto-5 shotgun, a firearm reflecting the Liège region's expertise in arms production.

would be spread out over the next few years. The amount was 24% higher than the previous year, partly because the order was larger but also because the prices charged by FN had increased.

During this period of beneficial commercial relations for both parties, the Japanese threat emerged: the Browning shotgun had been "slavishly" copied. John Val led an investigation and discovered that Miroku was responsible. Rather than get into a fight with the company, John Val Browning headed to Japan with André Laloux, Head of Service responsible for manufacturing FN's hunting rifles. Laloux was struck by the discipline of the Japanese workers. He observed that, while the guns that they made were not perfect, their shortcomings could apparently be easily rectified with training. John Val himself urged FN to consider manufacturing in Japan. The Board concluded this chapter in a matter of fact way by agreeing that "the issue will be studied closely".[173] On 11 October 1965, Browning again advanced his idea and suggested that FN should invest in production in Japan. The Board was not at all enthusiastic about this proposal.

The fact that John Val Browning was pressing to see agreements reached between FN and Miroku was no doubt linked to US concerns about the future of Belgium. A number of employer-employee conflicts[174] and the constant price rises of products following regular wage increases made the country seem less reliable. The Americans, who were becoming anxious, began to view Japan as a possible partner, as it was more disciplined and had fewer trade unions.

It was noted, on 13 December 1965, that: "Shocked by FN's price hikes due to increases in Belgian wages, J.V. Browning considered ordering from Miroku a single-shot shotgun first, then a boxlock side by side shotgun. However, he conceded some of the arguments that were set out against using Japanese industry and decided to defer any decision for the moment." Among the arguments that are said to have influenced him was FN's promise to supply Browning with a new firearm every year... However, in around 1966, Browning developed its first commercial links with Miroku.

In Europe, the signing of the Treaty of Rome in 1957 set up a common market between six countries (West Germany, Belgium, France, Italy, Luxembourg and the Netherlands) in which customs restrictions were to gradually disappear. A huge market was opening up for FN for the sale of its hunting products (and more generally for all its civilian products), breaking down the barriers between national markets which had become far too small.

The engraving workshop of the Fabrique Nationale d'Armes de Guerre, supervised by Félix Funcken, 1950s
In 1930, so four years after its creation, the engraving workshop employed 24 engravers. There were 180 in the 1960s, making this workshop the world's largest in this speciality.

THE HERITAGE OF
TRADITIONAL EXPERTISE

FN started producing hunting guns when the company was reorganised in 1896. This was a result of the need to diversify its range of products. Emerging from the Herstal factory then were sidelock shotguns and Anson[175] box-lock shotguns. The British Ansons quickly overtook the other models and became the standard. Their mechanical design was simpler. This meant fewer parts and mechanisation was possible, which in turn made manufacturing them less expensive.

As these guns came from the UK, they brought with them a British inventory of decorations for the gun. The inventory was drawn up based on the tastes of an aristocratic and conservative clientele and beautifully decorated guns of the time. The 19th century was all about rediscovering styles inherited from the past and particularly the neo-classical style. The standard for gun decoration work revolved around compositions using the acanthus leaf, which was often a feature of floral and hunting patterns. Engravers of the time were also inspired by the work of a publisher, Charles Claesen, who produced an inventory of patterns from the Renaissance to the Neoclassical age.

FN, which exported its products widely, followed this decorative approach – which contributed to its spread. The company's engravers had inherited know-how dating back four centuries. They skilfully practised the

'hammer and chisel' technique that was specific to the Liège region. Most often they produced steel engravings with 'English scroll' patterns, which were copied or lightly adapted.

John Moses Browning brought to FN the B25 over and under shotgun and the Auto-5 semi-automatic shotgun. These were radically different designs from the company's previous products. The guns were more angular. So the question quickly arose as to how to adapt the engraving to these new contours, so that they could feature aesthetic elements of Art Nouveau and Art Deco, which neatly matched the new designs.

FN was eager to maintain high-quality craftsmanship and to use its resources to improve its own products.

In 1926, it therefore founded an engraving workshop within the factory, managed by Félix Funken. This master engraver started the company's reflection on how to modernise the engraving process. One of his signature works was undoubtedly the hunting gun (a Browning B25 over and under shotgun) that FN displayed in 1939 at the Exposition de l'Eau (Liège, 1939). Thanks to its Art Deco style and its very modernist lines, this gun stood out like a work of art from previous engraved products.

Engraving drawing for the B25 D5 Model action frame, by the master engraver André Watrin, 1973

Despite their success, these trials ultimately did not define FN's decorative patterns, as they were not marketed. Customers' tastes were still highly conservative. For the B25, various decorative patterns were created, graded from A to D, which corresponded to increasingly complex engravings. For example, the B25 A1 only included a light border strip of steel-engraved acanthus leaves while the B25 B also depicted a little hunting scene. For higher grade models, either the steel engraving took up more space or techniques that were more difficult to implement were used: hollow-bottom, chasing or using inlays.

This range of styles, which defined the B25 and Auto-5 range, in turn influenced engraving worldwide via exports. Commercial successes helped to spread decorative styles. Customers' tastes determined this decoration, which has remained mainly traditional to the present day. The engraving process had barely changed since the 19th century, even though the techniques had evolved. For example, laser engraving became a fairly common industrial tool in the 1980s.

Through its commercial successes, FN thus had a major influence on contemporary guns manufacturing, technically of course as well as aesthetically. In the latter area, FN Browning rose to have just as much influence as the best-known firms. Its B25 and Auto-5 guns, with all the variants arising from technological improvements, became benchmarks for decorating guns.

When Félix Funken retired in 1960, he left behind him a department that had made a considerable contribution to the reputation of FN's hunting guns. FN was a place where 180 highly specialised male and female workers were employed. They had often been trained by the master engraver and by the *École d'armurerie de la Ville de Liège*.[176]

André Watrin and then Louis Vrancken, both master engravers trained by Félix Funken, continued on this path of artisan excellence up until the 1980s. There were many female engravers involved in this adventure too. Notable among them were Leah Van Laar (who became head engraver), Antoinette Brigante, Jeannine Vanderspiegel, Sophie Purgal, Jeannine Pirotte, Lyson Corombelle and Marie-Louise Magis. In November 1960, the latter's design was preferred to that of her masters, Félix Funken and Louis Vrancken, by Val Browning himself.

The master engraver Félix Funken at the engraving school completing the engraving of the SA 22 rifle receiver, 1950s

End of a tradition

MOTORCYCLES: EXCELLENCE ON DISPLAY

Since the beginning of the 20th century, FN had been producing motorcycles in a highly competitive environment both nationally and abroad. This forced the company to continually improve and renew its products, in addition to adapting them to enthusiastic customers who were well aware of competing innovations.

Just after the war, in 1946, FN launched the series XIII motorcycles. They were technically far superior to those produced before the war with an original fork in two parts 'with a pulled wheel', a rubber suspension system for comfort that was unmatched until then on this type of machine. Despite these obvious qualities, their appearance attracted some criticism. The front fork, which was deemed unattractive, was modified in 1949, yet maintained the same principle. The range was enhanced over the years: the pulled wheel fork or telescopic fork, model XIII RT (1951) and, in 1953, the pulled wheel fork was finally abandoned in favour of the telescopic fork.

FN continued to demonstrate the excellence of its 'motorcycle' products to its customers. It took part, as it had before the war, in a number of internationally renowned

and local competitions: Liège-Milan-Liège endurance rallies, the Algiers motocross and the Championship of France motocross, as well as more modest competitions of which there were blow-by-blow accounts in local newspapers. FN took every opportunity to talk about its machines or its official riders, such as Auguste Mingels, champion of Europe in 1953 and 1954, or Victor Leloup, who would take victory after victory for FN before he stopped competing in 1956.

As before the war, FN protected its image. It organised and encouraged the organisation of long-distance rallies with fearsome reputations, in order to show off the endurance qualities of its machines.

The rally that attracted most media attention was no doubt the Brussels-Kamina rally (9 December 1950 to 15 February 1951). It was an initiative of the Belgian government that wanted to promote its new Kamina air base in southern (then Belgian) Congo and to stress that it was a legitimate part of Belgian territory. Seven paratrooper officers rode from Brussels to Kamina on mass-produced 450 type XIII FN machines, purchased by the national defence ministry. No accompanying vehicle was

foreseen. The maintenance and repairs had to be carried out by the riders. It was clearly a success for FN and the government.

Mr and Mrs Cassiers, from Antwerp, set themselves another challenge: to tour around the Mediterranean, with their camping equipment, on a 350 cm³ FN motorcycle with side valves. In the spring of 1954, they covered a distance of 18,000 kilometres, to the company's great satisfaction: "In spite of the tough conditions in which the tour was undertaken, the motorcycle never showed the slightest weakness and returned to the country without having required the slightest repair. With an FN motorcycle, any trip can be carried out with total security."[177]

For their part, Mr and Mrs Dupont left Brussels to travel to South Africa in 1957 on a 250 cm³ FN-M22, to which a trailer carrying 200 kg of luggage was attached. They took delivery of their specially customised vehicle in FN's courtyard, before setting out on their expedition. They covered 43,000 kilometres in travelling to Cape Town and back. The motorcycle was equal to the challenges of the often dangerous tracks, sandstorms and wild animals.

At the Salon de Bruxelles [Brussels automobile sh 1956, FN presented the Royal Nord moped, whic designed to be comfortable and very well equ There were two versions, the Standard and Luxe. 'sports' model completed the range soon after. models would follow, but in 1964 the factory st manufacturing two-wheeled vehicles.

Miss Belgium posing on a Princess scooter, 1958
Between 1958 and 1967, Fabrique Nationale d'Armes de Guerre produced one range of 49.4 cc scooters with modern lines.
The Princess scooter was the luxury model and had three gears, leg guards, an attractive streamlined look and a two-seater saddle.

FN MEDIUM-ENGINED MOTORCYCLES AND SCOOTERS

From 1953, FN's goal was to meet the demands of younger and less well-to-do customers seeking affordable ways of travelling. Along came the M22, from 175 cc to 250 cc, in December 1952. The model emerged from cooperation with the famous Küchen brothers, Richard and Xaver, in Germany, who designed its engine. This was an opportunity for FN to enter into and compete on the markets of Gillet (whose models evolved from 100 cc to 250 cc) and of Saroléa (known for its 125 cc Oiseau bleu). Shortly afterwards, a sales agreement was reached with Saroléa for this model under the twin FN/Saroléa label, the aim being to reduce production costs and to face competition from abroad.

The 'trial' version of this vehicle, which was then produced under the M22 TT ('Twin Trial') name, differed from the 'general public' version by having no rear suspension, raised exhaust pipes and a reduced turning circle. The TT distinguished itself in many motocross competitions, as ridden by Auguste Mingels.

In 1959, FN started manufacturing scooters to meet public demand[178]. Each of the many ensuing models was better and more powerful: the Princess (50 cc) scooter in 1960, the Fabrina scooter (1960) and the 'scooterette' [mini scooter] (75 cc) in 1962. The advertising for these products was mainly geared around female models, including Miss Belgium in 1958. It clearly aimed to show how easy these very popular machines were to ride, and that they offered their owners freedom and independence.

It is instructive to look at the ways the company communicated with its customers in the 1950s. Before 1914, the brand sought to show how serious it was through illustrations of domineering women. In the period between the wars, posters took on a futurist feel, indicating a shift in approach. Illustrations now evoked movement as well the machines and their riders. In complete contrast, buoyed by the optimism of the Trente Glorieuses [the thirty years following the Second World War], the posters and visuals more generally (posters, pamphlets, brochures, etc.) introduced real life, men (and above all women), colour and light into the advertising approach. This was often the work of graphic designers hired by FN, the same ones employed in the inter-war years, such as Pol-François Mathieu and Albert Fromenteau. Young, smiling and energetic people – among them young women – now embodied 'modern' life. They intentionally sent a simple, clear and often repeated message: "You too!" The aim being to create a direct link with potential customers.

THE END OF FN'S MOTORCYCLE ERA

This apparent optimism and creativity were not enough to save the company's two-wheeler business. Production stopped, due to several factors at the turn of the 1960s. Firstly, even if this was never overtly announced, FN was losing interest in motorcycles and scooters. It was increasingly focused on plane engines, nuclear engineering and even farming equipment.

In 1963, during the General Assembly of 24 October, mopeds were only mentioned in two lines: "Our sales continue to be affected by the decline in 'two wheels' to the benefit of 'four wheels' in our country." This already sounded somewhat like a funeral speech, which was itself pronounced at the General Assembly of 1965: *As motorcycle manufacturers since the beginning of the century, it is not without regret that we have finally decided to abandon making the moped. In reality, in the two-wheeler business, we never succeeded after the war in achieving the minimum quantities that could have made our production economically viable. With Saroléa, Gillet and FN, the production of motorcycles was in the past prosperous at Herstal and we have to admit that there is a link between the wage levels in the Liège region and the business moving to other parts of the country where the remuneration is lower.*

Wages surely played a role in this decision to halt motorcycle production, though they were certainly not the only reason. Other factors contributing to this at FN included the production methods of the Pré-Madame factory, the sometimes risky technical and commercial choices, as well as a company culture that did not prioritise leisure goods. The sector was already on its last legs in Belgium. In 1950, the government directed the three competitors – Gillet, Saroléa and FN – to reach a commercial and technical agreement in exchange for maintaining a quota system for imported motorcycles.[179] Difficulties started to emerge with Gillet in December and soon after with Saroléa. Gillet Herstal folded in 1959 and Saroléa in 1963.

LAST TRIALS WITH MILITARY VEHICLES

After the end of the war, FN took up manufacturing trucks, with a model derived from pre-war advanced cabins. Responding to a call for tenders from the army, FN partnered with the firms Miesse and Brossel, using the name Entente Inter Constructeurs, to develop a four-wheel model which was then marketed in 1950. The Belgian Army bought 4,000 of them between 1951 and 1953. It was a real success story. A version of this truck, known as the Ardennes, was first made in 1958. It was highly robust but did not achieve the expected success: the Belgian Army bought only a few more than 200.

An innovative vehicle for the army, the AS24, was exhibited for the first time at the Salon de Bruxelles in 1960. It was a tricycle that could transport four soldiers carrying firearms, sitting in a row. It was unique because it was partially foldable and could be dropped by parachute. The trike weighed 200 kg and could carry a payload of 350 kg. The AS24 could be customised for different uses: transportation of injured people, as a snowplough, carrying a radio set or an electric generator set, etc. More than 450 of them were sold to the Belgian, French and Peruvian armies and a very small number to the US, South Africa, Israel, Qatar and a few other countries.

For these three-wheeler vehicles, FN once again endeavoured to use its knowledge of a technology acquired over several successive generations of vehicles. The 1,000 cc Tricar FN 12 T3 for passengers, designed for the army shortly before the Second World War, was transformed into a civilian product just afterwards (the commercial T8 Tricar) and most likely inspired the military AS24.

In 1967, the Board announced it was "stopping the production of vehicles, thus ending an uninterrupted series of over 65 years of often successful development".[180]

PRECAUTION AND DIVERSIFICATION

FN had set about making various high-added value products. Yet it continued making simpler products, knowing from experience that these could ensure the survival of the company and its staff in the event of a major crisis. Europe's Common Market opened up new prospects[181] for the more diversified farming equipment that FN manufactured, such as milking machines, refrigerated cabinets, and deep freezers.

The takeover of the 'industrial textile machines' section of the Société Belge d'Optique et d'Instruments de Précision de Gand in the early 1960s was difficult at first. However, it quickly found its feet and sales took off from 1962. FN innovated by developing a small circular knitting machine with four chutes. In 1963, FN ecstatically declared: "Our 'Cotton Pull Over' business is currently experiencing success that we wouldn't have dared hope for, both in Belgium and abroad, with 71 machines already sold."[182]

◀◆ **Air drop of the Tricar AS 24, 1959**
AS 24 is the acronym for 'Airborne, Straussler', the patent holder; 24 stands for the motorcycle's engine, the M24. Developed in 1958, the machine had a 243.5 cc two-cylinder and two-stroke monobloc engine, with turbine-driven air cooling and a four-speed sprocket and chain transmission. The Tricar could be put into operation in less than a minute and had a top speed of around 60 km/h for a range of 100 to 200 km, depending on the terrain.

◆▶ **Assembly of knitting machines in Ghent, 1964**
In 1974, the knitting machines division was concentrated in Bruges, but it would be abandoned four years later due to strong competition.

Boeing turbostarter, mid-1960s
FN-Boeing Turbines was a company started in Herstal in 1964, to produce the Boeing Model 502-12B aircraft compressed-air starter. This compact gas turbine-driven compressor provided a source of air to start jet engines or was used in other industrial applications requiring high volumes of low-pressure air.
The turbostarter had two sections: the gas producer, i.e. a single-stage axial flow turbine with two interconnected combustion chambers, and the compressor outlet, which included a second axial flow turbine coupled with a single inlet centrifugal compressor, and an air flow controller.

Redefining the fundamentals worldwide

ENTERPRISING FN

In 1964, FN turned 75 and honoured this with a major commemorative event on 27 June, attended by the Brownings, current ministers and Auguste Francotte, who wrote an official history of FN.[183] On that day, the company could take stock of the long road travelled since 3 July 1889. The evolution had been impressive, starting with the initial order of 150,000 Mausers to the FAL which would sell in the millions; from the chainless bicycle for the gendarmerie to the AS24 parachutable tricycle and the first four-wheel drive trucks; from 'sparklets' to plane engines, from steam engines to nuclear energy; from the cartridges that the company with no firearms experience first produced in 1891 to the manufacture of NATO ammunition; and from a factory where 1,000 workers worked to an industrial giant employing 13,000 people in 1964.

FN was now a major export industry. It no longer needed to hark back to the good old days and could redefine its core business on a global scale.

Armaments came first:

The company owes its foundations in 1889 to this division and the ongoing dynamism that it has shown, despite the handicap of two wars which interrupted its activities, is remarkable. What are the reasons for this? We will cite:

- *The skills of a region where the manufacture of weapons has always been honoured;*
- *FN's long partnership with the Browning family, attracted to Liège by the reputation of its arms manufacturers;*
- *The longevity of its products, some of which date back to the beginning of the century;*
- *The production at Herstal of the two components of shooting: the gun and the cartridge;*
- *The advanced mass production methods employed at FN and the constant improvements in productivity;*
- *Finally, bringing together two activities, civilian and military, in the same workshops. These were technically similar although commercially different.*[184]

This introduction underlines for us the key qualities of the 'Weapons' Division of FN – qualities introduced by its management: historical legitimacy, consistency of production, innovation, rationalisation of production, and productivity. However this introductory text also mainly served to reaffirm that the arms manufacturing sector would remain dominant in the company's business.

Gift of a Browning over and under shotgun to the President of the French Republic Georges Pompidou, from Othon Drechsel, Managing Director of the Fabrique Nationale d'Armes de Guerre, 1971

The reminder was probably necessary: the rest of the report notably covered the high-technology products – underway or simply being explored – in which the Pré-Madame factory was involved: J79 turbojet engine, Tyne turboprop, Dart turboprop, Boeing 553 turbine, 550 turbine, Blue Streak rocket and Hawk guided missiles, as well as nuclear engineering.

Still on the lookout for weapons of war or ammunition with a future, FN was quick to adopt production processes that were new or simply variations of existing products, often in partnership with seasoned partners. For instance, in 1957, FN built various models of rifle grenades developed by the Société Technique de Recherches Industrielles et Mécaniques (STRIM) of Paris. Subsequently, it began research on rockets and a test bench was set up in Zutendaal, in Limburg.

In 1936, FN had built anti-aircraft artillery pieces developed by the Sweden's Bofors Aktiebolaget.[185] In the aftermath of the Second World War, this firm invented an even more powerful (57 mm calibre) anti-aircraft gun complete with a remote control mechanism. It gave FN a licence to produce this highly complex gun for the Belgian Army (1953-1956).

Besides diversifying its products, FN often acquired holdings, by joining the capital of certain companies. The aim was not to make profits quickly but either to provide help setting up a company with promising projects or to support that company in developing new ones, and in so doing to extend FN's own operations.

Consequently, in 1958, FN created the public limited company Métallurgie et Mécanique Nucléaires (MMN) with the Société Générale Métallurgique de Hoboken (SGMH). The factory was set up in Dessel, close to the Mol research centre. The goal was to get a foot in the door of this new form of energy, nuclear energy. Success came quickly to the company, though it was noted in 1971 that there was still insufficient commercial production. Nevertheless, FN was determined to be active in the leading sector of mechanical manufacturing for nuclear power plants – which it did for some 15 years.

On 6 April 1964, FN created the US company 'FN-Boeing Turbines' with the Boeing Company of Seattle. Established in Herstal, this company would build, sell and maintain small gas turbines. Its aim was to manufacture 250 to 500 horsepower turbogenerators (which were popular at the time), for aircraft engine starters and

Test of missile launchers at the remote controlled unit of the Fabrique Nationale d'Armes de Guerre, 1965
At the time, FN worked with ACEC on producing the electric generator and certain parts of the launch ramp of the ground-to-air Hawk missile.
The MIM-23 Hawk – an acronym for 'homing all the way killer' – was a medium-altitude ground-to-air guided missile developed in the 1950s by the US firm Raytheon Company. With a range of 50 km and a top speed of Mach 2.4, it could intercept enemy planes and ballistic missiles..

for the Swedish 'S' tank. This company was dissolved in 1968.

A list sent to shareholders in October 1972 highlighted the main companies in which FN had an interest and the shareholding that it had in their capital:

Companies distributing FN products

Browning, Morgan, United States	3% of the capital
Browning, Montréal, Canada	30 %
S.A. Cartoucherie Française, Paris	13 %
S.A. Fusi et Co, Milan	33 %
S.A. Schroeder Frères, Liège	33 %

Companies related to FN's industrial activity

S.A. Métallurgie et Mécaniques Nucléaires M.M.N.	19.8 %
S.A. Belgonucléaire	0.85 %
S.A. Établissements Lachaussée, Ans	3.5 %
S.A. Fokker	3.1 %
S.A. Dumont-Sclaigneaux	37.5 %

The financial results posted by the supply companies were not surprising and were more or less repeated from one year to the next: Browning US and Canada were at the forefront whereas Fusi and the Cartoucherie Française were struggling in their respective markets; Schroeder, FN's general agency, "is working pretty well and provides reasonable profitability".

We referred earlier to the disappointing results of MMN. Like all nuclear energy companies in Belgium, Belgonucléaire also suffered due to a lack of government support. Les Établissements Lachaussée, originally a family business and a specialist in cartridge machines, worked closely with FN. FN's interest in the Fokker company dated back to 1966, when the Dutch aeronautical equipment manufacturer bought up SABCA (Société Anonyme Belge de Constructions Aéronautiques). FN had participated in the latter's foundation in 1921. Fokker's results surpassed all expectations and a merger with a German company was envisaged. FN and the Société Générale de Belgique took out a majority share in the company Dumont-Sclaigneaux, so as to continue the mechanical engineering workshop activities of Société G. Dumont when it was liquidated.

Ammunition factory's machining and revision section, mid-1950s
In late 1965, the Fabrique Nationale d'Armes de Guerre employed some 11,000 workers, including 3,500 women. Six hundred female employees worked in the ammunition factory, 200 in the Engines Division, 350 on the assembly of firearms, 1,200 in the great hall and the small hall for machining, 1,000 in the Mauser workshops and 200 in general services. Most of these workers were categorised as 'specialised labour'.

KEEPING PACE WITH SOCIAL CHANGE

In the late 1950s and 1960s, the industrial world saw an upheaval comparable in scope and impact to the arrival of mechanisation in the 19th century. It became a world of specialists driven by technical and technological innovation, influenced by new sociological approaches. They were the witnesses and sometimes the participants in the trend towards more financialisation[186] of the economy.

This was the new world in which Fabrique Nationale d'Armes de Guerre would once again need to evolve. Starting in the 1950s, minutes of the Board of Directors' meetings reported, at almost every monthly meeting, on "the workers' question", "personnel problems", negotiations, consultations with the unions, and sometimes work stoppages: the size of the company (13,000 employees including 11,000 production workers) lay behind many of these difficulties.

Apart from these conflicts typically found in big companies, FN experienced an unusual dispute in 1966. It was very long, with work halted for three months, and above all its form and impact in Europe were unprecedented. The dispute began in February and was initiated by the factory's female workers, who demanded the same pay as men for the same work. The obligation imposed by the Treaty of Rome on all member countries to end wage inequalities between men and women was set to come into force by 31 December 1964 at the latest. But the

discussions between employers, authorities and unions dragged on.

French newspaper *Le Monde* reported every move of this dispute, explaining the causes and consequences. Beyond a straightforward demand for equal pay, this strike reflected the new aspirations of female workers, in particular that they should no longer be restricted to secondary positions.

In the first few weeks, the women strikers came up against a widespread inability to grasp what was happening. Their fellow workers, the unions – socialist or christian – as well as the company's managers and directors failed to understand what lay behind this strike. The dispute ended in May, after the women workers had partially won their case.

FN has often justifiably believed that it does much more for the welfare of its employees than other businesses. It was quick to point out that it provided substantial assistance to its employees or workers who want to buy their own home. The company could also highlight its end-of-year bonus, ensuring that staff could share in the company's success. Plus it could refer to measures to limit redundancies as far as possible when the volume of work decreased, though this is of course a way to keep qualified staff for better times.

Yet the disputes continued. For the socialist and christian trade unions, FN has often been used as a sort of testing ground for their demands. It's a crown jewel of

9 February 1966, the Fabrique Nationale women go on strike

industry in Liège, Wallonia and Belgium, thus making the company a prime target.

Trade unions sometimes struggle to fully appreciate the economic changes that are taking place. Rather than imposing anything, the company wanted to convince. In 1979, the Works Council of FN[187] was invited by management to join a study trip to the United States, which included visits to the Pratt and Whitney, Browning and Ford plants. For management, the aim was to introduce the union representatives to the realities of the industrial world. The resulting conclusion was that "this trip was very useful for raising awareness of the competitive conditions facing the company as compared with US companies, along with the importance of the research and investment efforts required". The initiative was somewhat limited, and probably would not have been successful in the long run. But the innovative approach was still instructive.

Although the unions were not responsive to the demands of modern industry, FN had learned from its past experiences. In 1974, a new strike by female workers was quickly resolved: the company had no intention of prolonging a dispute that could damage its image. The female workers were successful in their demands, which were not simply wage-related. This strike's main achievements were the abolition of piecework, reclassification of jobs, promotion of women, improvement of workplaces and, above all, permanent consultation on these matters.

On 24 October 1974, five joint commissions were set up: 'Wages and Reclassifications', 'Promotion of Women', 'Daily Work', 'Working Conditions' and 'Information Problems'. The new measures arising from this were applied not only in redeveloping Herstal's old production facilities, but also in the new factories that FN built in Belgium and abroad.

This is perhaps the highlight of the Board's report to the shareholders' meeting of 27 October 1977, underlining the complex role of FN:

The company must achieve both economic and social progress. In publishing this report, FN wishes to inform its shareholders, employees, customers, suppliers and, more generally, all those interested in its activities, of the way in which it is fulfilling this dual responsibility.

In those years, these serious difficulties slowly weakened FN. However, the company was more affected by certain structural elements in the Belgian economy.

A BELGIAN STRUCTURAL PROBLEM

Between the late 1950s and early 1970s, inflation was a constant. Although still at relatively low levels, inflation necessitated regular wage increases, which of course only made FN's products more expensive on the international markets.

Government measures in social affairs sought to ensure a better standard of living for the workers and to reduce their workload. However, these measures also seriously affected the company's activities.

In 1964, according to the company's own estimates as presented to the General Assembly, the payroll (wages and social security) had risen by 94.3% since 31 December 1961. This figure should no doubt be interpreted with some caution, because over those years, FN had notably hired growing numbers of qualified managers and engineers with high salaries. Thus the workers' salaries were not the only cause of this rise.

At the same time, working hours were decreasing. They had been fixed at 48 hours in 1921, when the State first became involved in this field, and were maintained at that level during the war. From 1955, businesses were allowed to reduce this time to 45 hours a week.[188] Until the mid-1970s, working hours were constantly reduced and the 40-hour working week became the rule, prior to being reduced to 38 hours a week in 1978 and to 36 hours in 1981. Paid holidays also increased, rising from one week just after the war to four weeks in 1975.

As management and directors pointed out at every opportunity, it was impossible to increase productivity to levels that would be required by increases in wages and decreases in working hours:

It goes without saying that we are powerless to compensate for increases of this magnitude by an increase in work pace or productivity. If we want to avoid working at a loss or not being able to invest - which is another way of being forced out of business - we have no other solution than to increase our selling prices, with all the risks that this entails in the presence of increasingly fierce international competition.[189]

This was the crux of the problem: FN was an exporting industry, with the domestic market in Belgium only accounting for 11% of its sales in 1969. In 1973, FN's customers from 117 countries, all categories combined, were from outside the Common Market.

We have already seen how much the regular increases in sales prices had damaged relations with Browning, to the point of pushing the American firm to seek suppliers capable of manufacturing at lower cost. This was evidently the case for all the company's sectors, with the less automated ones becoming more vulnerable.

The company's turnover was still thriving, reaching nearly

Royal Air Force Hawker Hunter flying over troops from the 45 Commando Royal Marines in the Radfan region, in Yemen, 1967
British troops in those days carried small arms that were standardised on FN products: the FAL assault rifle (or L1A1) and the MAG 58 machine gun (or L7 GPMG). The UK had provided the Engines Division of the Fabrique Nationale d'Armes de Guerre with the technology needed to produce engines for fighter planes. Having proven its capabilities through production of Derwent V engines, FN was tasked with manufacturing the new Rolls-Royce Avon engine, for the Hawker Hunter aircraft of the Belgian and Dutch air forces. The Avon was higher performance, had an axial compressor (replacing the former centrifugal compressor), and required many parts and more complex tooling.

three billion Belgian francs in 1965. Yet this indication of good economic health was misleading: the wage bill was becoming ever larger, at around 60%, and this considerably reduced the company's self-financing capacity.

KEEN INTEREST IN TECHNOLOGIES

From the end of the post-war period until the 1970s, FN started manufacturing products that were seemingly very different from its core business. When the company was invited in 1948 to participate in the manufacturing of Rolls-Royce engines, it reacted cautiously. However it seized the opportunity, equipped itself accordingly, and trained its managers and workers in new high-tech practices. Technical and commercial success ensued, FN made a name for itself worldwide and, now established in this new field, it was able to conclude a series of contracts.

Different technologies were needed for each engine or missile, so each new order called for the purchase of new

machines and the retraining of personnel. FN accepted these obligations, because it was seeking profitability in the short term, and more importantly was eager to learn. The company wanted to learn how to use new technologies, new materials and new machines – in other words, it wanted to acquire tomorrow's know-how. The company's future would show that it was right to follow this path.

FN approached these new fields as it had always done: with caution and prudence. FN was well aware that others knew more and did better in areas that were not yet its own, so it was prepared to follow a learning curve. During initial production of the Derwent Rolls-Royce engines, parts from the British supplier were first assembled in Herstal. The company then learned how to make these parts and started to assemble them with other parts from England. When producing the 81st engine, it was already familiar with the techniques. The parts were perfect and the engine was now a true FN product. Once this stage had been completed, other increasingly

complex engines could be manufactured in Herstal, including the J79 by General Electric.

This keen interest in technologies would take FN even further. The company manufactured – under American licence for Western armies – Hawk surface-to-air missiles capable of destroying a target flying at an altitude of 15,000 metres. Moreover it participated, albeit on a modest scale, in the European programme ELDO.[190]

Here we should perhaps read the words of Mr Denning Pearson, Managing Director of Rolls-Royce, in his letter to René Laloux in 1960: "that no licensee had offered them a friendlier relationship than the one they have with FN and expressed the hope of resuming at a later date the business relations that had been so harmoniously established with his company".[191]

Fabrique Nationale d'Armes de Guerre becomes Fabrique Nationale Herstal

On 10 September 1971, the Board of Directors met to ratify a decision to be submitted to the Extraordinary General Assembly convened one month later: the proposal was to remove the words 'armes de guerre' [weapons of war] from the company's name. All the directors agreed, apart from René Laloux,[192] Honorary Chairman of the Board and a key figure at FN: he thought this decision was "regrettable".

Unconcerned about the tautology, the talk initially focused on keeping the name 'Fabrique Nationale FN', because these two names were well known. In the end, after two sales managers contributed their views, the Board settled on the name 'Fabrique Nationale Herstal'. This name change was surely driven by the period's hostility towards everything that referred to weapons and war. The Vietnam War, which ended a few years later, provoked large-scale demonstrations which crystallised Western opposition to any armed and deadly conflict.

Yet this decision was not simply a semantic and commercial adaptation to contemporary preferences. It also highlighted a real desire to turn the page, to mark progress, to show ambition, to open up to the industrial world as it was and to highlight this without the constant reminder of 'the first core business', which dated back to 1889.

This change of name accompanied a process to reorient FN's activities, integrating the company into a global economic and industrial network and modifying its internal structure. Over a decade or so, these three deliberate and simultaneous steps transformed the very essence of FN. They were interdependent and could be readily explained by reference to one another. FN harnessed the new means made available to companies by management sciences, thus making these steps feasible and effective.

In the late 1960s, under the guidance of Michel Vandestrick, then Secretary General, FN set up an analytical accounting system. This rigorous method for analysing costs internally enabled short-term cost control and therefore a rapid reorientation of an industrial or commercial process if necessary. It gave the company a true dashboard.

Bringing in external consulting firms, formerly distrusted in the industrial world, also signalled an evolution in the company. In 1974, the Board had suddenly rejected the advice of a Swiss company for fear "that the advisors would interfere and eventually replace the managers". Several months later, the scepticism and distrust had vanished. FN drew on the expertise of McKinsey consultants to define the organisational procedures for

In 1964, Fabrique Nationale d'Armes de Guerre employed 13,000 people.
This was the highest number in its history, turning the company into a real town within the town.

manufacturing F100 engines. Between 1970 and 1980, FN began opening up to modern management tools. Even the Board, on the lookout for new ways to understand and make decisions, opted on 12 July 1976 to try out the 'SWOT matrix'. The goal was to assess the Strengths, Weaknesses, Opportunities, and Threats of a project to relocate some of the production somewhere other than the Herstal factory.

This use of modern management tools was significant. FN, in this field and others – as noted above, was marking its intention to enter the emerging new world.

Though the need for modernisation was acknowledged, it still gave rise to questions within the company. Defenders of the 'old guard', led by René Laloux, opposed modernisation. The debate continued about dropping the reference to weapons of war, as we have seen, plus the outsourcing of certain production processes, either by subcontracting or by creating new workshops "outside the walls" of the Herstal factory. There was also lively yet discreet discussion about the policy of participating in the capital of complementary companies. Two opposing philosophies were at play here. They concerned the traditions centred on Herstal versus a form of modernity linked to opening up to the world. The world of arms manufacturers set against that of new production

methods; and a passion for work done well contrasted with economic and financial necessity. The proponents of modernity clearly won the argument, bringing a change of paradigm and shifting the company's centre of gravity.

INTERNAL OR EXTERNAL?

FN often evaluated the best ways to physically expand its workshops, in order to support its expanding activities. The most conservative directors were against the opening of new sites, preferring improvements to the Herstal factory. René Laloux notably opposed moving the 'Weapons' Division to Harzé, some 30 km away, in 1974. His opinion was ignored, as colleagues were tempted by the prospect of smoother production, as well as moving the industry to a potentially new workforce.

The same question arose over the manufacturing of engines for the F-16 fighter plane. This time even more was at stake, as the investment required was huge against the backdrop of a financially difficult period for FN. Ultimately the 'External' decision was made (outside Herstal) in September 1976, despite the opposition of René Laloux.

Internal restructuring

Until the mid-1960s, FN had never worried much about its internal structure. One management team was in charge of three large and adjacent divisions (Weapons, Ammunition, Engines), each benefiting from the shared services of the laboratories, the foundry, the forge, the heat treatment department as well as the construction and maintenance department.

FN had been conceived as a typical 19th-century company. It evolved over time, but the industrial, economic and social changes of the early 1970s prompted it to implement a systems-based organisation. In other words, the company would now be seen as a complex set of interactions and no longer as a collection of parts that were disconnected from the functioning and overall activity of the entire company.

The organisation chart published in 1978, following several phases of internal restructuring since 1974,[193] underlined a clear desire to improve the efficiency of the structure of what was now a group employing an average of 10,000 people at the time.

Under the general management, there were four sectoral managements with well-defined competences: personnel management, administrative management, financial management, and the management of planning and equipment. The last of these was responsible notably for the planning of the central laboratory, for the coordination of research and development, and for equipment and facilities.

The activities of the divisions, under the leadership of the various managements, were also carefully described:
- FN Sports Division: ammunition and guns for hunting and shooting, golf clubs, tennis rackets, fishing rods, archery, sports promotion;
- Defence and Security Division: defence ammunition and firearms, security equipment, engineering;
- Weapons Series Division: mass production of sports and defence guns;
- Engines Division: turbojet engines;
- FN Formétal Division: precision forging, investment casting, heat and thermochemical treatments, surface treatment;
- The Industrial Equipment and Services Division comprised most of the new products of FN Herstal developed to ensure the diversity of the company's production (plasticising machines, aerial work platforms, waste collection vehicles, hydraulic equipment, operation and maintenance of public and industrial equipment, etc.) as well as everything related to the research and development of industrial mechanisation, industrial cooperation and technology transfer.

When reading this organisation chart, one is struck by the company's remarkable diversification. Nothing was apparently beyond its sphere of activity, so long as there was a connection with mechanics and the new technologies that now surrounded this field. FN was strongly involved in both the economy's secondary sector (manufacturing) and the tertiary sector (service activities).

It's also noteworthy that in the past FN was organised around three large factories (firearms workshops; cartridge factory; motorised equipment at Pré-Madame), to which a complementary activity was sometimes added (e.g. 'sparklets' at the cartridge factory until 1954, or post-war agricultural production at the Pré-Madame factory to use the presses that were then under-utilised). Yet in just a few years, it had established production centres defined by their customer(s): governments for the Defence and Security Division, for example, or sports enthusiasts and hunters for FN Sports, and local authorities or large companies for industrial equipment and services.

In 1974, the General Assembly was reminded that the Weapons Division employed 4,300 people and had 5,000 machines and that this division was the centre of production for guns sold by the FN Sports Division as well as the Defence and Security Division. The main objectives set for it were increased production, higher productivity and improved working conditions. This drive to streamline things, which led to the emergence of a technical division tasked with implementing civilian and military products, was a real change of paradigm for the company. Up until then, it had carefully kept the two activities separate, even though they were very similar at the industrial level. This place for interaction for civilian and military could only be beneficial for both sectors.

FABRIQUE NATIONALE HERSTAL S.A.

In 1972, 'Fabrique Nationale d'Armes de Guerre' became 'Fabrique Nationale Herstal'

FN stand at the Foire internationale de Liège [International Fair of Liège], 1971

Browning catalogue, 1980
For the FN Group, the equipment and services for practising sport and leisure activities were a booming sector. Research and development programmes aimed to capture and exploit the new market trends in areas as varied as hunting and shooting firearms; hunting bows and leisure, hunting knives; fibre glass, graphite or boron fishing rods for all disciplines; Browning tennis rackets with a honeycomb structure; golf clubs; and competition windsurfing boards.

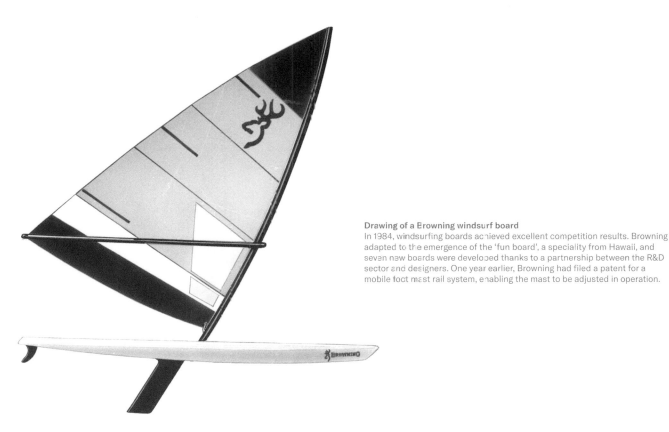

Drawing of a Browning windsurf board
In 1984, windsurfing boards achieved excellent competition results. Browning adapted to the emergence of the 'fun board', a speciality from Hawaii, and seven new boards were developed thanks to a partnership between the R&D sector and designers. One year earlier, Browning had filed a patent for a mobile foot mast rail system, enabling the mast to be adjusted in operation.

A DYNAMIC DIVISION: FN SPORTS

This division showed off FN's ability to quickly integrate others after the takeover of Browning in 1977.[194] In 1978, FN founded Browning France and took control of its old ally, the Fusi company in Milan. It now had a network of subsidiaries covering five major markets with a total population of 350 million (France, Italy, Belgium, United States, Canada), and this network was gradually absorbed into FN's marketing and sales departments, which were strengthened in the process. The research and development potential of FN Sports was increased, with the hiring of researchers and technicians.

The traditional Superposed line continued, with new variants, such as the 'Spécial Parcours de Chasse' in 1974 and the 'Super Trap' in 1976, thus renewing the success of these classics in high-level competitions. At the same time, some technological advances in the US would lead to the development of a new shotgun: the Citori. Some gun models were abandoned. The manufacture of others was entrusted to foreign countries (the BAR to Portugal), but new firearms especially were launched and these were better adapted to the market. FN's firearms production then tapped into the great strength of its network of partners and subcontractors. With Browning US and its own suppliers, FN with its partner Beretta or occasional associates, the company successfully combined the contributions of each into new products. These multiple inputs were found in the new models of .22 pistols: such as the 'Challenger III' (1982) and the double-action pistol, 'DA 140', which was launched in 1978. In the same year,

various .22 and large-calibre rifles were launched, and even a semi-automatic shotgun, the B80.

For FN, the takeover of Browning meant continuing the story that had begun with John Moses in 1897 and was continued under Val and Bruce. The expanded range of hunting guns had to continue with the same intensity and same inventiveness. At the same time, it was vital to reduce manufacturing costs, which was a key consideration.

For hunting ammunition, FN developed the 'Légia Star FN' in the 1950s. The cardboard tube containing the pellets was sensitive to humidity, so it was replaced by a more reliable metal tube. FN kept on improving its performance and developed a high-power cartridge: the Légia Star High Speed. Featuring a plastic wad shaped like a bowl, its reputation spread throughout Europe, helped by the gradual creation of the Common Market.

Besides these traditional products, the company took a key step forward in its programme to diversify activities, through internal growth and acquisitions. Browning's golf clubs were gradually gaining worldwide recognition, and the FN Browning range now offered a full range of irons, woods, putters, bags and golf carts. Two tennis racket models were successfully presented at the Cologne Trade Fair, before going into industrial production in 1979. The French company LERC was a specialist in technologies for processing composite materials. FN took a majority stake in this company at the end of 1978, because it was convinced this know-how would help with the future development of sports equipment.

Heavy Machine Gun Pod, early 1980s
Fitted with a .50 calibre M3P machine gun, the Heavy Machine Gun Pod (HMP) can be mounted under the fuselage or the wings of a subsonic plane or a helicopter. The aerodynamic streamlined design allows the continuous feed of the cartridge strips without the risk of jamming, as well as the collection of links and cases. This design also facilitated the air inlet for cooling and evacuation of the powder residues. To make the machine gun compatible with the structure, several elements were added to the gun: a solenoid sensor assembly, making it possible to adjust and fire shots from the cockpit and to see the number of rounds fired; a back block to absorb the recoil, since the rate of fire is around 1,000 rounds per minute, and a flash suppressor. The HMP is 1.9 m long and weighs 120 kg fully equipped. It is fitted on an elastic cradle that absorbs the recoil forces and it is fixed by NATO-approved attachments.

SUCCESSES OF THE DEFENCE AND SECURITY DIVISION

At the end of the 1970s, FN could rely on easy sales of firearms that had become mainstays in military arsenals: the FAL, the MAG, and the HP among others. Rather than rest on its laurels, the company continually improved its manufacturing processes and kept innovating by introducing new products.

A multi-purpose weapon

The MAG's versatility meant that it could be used in all kinds of combat. To ensure its effectiveness in every situation and to be useful in all kinds of strategies, FN added optional elements. So the gun could be used on the ground, but also when fitted to land, air and naval combat vehicles in increasingly sophisticated configurations. It was an opportunity for FN to introduce new know-how, which would became one of the hallmark capabilities of the Herstal Group in the years ahead. In 1978, the company released the 'Pod' – a new product that was destined to be a great success. This is a pod that can be mounted on an airplane or a light helicopter, equipped with one or two machine guns remotely controlled by the pilot and intended for close air support. FN has continued to adapt these systems to new needs. For example, the 1980s saw development of the 'Twin MAG Pod'. These have been perfected and redesigned right

through to today's Pods, which are digitally controlled and can be adapted to different combat situations. FN's periodical, *Le Journal FN*, gave the following account of this invention in January 1980.

The Twin MAG Pod: an original product ahead of its time
To expand the commercial range of our weapons, we're striving to diversify their use as much as possible. We have thus been led to adapt the MAG to different vehicles and, in particular, US Army tanks, which has enhanced their great reputation. There was a further step to be made or, if you like, a leap forward to be made: giving the MAG wings, i.e. adapting it to planes and helicopters. Today, that is now a reality. The Service Recherche et Développement de la Branche Défense et Sécurité [the Defence and Security Branch's Research and Development Service] has designed and developed a 'POD' or streamlined container, which is worthy of the star status that MAG has come to enjoy. The POD contains a pair of twinned MAGs that are positioned on elastic cradles, which ensure excellent levels of accuracy and eliminate the shocks and vibrations that could harm the aircraft carrying it.
It includes a mechanism that allows the machine guns to be rearmed during the flight and makes them secure on take-off and landing. [...] [The Twin Mag Pod] gives access to a completely new commercial sector for us, that of aeronautical weapons.

Combat helicopter equipped with Twin Mag Pods, 1982
The Twin Mag Pod (TMP) is fitted with two MAG machine guns that are made especially lighter for airborne operations. The TMP has a capacity of 1,000 rounds and can sustain a high rate of fire of 1,500 to 2,000 rounds per minute.

FN-BROWNING
Heavy Machine Gun
M2 HB-QCB Model

M2 or M9 detachable link belt

Feed mechanism

System
Automatic machine gun, operating by a short
recoil of the barrel

**Chromed barrel
or resistant alloy**

114.3 cm
Barrel length

Quick Change Barrel

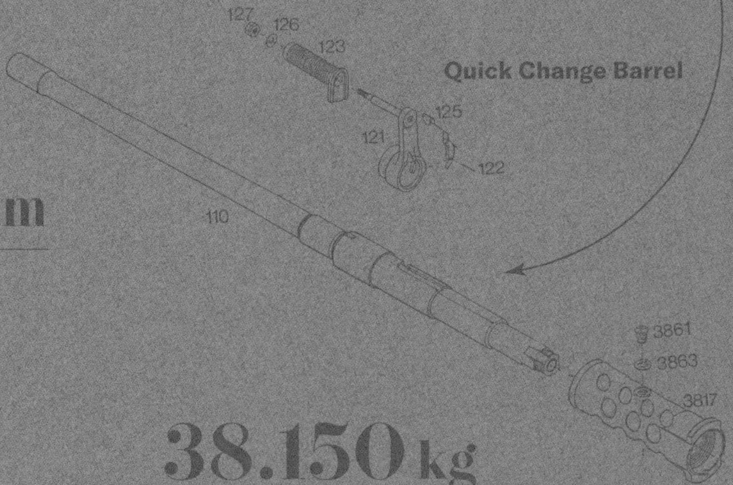

127 126 123
125
121 122
110
3861
3863
3817

165.6 cm
Overall length

485 / 635 rpm
Firing rate

38.150 kg
Weight

**Heavy Machine Gun, FN-Browning,
M2 HB-QCB Model**

In 1921, John Moses Browning developed a .50 calibre
machine gun, with short barrel recoil, cooled by water or
by air, to meet the US Army's request for a heavy machine
gun with armour-piercing capabilities. Based on the
design of the .30 calibre machine gun of his invention,
this machine gun nevertheless had to be able to absorb
five times more energy, which called for the addition of a
buffer. In the 1930s, the gun was further developed and
fitted with a heavy barrel, to enable sustained firing. This
led to the new name: M2 HB (Heavy Barrel).
Its firepower gave the US forces a considerable advantage during the Second World War, forging the machine
gun's reputation, nicknamed 'Ma Deuce' by soldiers.
Sales of .50 machine guns started in Herstal in the
1930s. After the war, the MAG became more popular
than its predecessor, but there was renewed interest in
the M2 HB in the 1970s. FN then modified the aviation
type .50 that it had bought in order to adapt them to the
army's needs. The aim was to reduce its weight, improve
its firing rate and its durability, but above all to include
the possibility of quickly changing its barrel. In 1978,
FN bought the rights to develop, produce and sell the
Quick Change Barrel QCB system developed by Mack
Gwinn Jr, a former member of the US special forces in
Vietnam, who developed various technologies to improve
the operation of several guns. The Quick Change Barrel

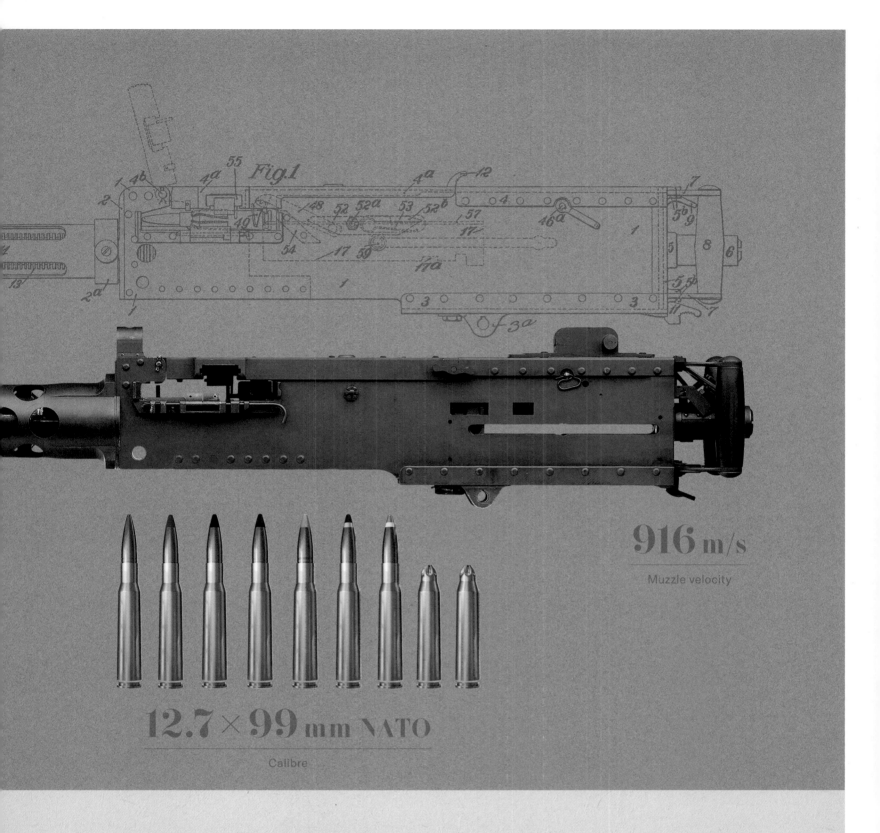

916 m/s

Muzzle velocity

12.7 × 99 mm NATO

Calibre

(QCB) system made it easier to replace a barrel which had overheated during long periods of sustained firing. Supported by this technological advance and improvements made, FN was able to supply .50 machine guns in different versions: army (with tripod), for mounting on vehicles or mounting on aircraft and helicopters.

Le Journal de la FN [FN's newspaper] from October 1978 ran the headline 'The return of the .50' and highlighted the commercial dimension of this evolution:

For FN and the Security and Defence Development division in particular, the relaunch of the .50 means an expansion in the range of weapons available, the possibility to reach new clients, a contribution to maintaining employment levels, an increase in turnover and also the prospect of selling ammunition of this calibre that FN has in fact never stopped producing.

Forty years on, the company's commercial success speaks volumes. The strength of FN was already, back then and today still is, its constant eagerness to adapt its arms production to the new strategies and changing nature of combat, as well as to incorporate or invent new approaches and new technologies.

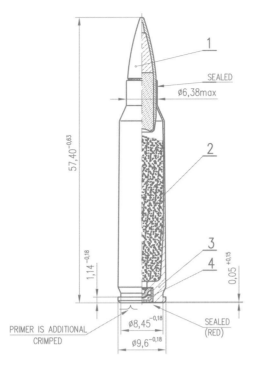

1

SEALED
ø6,38max

2

3

4

57,40⁻⁰·⁶³

1,14⁻⁰·¹⁸

0,05⁺⁰·¹⁵

PRIMER IS ADDITIONAL
CRIMPED

ø8,45⁻⁰·¹⁸

ø9,6⁻⁰·¹⁸

SEALED
(RED)

In 1967, NATO proposed to set a new standard for the 5.56 mm calibre. After numerous tests, Fabrique Nationale Herstal's 5.56×45 mm SS109 ammunition won against the rival American M193 ammunition. The SS109 Belgian cartridge uses a 62 grain projectile rather than the 55 grain projectile used by the M193, and it has a hardened steel penetrating core, which allows deeper penetration from a distance.

The stimulus of competition

In some situations, the 7.62 mm NATO ammunition turned out to be too powerful (leading in particular to a considerable recoil from light arms) while being too heavy to be carried in large quantities in the new theatres of combat. These included fighting constantly at close quarters, demanding a high level of mobility from combatants. Good examples include Vietnam, a key conflict of the time, and guerrilla warfare in Latin America.

So the hunt was on to develop a less powerful type of ammunition, suitable for typical firing distances, thus enabling soldiers to carry more cartridges during operations. FN developed this ammunition, standardised with NATO in 1980 as 'NATO 5.56×45', or SS109, as had been the 7.62 × 51 mm ammunition.

During this same period, Western powers sought to provide their troops with a versatile firearm offering a range, precision and power equivalent to that of the AK-47, developed by the Soviets.

From the early 1960s, the US, which was widely involved in Asian and South American conflicts, developed the M-16 rifle. This was novel because it used small-calibre ammunition, which was however slightly larger than 5.56 ammunition.

FN had no option but to react to the real success of the M-16. From the mid-1960s, it did so by conducting studies of a gun capable of firing the new 5.56 mm ammunition.

The Carabine Automatique Légère [Light Automatic Carbine] – the CAL – emerged from this research and it entered production in 1968. This gun was a masterpiece of arms engineering, but proved to be too complex, fragile and expensive. It was replaced in the mid-1970s by the FNC (Fabrique Nationale Carabine), and quickly recognised as a reliable, efficient and ergonomic gun.

The FNC's mechanism was a variation of the tried and tested AK-47. However, it was adapted and refined by the FN engineers, with the aim of simplification and cost reduction without loss of reliability. These included use of pressed and welded sheet metal for the casing, aluminium alloy for the whole receiver and a micro-fusion technique for the trigger mechanism and other steel parts. The gun enjoyed less success than the FAL, although in subsequent years the FNC equipped the Belgian, Latvian, Lebanese, Moroccan, Mongolian, Nigerian, Salvadorian, Ceylonese, Tongan and Venezuelan armies.

↪ The **FN CAL (Light Automatic Carbine) carbine** was produced to complement the FN FAL, for times when the action or terrain called for a lighter yet still effective gun. Chambered for 5.56×45mm calibre, it was a locked and gas-operated firearm. It made use of several solutions on the FAL, such as the placing of the gas piston above the barrel, adjustments enabling the gun to be disassembled, plus use of plastic for the stock, the guard handle and the handguard lining. Innovations included the sheet metal manufacture for the receiver and the trigger guard, making the gun lighter, as well as simplification of the barrel breeching, a rotating bolt stop, and burst options that allowed a single shot, a three-round burst or full automatic fire. This gun could also be fitted with a fold-out stock and a 40mm grenade-launcher.

↪ ↪ The **FNC M3** built on the improvements introduced with the CAL, thanks to simplification of certain key functions. It was also fitted with an M16 A1 type of 30-round magazine, interchangeable with the FN MINIMI machine gun. This meant that Fabrique Nationale Herstal was marketing a true family of light guns that used SS109 NATO ammunition – which the company had itself designed.

The FN MINIMI Machine Gun being tested in adverse conditions
Firearms tested by Fabrique Nationale Herstal must be able to handle the worst climate conditions present in various theatres of operation, including Arctic conditions.

THE MINIMI

Around the same time, FN started developing a light machine gun, which would be the "mini-machine gun", known as the MINIMI. The MINIMI was not really a new gun, as the United States had already designed an individual small-calibre infantry gun; but there was no standard military issue automatic gun.

The MINIMI is a gas-operated machine gun designed to fire 5.56 mm NATO rounds. This ammunition is less powerful than that typically used in machine guns, but it is smaller, enabling a more mobile gun that can be handled by a single person. The MINIMI has a high rate of fire: it can be used as a support, either in defence or in attack. It was designed to support the FN MAG, not to replace it. A squad (a small military unit of about 10 soldiers) is often equipped with one MAG and two MINIMI for support.

Production of the MINIMI began in 1974. Adopted by the United States in 1980, the gun is manufactured in America – in line with US legislation – by the subsidiary FN Manufacturing (FNM) under the name M-249.

The Defence and Security Division was warmly congratulated by the Board on 22 September 1980. FN learned at the same time about the success of the MINIMI in the United States during comparative tests on four guns (two American, one German, plus the MINIMI) as well as the adoption by NATO and the United States of NATO 5.56 ammunition. The Board underlined the importance of having ammunition and gunsmith specialists within the same company.

America's infantry to be equipped with Belgian machine guns

Brussels. – The American army's new machine gun will be Belgian. Following tests lasting 18 months, the US Department of Defense has decided to equip America's infantry with the calibre 5.56 MINIMI. The Belgian machine gun faced three competitors, two of which are made in America.

The news was announced by Fabrique nationale (FN) d'armes de guerre de Herstal, in a press release from Liège on Monday 22 September. The advantages of the MINIMI over previous machine guns are: a lighter weight, the NATO 5.56 calibre and the possibility of being loaded with the magazine of the American M-16 rifle. It is a support gun for infantrymen.

The American army is expected to acquire several tens of thousands of the MINIMI, which are to be mass-produced by FN's American subsidiary, Manufacturing of Columbia, in South Carolina.

This is the second time in two years that FN has won over its competitors on the American market. In 1979, the United States chose the MAG-58 Belgian machine gun for battle tanks. Twenty-six thousand units of this gun were ordered and produced in South Carolina by FN's subsidiary. These contracts are also a trade-off for Belgium's purchase of the American F-16 fighter jet.

Le Monde, 24 September 1980

FN MINIMI
Light Machine Gun
MK1 Model

5.56 × 45 mm NATO
7.62 × 51 mm NATO

Calibre

System
Automatic machine gun, gas-operated
open-bolt operation

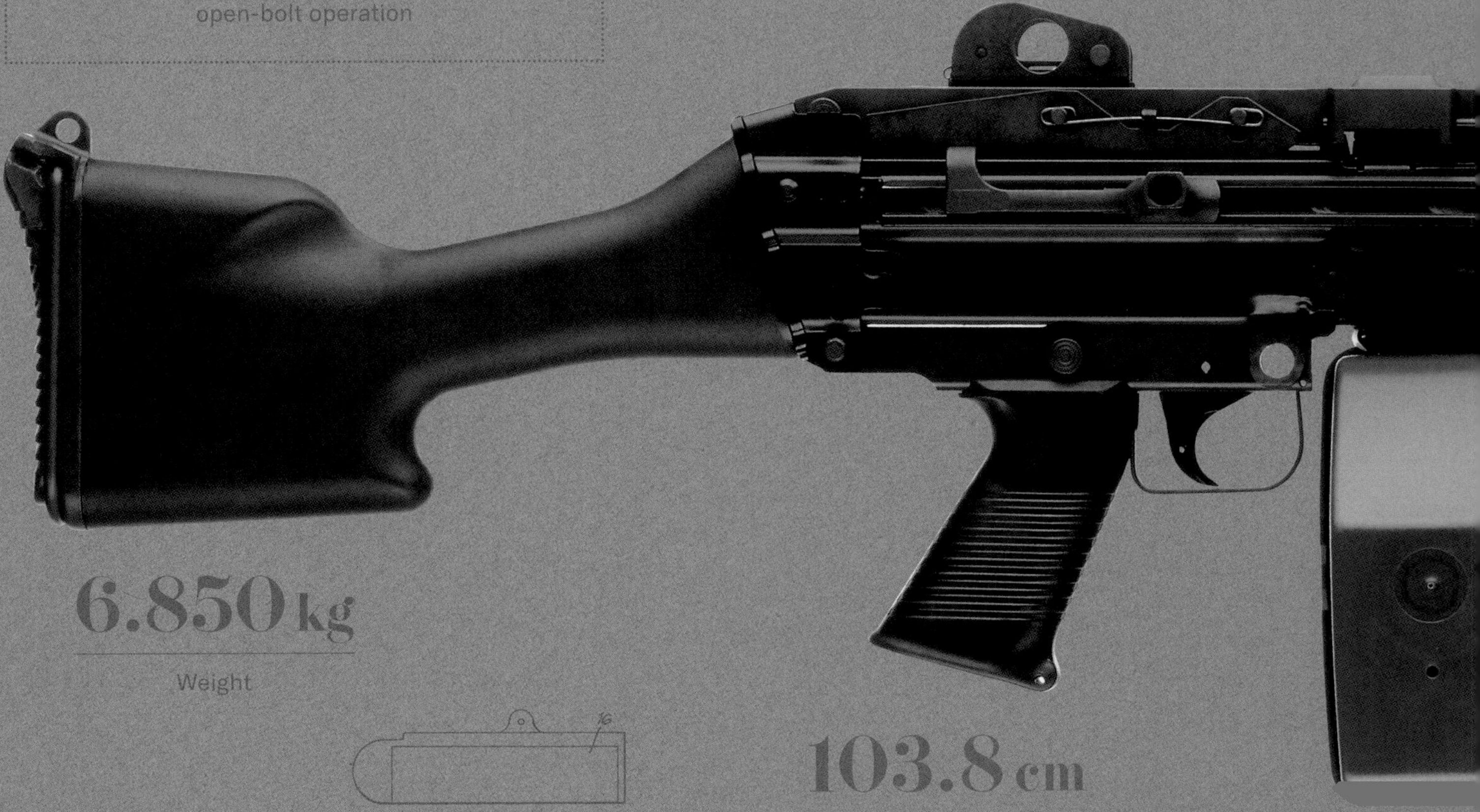

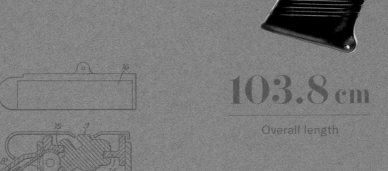

6.850 kg

Weight

103.8 cm

Overall length

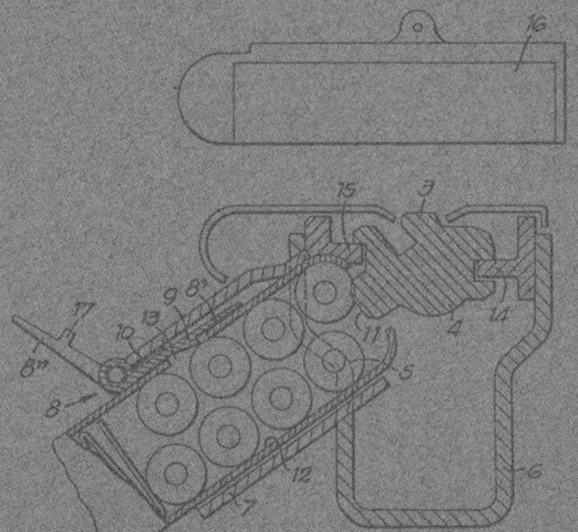

Research into the **FN MINIMI** began in 1971, alongside that on 5.56×45 mm calibre ammunition, to respond to the concept of a Squad Automatic Weapon. On 19 December 1981, after numerous tests in adverse conditions, the US Army chose the FN MINIMI for the last phase of tests, focusing on ergonomics and adapting it to the SS109 ammunition chosen by NATO. Mass production began at the end of 1981 and, from February 1982, the machine gun was incorporated into the basic equipment of the US Army under the name M249. Since it was adopted by the leading NATO country, its reputation was forged worldwide.

The FN MINIMI was at that time the only machine gun in the world offering a triple feed: belt, box of 200 cartridges on links, or the M16 or FNC-type magazine of 30 cartridges. It is also equipped with a QCB system and operates in open bolt and is gas-operated through a small hole in the wall of the barrel. The open bolt, guided

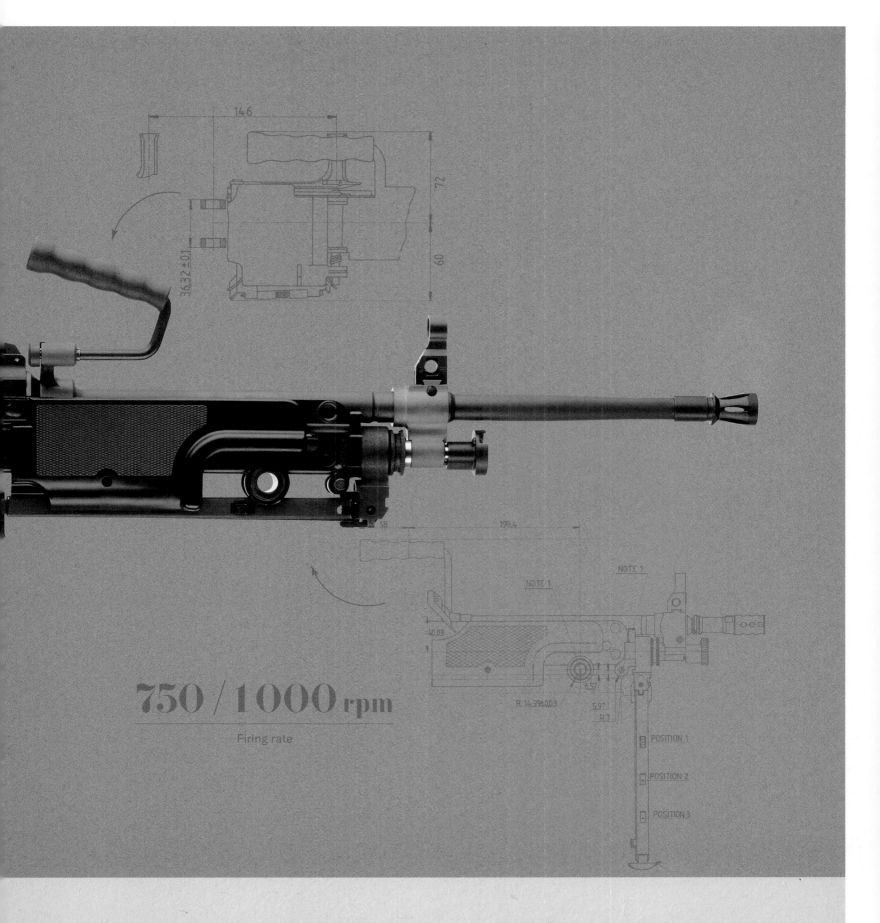

146

72

60

36,32 ±0,1

750 / 1000 rpm

Firing rate

58

199,4

NOTE 1

NOTE 1

NOTE 1

40,03

R 14,39±0,03

6,57

5,97

R 7

POSITION 1

POSITION 2

POSITION 3

by two connected longitudinal rails, increases cooling by circulating air through the barrel and prevents 'cook off'. A two-position gas inlet regulator enables shooting in normal conditions without unnecessarily stressing the moving parts while maintaining a power reserve for use in adverse conditions. The gun is fitted with an anti-runaway mechanism, to prevent a traditional shortcoming of machine guns that arises when the moving parts do not move back enough to be caught by the sear.

As a result they move forward again under the pressure of the recoil spring, which can lead to an uncontrolled burst of fire. Finally, the principle of rotatory locking in the barrel end has enabled the construction of a lightweight receiver made of pressed steel.

THE ENGINES DIVISION – FOR CIVILIAN AND MILITARY AEROSPACE

Starting in 1974, FN invested extensively in a project to replace the Belgian Air Force's F-104 G jet fighters. It negotiated joint production pre-agreements with different engine manufacturers that could potentially be involved in this particularly intense competition.[195]

A decision was made to select the F-16 in 1975:

The 'Engines Division' saw enhanced prospects for future growth. The experience and technical know-how acquired through previous programmes with different licence holders and via different forms of international cooperation today helped us gain a greater stake in the joint production of Pratt & Whitney's F-100 engine [...]. All the engines intended for planes from the European consortium will be fitted and tested in our Herstal and Liers facilities, while our production workshops will manufacture a selection of the parts and mechanisms for the European engines for F-16s bought by third countries as well as for a proportion of engines [...] destined for the US.

The size of this huge programme [...] will create additional highly skilled jobs. [It] will considerably increase the division's technological potential and knowledge.

F16 fighters flying at night
In June 1975, the Belgian, Danish, Dutch and Norwegian air forces selected the General Dynamics F-16 to modernise their fleets, just as the US had done a few months earlier. Dubbed the 'Swing Force Fighter' for its capacity to adapt to air combat missions and to air-to-ground attacks, the F-16 offers enhanced combat capacities and manoeuvrability thanks to its high-thrust engine, a low wing load, high resistance to gravitational force and a simplified design.

Major investments will need to be granted to adapt our equipment and installations to the specific needs and to the new techniques of the programme and to meet the demands for competitiveness, which is much more intense than in the past, given the levels of our wage costs.[196]

The distribution of the work was set out in a diagram during the General Assembly of 1976: the fan and high-pressure chamber were manufactured by FN; the low-pressure turbine by Norway; the post-combustion nozzle by the Netherlands and the gear box by Denmark. The fitting and the final tests were also carried out at FN.

The projected budget was a total expenditure of 28 billion Belgian francs (manufacture, assembly, tests, purchase of materials, raw and finished parts, etc.). Investments in machines and manufacturing equipment and testing equipment amounted to 1.15 billion Belgian francs and the construction of a new factory came to 1.2 billion Belgian francs. Two thousand people would need to be recruited for the mass production phase.[197] Production was scheduled for 1978 to 1984.

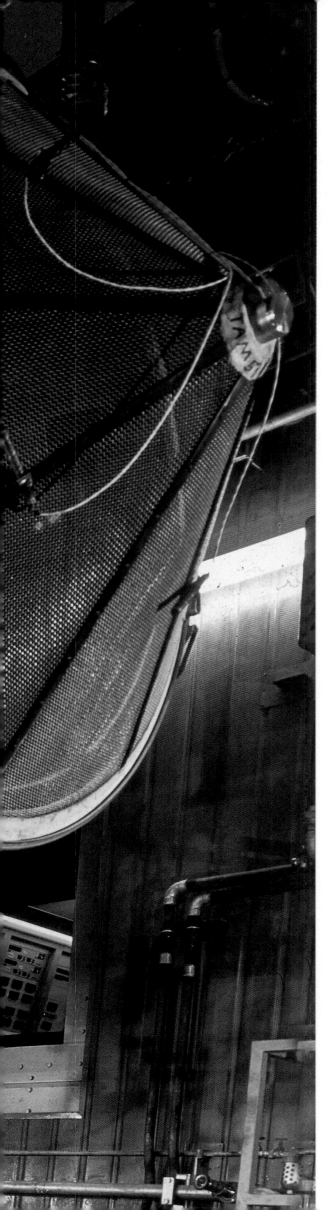

The new factory was located in the industrial zone of Hauts-Sarts, near Liège. This new factory was completed with the construction and fitting of new test benches in the Liers test centre.

The first F-16 was delivered to the Belgian Air Force on 26 January 1979. On that date, FN underlined that the "contract of the century" was not limited to the manufacture of modules, or the assembly and testing of engines. Notably, FN contributed to the creation of SONACA,[198] which was responsible for assembling the central part of the F-16's fuselage, by buying a quarter of its capital. That highlighted the importance of this programme for Herstal's aeronautics business. In 1979, it was estimated that more than 2,000 F-16s would be sold and perhaps even more, and that the plane would remain in service until the year 2000. The FN periodical, *Le Journal de la FN*, ran the following headline in March 1979: 'Bon pour le service jusqu'en 2000' ['Fit to be in service until 2000']. In 2020, the F-16 was the world's best-selling fighter plane, accounting for 16% of all fighter planes in service.[199]

The company was also keen to get involved in civil aeronautics. Air transport was growing strongly and had become a highly profitable industry. So by being involved in civil aviation, the company could sidestep the cyclical nature of military orders.

In 1971, the French company SNECMA and the American company General Electric signed a cooperation agreement to jointly manufacture 10 tonne thrust engines, the CFM56, for the re-engining of civilian and military aircraft. FN had already worked for a few years beforehand (1968) with SNECMA on the military ATAR 9C engine, so it made sense to call on FN again.[200] This time FN would be involved in a cooperation and not just as a subcontractor. FN's Engines Division was the joint developer of the programme and took part in the research, which again gave it access to cutting-edge technologies.

PW F100 engine on the test bench at Liers, 1980s
The Liers fort was a military structure erected in the late 1880s to defend Liège. Thanks to its isolated and secure position, it was ideal for setting up a test bench: this was fitted in the 1950s and completely renovated in the early 1980s. At that time, it offered four benches for FN Engines, enabling tests of over 20 tonnes of thrust with post-combustion and four benches for small engines. Complete with computerised systems for automatic and processing of measurements during tests, the Liers test bench could be used for mass production testing as well as maintenance and development.

Assembly of JT9-D at the Pratt & Whitney factory, 1975
FN Engines became involved in civil aviation with General Electric-Snecma's CFM56 engine in 1972. In 1980, FN Engines partnered with Pratt & Whitney to produce parts for the JT9-D engine for the Boeing 747 and the Airbus A300.

The Boeing 747, America's mass-produced jet airliner and cargo plane, making its maiden flight, February 1969

The CFM56 would become the best-selling aircraft engine in the world, being fitted in 33,000 civilian planes, and notably those made by Boeing and Airbus. The timing was perfect. Since the early 1970s, civil air transport has doubled in size every 15 years and the medium-haul carriers have contributed most to this expansion. Thanks to its reliability and relatively low-cost maintenance, the CFM56 engine has also contributed strongly to this growth.

This programme enabled the Engines Division to fill its order book in the civilian sector. Above all, it helped the division to take its place among the elite companies in aeronautical industrial cooperation.

Civilian air traffic continued to grow and, from the end of the 1960s, there was a move towards manufacturing big planes, the Jumbo Jets. The first Boeing 747 had a capacity that was two and a half times greater than the 707. FN benefited from this trend. In 1983, it was sub-contracted to produce parts and assemblies for the Pratt & Whitney JT9D, variant 7R4, for the re-engining of Boeing 747 and 767-200 aircraft as well as the Airbus A300 and A310.

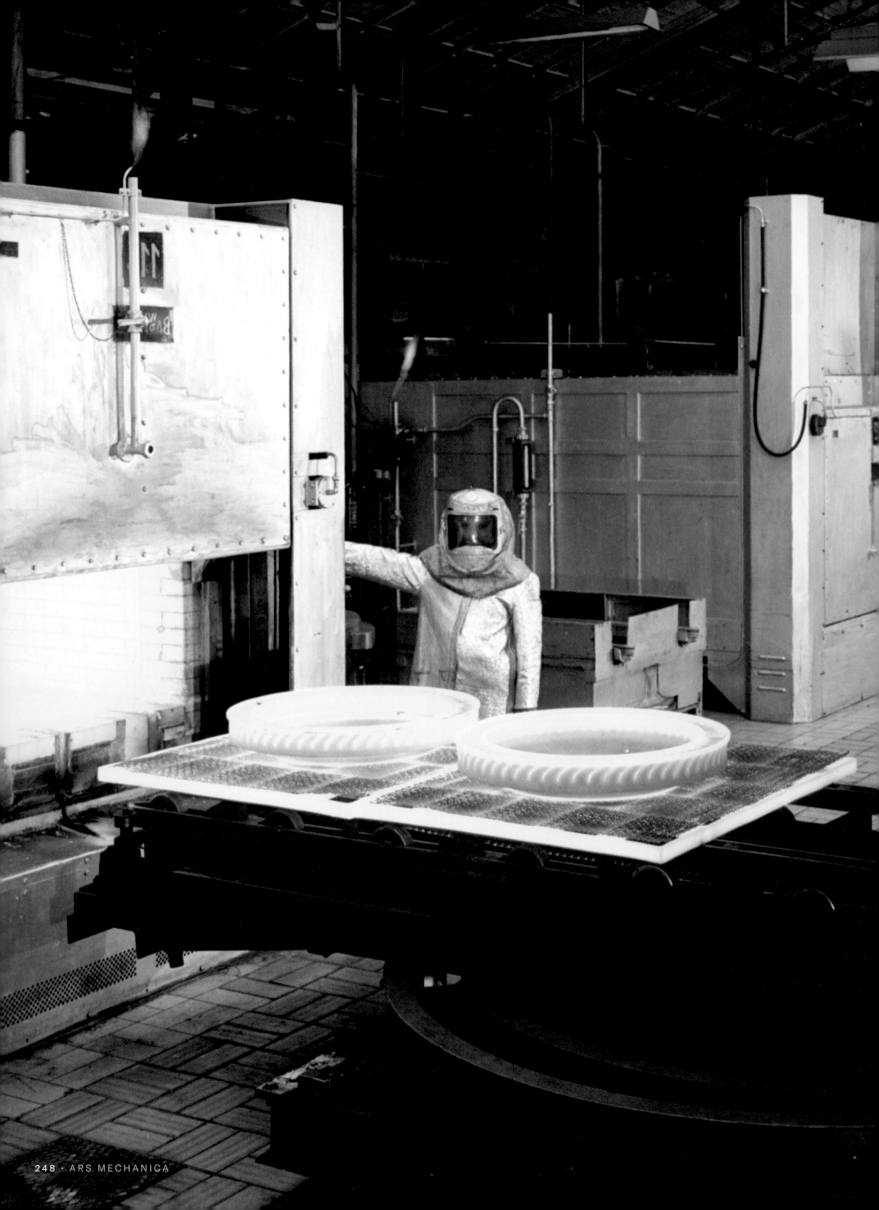

FORMÉTAL:
HOW A COMPANY DEPARTMENT
REFOCUSED ON EXCELLENCE

Central Production Services were restructured in 1975, on the initiative of McKinsey consultants. They were able to estimate the cost of these services at 700 million Belgian francs per year, money that could be used to better advantage by making it available to other companies. A market study was therefore launched on how to develop industrial and management services for large facilities, whether for the private or public sector. The study revealed that this type of service could be developed in companies for the maintenance of machine tools or the management of fluids. This was an opportunity for the repurposing of the general services, provided that a partner with the necessary experience was involved. This is why Cofreth, a subsidiary of the Compagnie Lyonnaise des Eaux, was immediately contacted. An innovative joint venture was devised, in which the maintenance of the FN Engines Division would be a testing ground for this new activity for FN. This activity would be carried out by spare FN personnel.

It was soon realised, after some initial contacts had been made, that other companies as well as the Belgian Army, could be interested in calling on technical management by a company that was external to some of their own facilities.[201]

FN's former general services thus became a separate division of the company. It was no longer solely responsible for supplying the company's other divisions, but instead opened up to the outside world. Major investments were quickly made, which gradually led to the acquisition of cutting-edge technologies for the production, shaping, processing and use of metallic materials. The General Assembly of 27 October 1977 was advised that "Formétal already enjoys a strong position in the aerospace and energy markets, which require the use of advanced technologies and very strict quality requirements."[202]

A new workshop was purchased and fitted out, becoming operational in July 1978, and specialised in making pieces for the aerospace and nuclear markets. A licence contract was concluded with the company Thomson Ramo Wooldridge (United States, a supplier to Pratt & Whitney) for precision castings for the F-100 engines.

To meet the demands of the aeronautical industry, a leading sector but not the only one, a new American-designed investment casting workshop was built in 1977 in the industrial zone of Hauts-Sarts, in Herstal. 'FMP Aéro' enabled FN to master state-of-the-art techniques, giving the company swiftly acknowledged know-how in turbine and compressor blades, further enhanced by the precision forge's activities. However, the traditional foundry experienced a recession, which led to its definitive closure in the early 1980s.

The foundations for excellence in FN's products were laid by combining the foundry and laboratory. The two entities, both at the very forefront of technology, worked hand-in-hand and acquired a mastery of steel.

FN Formétal foundry for investment casting patterns, 1983
FN Formétal and the central laboratory developed alloys with exceptional properties in terms of their toughness, hardness and heat resistance.
All were essential to meet the demands for precision and high-quality manufacturing of structural parts for aeronautical equipment or the nuclear industry.

‡ Construction of escalators at FN Bruges, 1981

➥ Comet hydraulic lift, Place du Marché in Liège, 1981
Based in Namur, FN Comet specialised in manufacturing hydraulic lifts adapted to all kinds of vehicles and types of lifting (jib cranes, articulated or telescopic models). These lifts featured proximity detectors and electrohydraulic valves with progressive control.

MOVING FROM INDUSTRIAL EQUIPMENT AND SERVICES TO FN INDUSTRY

FN Industry grew out of the merger in 1979 of the Subcontracting, Engineering and Participation department with the Industrial Equipment and Services Division. This entity had initially combined the production of agricultural equipment, knitting machines and household appliances, though these could not survive the fierce competition from specialist manufacturers. However, FN Industry turned out to be a very effective tool for new business sectors, such as industrial oil hydraulics. FN Industry brought together small companies with highly diverse activities, enabling it to successfully launch into industrial cooperation, maintenance and industrial services, aerial work platforms, robotics and plastic waste recycling. This latter activity took off at FN

in 1974, when the company was on the hunt for emerging industrial fields. FN found itself leading this research, thanks to the plasticiser that melted and blended plastic waste into transformable granules with excellent yields, and other equipment that was new at the time, such as biodigesters for organic products resulting in their methanisation. Shanghai purchased equipment like this, using it to recycle the city's plastic waste and transforming it into millions of bicycle handles.

FN Industry's activities allowed FN to combine its expertise with technology and innovation and to launch itself at a very early stage into economically, commercially and societally promising developments.

FN evolves into an international group

As stated in a June 1979 report by the Board of Directors: *FN has recently created several subsidiaries and taken holdings in the capital of a number of Belgian and foreign companies. It has thus become an international group with an industrial focus, whose main objectives are:*

- *to control more closely the network that markets its range of sports products and to be in a better position to promote the sale of these products; this policy has taken practical shape through the purchase of the companies Schroeder and Browning and through the setting up of a European Browning network.*

- *to assert its presence in certain countries in order to maintain its foreign customer base, mainly governmental; the presence of the companies FN do Brasil and FN Manufacturing contributes to strengthening the commercial position of FN in Brazil and in the USA.*

- *to diversify its activities by completing the range of products and services offered to its customers: the creation of the company FN-Cofreth Services has enabled FN to enter the industrial service market and, following the acquisition of the company LERC, the "fishing rod" has been added to the list of sports goods marketed by FN.*

- *to create synergies within the group that will enhance the competitiveness of the companies that make it up; this is, in particular, the purpose of FN Viana.*

- *to collaborate on the development or the relaunch of Belgian companies whose activities correspond to its own areas of development; equity investments in the SONACA companies and, more recently, Belairbus and Raskin Liège, meet this objective.*

- *lastly, to have a structure that allows access to the international capital market, to secure the best financial conditions in international operations and to reduce exchange risks. These are the missions of the companies FN International and FN America.*[203]

In 1979, FN published an organisation chart distinguishing between 'Subsidiaries and Sub-subsidiaries'; 'Related companies'; 'Companies with which there is a shareholding link' and finally 'Other important shares and holdings'.

We can overlay on this diagram (on the right) – which shows the company's financial strength, or at least its capacity to raise funds to complete a horizontal and vertical consolidation of industrial activities – another functional diagram: this shows the increasingly varied products of FN in the 1970s. Aeronautics, shotguns, sports goods, nuclear power, new technologies, machine tools, munitions and weapons of war: FN took an interest in everything and was open to anything.

In 1972, one should remember, FN had a stake in several companies distributing its products.[204] Over time, it also acquired significant shareholdings in MMN, Belgonucléaire, with a view to diversifying its activities, and smaller holdings in the company Fokker and Les Établissements Lachaussée.

These holdings increased in number and complexity, with subsidiaries and sub-subsidiaries acquiring stakes in companies whose activities were either complementary or competitive with those of FN.

Just as it did between the two world wars, FN made use of its stakeholdings as a multifunctional tool. Between 1919 and 1939, this tool was used alternately to control the sale of its products in Belgium and abroad, to invest in promising technologies, to increase its profits and to secure supplies for the company. However, from the 1970s onwards, FN used this strategy on a completely different scale. Between 1970 and 1981, the company's 'portfolio' of holdings of all kinds rose from 100 million to almost 600 million in current Belgian francs. Even taking into account the period's inflation, this increase was remarkable. Another difference with the previous period was the exceptional sophistication of the shareholding operations. In September 1972, FN International [FNI], a holding company, was created to finance projects abroad. In January 1973, FNI oversaw the acquisition of a minority but significant stake in the Italian company Fabbrica d'Armi Pietro Beretta, which was coupled with Beretta's minority stake in the capital of FN.

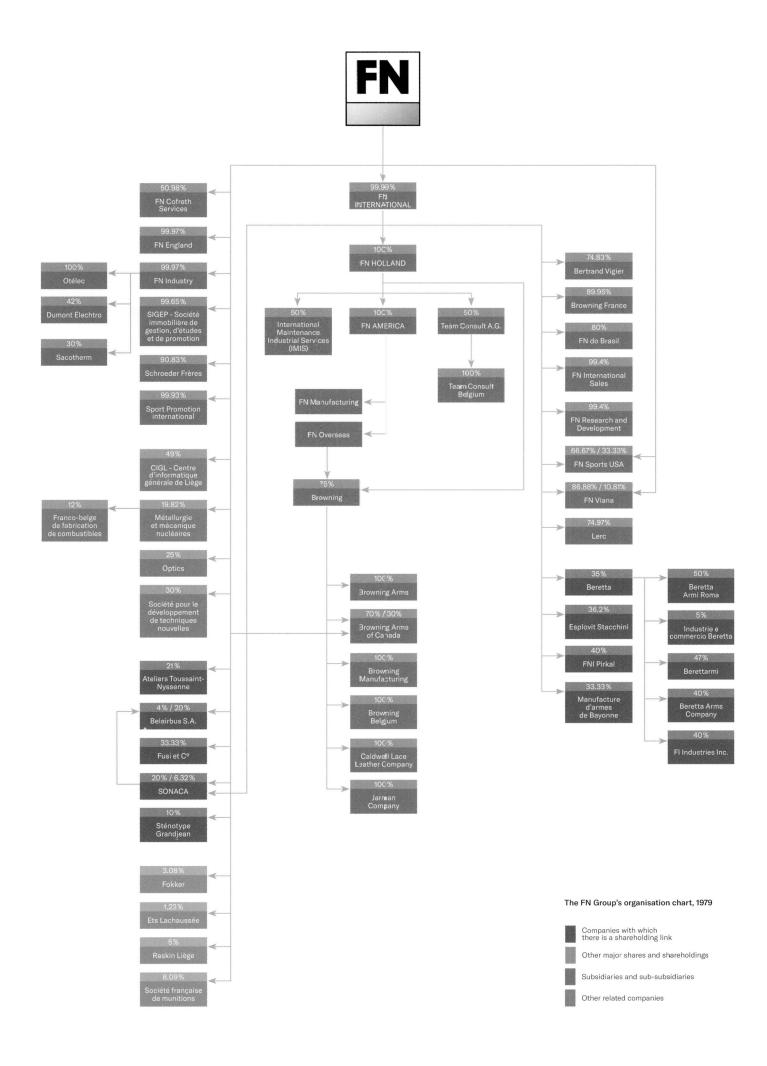

50.98%
FN Cofreth Services

99.97%
FN England

100%
Otélec

99.97%
FN Industry

42%
Dumont Élechtro

99.65%
SIGEP - Société immobilière de gestion, d'études et de promotion

30%
Sacotherm

90.83%
Schroeder Frères

99.93%
Sport Promotion international

49%
CIGL - Centre d'informatique générale de Liège

12%
Franco-belge de fabrication de combustibles

19.82%
Métallurgie et mécanique nucléaires

25%
Optics

30%
Société pour le développement de techniques nouvelles

21%
Ateliers Toussaint-Nyssenne

4% / 20%
Belairbus S.A.

33.33%
Fusi et Cᵉ

20% / 6.32%
SONACA

10%
Sténotype Grandjean

3.08%
Fokker

1.23%
Ets Lachaussée

5%
Raskin Liège

8.09%
Société française de munitions

99.99%
FN INTERNATIONAL

100%
FN HOLLAND

50%
International Maintenance Industrial Services (IMIS)

100%
FN AMERICA

50%
Team Consult A.G.

100%
Team Consult Belgium

FN Manufacturing

FN Overseas

75%
Browning

100%
Browning Arms

70% / 30%
Browning Arms of Canada

100%
Browning Manufacturing

100%
Browning Belgium

100%
Caldwell Lace Leather Company

100%
Jarman Company

74.83%
Bertrand Vigier

89.95%
Browning France

80%
FN do Brasil

99.4%
FN International Sales

99.4%
FN Research and Development

66.67% / 33.33%
FN Sports USA

86.88% / 10.81%
FN Viana

74.97%
Lerc

35%
Beretta

36.2%
Esplovit Stacchini

40%
FNI Pirkal

33.33%
Manufacture d'armes de Bayonne

50%
Beretta Armi Roma

5%
Industrie e commercio Beretta

47%
Berettarmi

40%
Beretta Arms Company

40%
FI Industries Inc.

The FN Group's organisation chart, 1979

Companies with which there is a shareholding link

Other major shares and shareholdings

Subsidiaries and sub-subsidiaries

Other related companies

Staff members of FN Viana, Portugal, 1983
In 1973, 84 people were employed for the assembly of hunting guns in a new factory at FN Viana.
By creating a subsidiary instead of subcontracting, FN Group maintains control of the concepts and
their industrialisation, as well as training staff who would gradually be required to play a major role in
the company's strategy.

SELECTION OF PORTUGAL

In April 1973, the industrial subsidiary FN Viana, Fábrica de Artigos de Caça was also created under the supervision of FNI. Eighty percent of the capital was purchased by FNI, 10% by FN and 10% by Portuguese entities.

The previous year, there was some concern about the company's interest in setting up an assembly shop for hunting guns outside Belgium. Europe and Asia had been considered, but ultimately Portugal turned out to be the most attractive country. It soon became clear that it was possible to recruit sufficiently skilled personnel, if they were well supervised and managed. It was also noted that salaries were three times lower there than in Belgium. This was the main challenge for the company: regular and significant salary increases in Belgium meant that its products were not very competitive on foreign markets.

The new company, incorporated under Portuguese law, was set up in September 1972. A plot of land north of Porto was quickly purchased and the first workers were hired the next month. FN clearly sought to retain control of the designs and their industrialisation as well as to keep the manufacture of complex hunting gun parts

for its own workers in Herstal, while outsourcing production under its own control and responsibility rather than as a subcontractor. In 1973, the Portuguese factory employed 84 people and technical training courses were held there, run by technicians from Herstal (engraving, polishing, surface treatments, assembly).

FN Viana's first achievement was the assembly of 1,200 Trombone rifles, using the parts sent by Herstal. The factory subsequently kept pace with FN production and was involved in developing numerous firearms and later Browning 'non-gun' products. Today it is part of the company's strategy and one of the Herstal Group's main factories.

MANUFACTURING IN BRAZIL

Brazil was a long-standing customer of FN and, since 1894, its successive governments had regularly procured from Herstal. As early as October 1976, the creation of a subsidiary was envisaged by the Board of Directors, which said: "supplies to the government [of Brazil] will no longer be possible at an on-site facility. The plan is to set up a machining workshop in the near future [...]".

Bernard Regout, director of the Defence and Security Division, 1987
Bernard Regout played a vital role in the modernisation phase of the FN Group and its expansion across the Atlantic. Inaugurated in 1981, the FN Manufacturing factory when running at full capacity employed 200 people for the assembly of FN MAG coaxial machine guns, adopted by the US in 1977 to equip the nation's M60 and M551 tanks. This task would later be extended to the production of army models and FN MINIMI light machine guns.

FN Manufacturing staff member, Columbia, c. 2000
By setting up a factory in Columbia, the FN Group benefited from a business-friendly environment, but more importantly a skilled workforce, whether in machining or assembly. The pride of the FNMI staff, working for a world-renowned group, would only grow over the decades.

In December of the same year, the Board said that FN risked losing this market, if it did not undertake local production.[205] In late December 1976, the company FN do Brazil was formed and production there started in January 1978. This was an advantageous operation for FN, as it enabled production at a lower cost and served as a key bridgehead for the South American market, with which FN was already highly familiar.

WHY MANUFACTURE ON AMERICAN SOIL?

In 1977,[206] it was noted that the manufacture and/or the assembly of guns in the USA was a path to explore.
In January 1977, FN concluded an agreement to supply the US Government with 10,000 MAG machine guns. The guns were manufactured at Herstal. It was immediately clear that the US Army's requirements would lead to more orders after this initial order. The Americans were keen to ensure the security of their military supplies. So the 1977 contract stipulated that the manufacturing permit would later be handed over to US manufacturers, enabling additional quantities of the M240, the American name of the MAG, to be produced in the United States.

FN obviously did not want to miss out on this significant market. So it decided to participate in the call for tenders launched by the US Government. It created a subsidiary under US law – FN Manufacturing – and, after a lengthy and difficult competitive tender, was tasked with supplying 16,000 M240s. Even before it had heard the tender's result, it had decided to set up a factory in South Carolina, near the capital Columbia, with the Engineering Department of FN given responsibility for designing the building. We also learn that, in April 1980, "the establishment of FN Manufacturing is ongoing and the planning approved by the US Army is respected". Production began in 1981 and continued until 1985. Most of the money needed for this new factory ($20 million) came from loans from the US.
The creation of FN Manufacturing provided a strong foothold for Herstal on the US market. The Columbia factory was a leading light for FN and could meet all the army's supply needs: M16 rifles (under licence), M249 light machine guns, M240 machine guns, and M2 machine guns, etc.

CONSOLIDATION OF THE ARMS INDUSTRY AND THE ACQUISITION OF BROWNING

FN Viana (Portugal), FN do Brasil, Beretta, Manufacture d'Armes de Bayonne, Esplovit Stachinni, FN Pirkal (Greece), Société Française de munitions, Schroeder Frères... for the arms sector alone, these were all companies where FN gained a foothold and sometimes took control in the 1970s.

The most significant of these acquisitions was certainly that of Browning in 1977. It was a financial operation that put an end, by means of a takeover bid, to a genuine industrial friendship that had been born on 17 July 1897, eighty years earlier. It should be remembered that this friendship had survived two wars and the death of Browning's founder John Moses. So this link had to be profitable for both parties. Each had to make substantial profits from this relationship, and that was more or less the case until the 1970s.

One possible reason for the breakdown of relations between Browning and FN was purely commercial – the steady increase in prices charged by FN was the first blow. Driven by the dizzying increases in salaries and social charges in Belgium, these prices became less and less competitive. The Browning company's margins shrank. For the first time, in the third quarter of 1973, Browning started losing money with the guns of FN.[207] The spiral had begun, Browning no longer supplied its products exclusively to Herstal, and the trust was broken. FN took advantage of the financial and legal difficulties[208] of Browning to launch its own operation, which some people considered unfriendly.

The reality was rather more complex. FN was faced with the risk of gradually losing its exclusive access in Herstal to Browning's supplies, but the problem was far more serious, as noted at the Board of Directors meeting on 11 July 1977:

The arrival of one or more new partners could be very detrimental, as the Browning brand and name are major assets for our company, which is by no means sufficiently protected in the event of new negotiations; it could, for example, lose the use of the Browning name after a certain period of notice.[209]

Furthermore, new owners could get together to get their firearms produced in the U.S.A., at more competitive prices. Because of the vital importance of the Browning distribution network to our production of hunting firearms, the time has come for FN to take a stand on its interest in this company. The legal department's advice is to negotiate with the Browning family, which is favourably disposed toward us, to buy back their shares based on $10 to $12, which represents a cost of about $300 million. However, since the family owns only 40% of the

HP Gold Classic Semi-Automatic Pistol, engraved with the image of John Moses Browning and the United States, 1983

shares, it is possible that a takeover bid for all of the shares will have to be made to give the other shareholders the same opportunity to sell at a price higher than the stock market price.

Having been informed of this situation, the directors had no hesitation in mandating the company's management "to let the Browning family know that they may consider FN as an intermediary to study the ways in which FN's presence could be expanded in the Browning company". Another equally important question immediately arose, about the ownership of the Browning name for products other than guns: this seemed to be legally less certain.

The takeover bid was launched within the very short period of time required by the new stock market information received in Herstal: it was necessary to act quickly. The General Assembly of shareholders was informed of this acquisition on 27 October 1977: "Based on a previous stake of 4%, FN now owned 71% of Browning shares. The acquisition by FN was made at $13 per share".

This was followed by a concise overview of the Browning issue:

Browning's business is the development, import, manufacture and sale of sporting goods: hunting and sporting guns and accessories, hunting knives, pocket knives, hunting and camping clothing and equipment, hunting and shooting archery and accessories, fishing rods, reels and accessories, golf club carts, and leather goods. [Browning and its six subsidiaries[210] employ around 750 people. Only the bows, golf carts, fishing rods and leather goods are manufactured by Browning and its subsidiaries. [...] Until around 1970, FN produced almost all the guns sold by Browning. These represented 80% of the total production of Herstal of hunting and sporting guns. Since then, FN's competitive standing in the US market has deteriorated significantly. As a result, the selling prices of FN firearms have, over five years, increased by 115% in the US.

This detailed communication ended with a few figures. Browning had a distribution network of 10,000 stores in the 50 US States, together with 1,500 distributors in Canada and 200 after-sales service stations. FN was therefore undertaking an operation that simultaneously responded to commercial, industrial and financial imperatives. At that time, FN was not seeking to make or not make a 'good deal', but to respond as well as possible to the market requirements.

B25 Over and Under Shotgun, 1980
Signed Vrancken, this commemorative engraving was made with the image of Utah, a State that saw the birth and development of the Browning company.

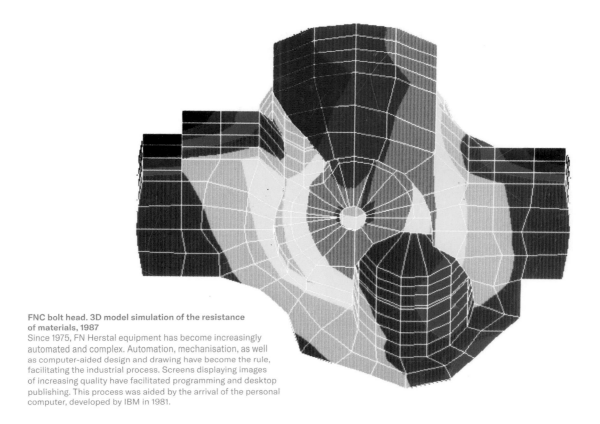

FNC bolt head. 3D model simulation of the resistance of materials, 1987
Since 1975, FN Herstal equipment has become increasingly automated and complex. Automation, mechanisation, as well as computer-aided design and drawing have become the rule, facilitating the industrial process. Screens displaying images of increasing quality have facilitated programming and desktop publishing. This process was aided by the arrival of the personal computer, developed by IBM in 1981.

Growth and progress

The company's economic progress in the 1970s and 1980s was all about growth, in terms of production levels, sales and the dividend. These three factors (and a few others) would of course influence each other and lead to them all increasing thanks to a spillover effect.

FN's turnover[211] grew in a fairly regular and moderate way in the late 1960s until the middle of the 1970s, from four billion Belgian francs in 1970 to six billion in 1975. It then soared to 22 billion in 1981 and continued to rise to 24 billion in 1985.

This dramatic increase could lead one to believe that FN was making great 'progress'. But it should be viewed in context. The wage bill (salaries and social security charges combined), for example, represented 70% of turnover in 1970, 80% in 1974, though 'only' 49% in 1980. This figure subsequently fell to 40%, resulting mainly from a government policy curbing social costs introduced in 1981. Another concern was the steady and significant increase in financial costs (debt servicing) and insufficient self-financing capacity.

The company's long-term debts exceeded its equity. Earnings were once again positive and reached their peak in 1977, when the company made a profit of 200 million Belgian francs. This was an opportunity for the Board of Directors to quote impressive figures at the 1978 General Assembly:

The annual turnover per employed person is currently 1.26 million Belgian francs. In the previous year, it was 1.07 million. 6,435 customers honoured FN with their orders [...]. 30% of the turnover was generated with industrialised countries and 70% with industrialising countries [...]. As of 31 December 1978, FN employed 9,405 people, a reduction of 373 [...]. FN employs about 3% of the metal manufacturing industry's workforce. In Belgium, the number of people who directly depend on FN is estimated at more than 20,000 [...]. In 1977, it contributed 0.27% to Belgium's gross domestic product and accounted for 3.47% of the value added by the entire metal manufacturing sector, excluding shipbuilding [...]. Through its exports, FN makes a positive contribution to Belgium's trade balance: it exports 4.5 times more than it imports.

This triumphant announcement also detailed the highlights of each division over the past year. They included delivery of the first MAGs to the US Army, start-up of the plant in Brazil, testing of the MINIMI and the FNC rifle by NATO, commissioning of the new Engines Division, new test stands for jet engines, delivery of the first F100 engine, establishment of a new investment casting foundry, acquisition of a stake in SONACA (Société Nationale de Constructions Aérospatiales), takeover of Browning, takeover of the French company LERC, launch of new golf clubs, creation of FN Manufacturing to produce guns on American soil, conclusion of an agreement with the company Cofreth to provide industrial services, creation of FN Industry, and so on.

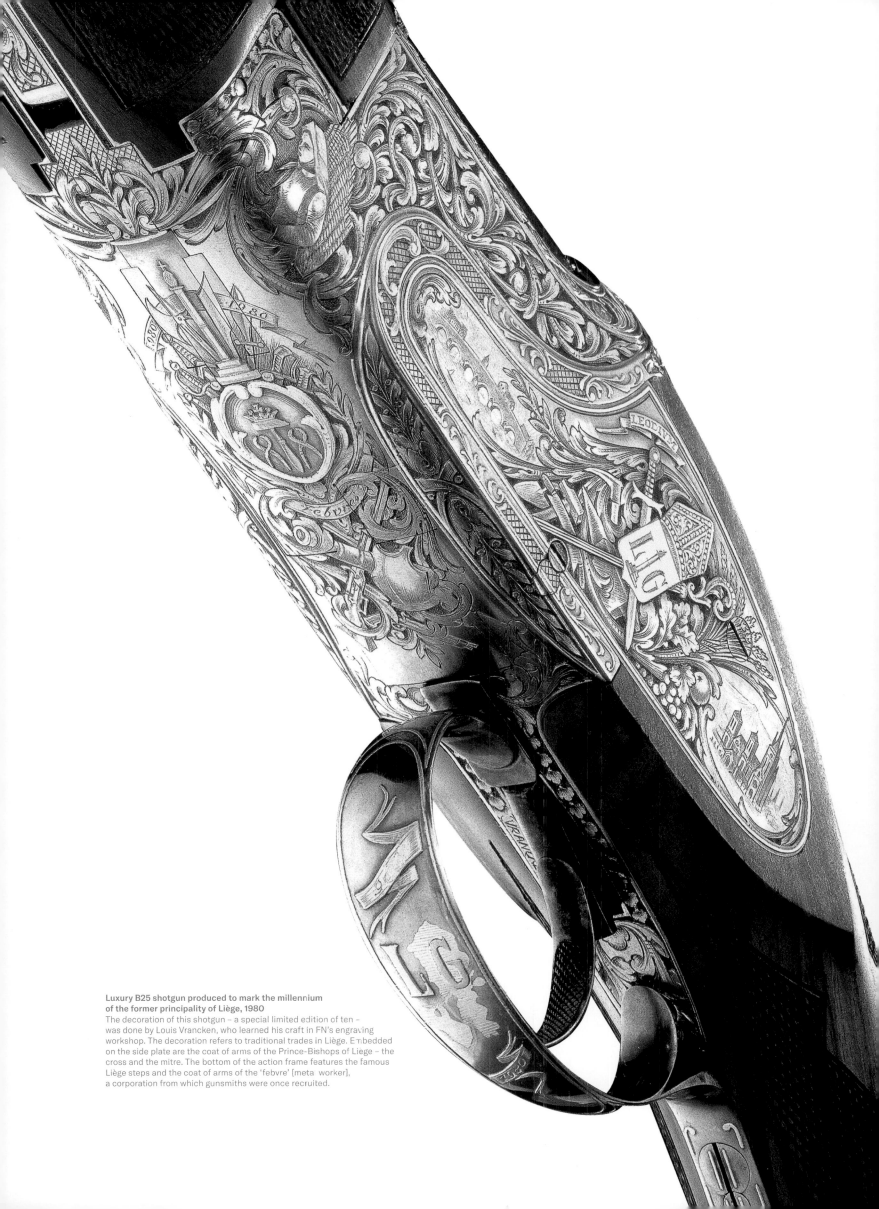

**Luxury B25 shotgun produced to mark the millennium
of the former principality of Liège, 1980**
The decoration of this shotgun – a special limited edition of ten –
was done by Louis Vrancken, who learned his craft in FN's engraving
workshop. The decoration refers to traditional trades in Liège. Embedded
on the side plate are the coat of arms of the Prince-Bishops of Liège – the
cross and the mitre. The bottom of the action frame features the famous
Liège steps and the coat of arms of the 'febvre' [metal worker],
a corporation from which gunsmiths were once recruited.

Space-bound

In concluding this period, which began in 1945 and ended at the beginning of the 1980s, one should highlight the involvement of the FN Group in the space sector. It had already embraced this in 1964, in collaboration with Rolls-Royce for manufacturing parts for the first stage of the Blue Streak rocket, under the ELDO (European Launcher Development Organisation) programme. Due to the low volumes and use of sub-contractors for the work, there was little enthusiasm for what amounted to a timid entry into the realm of space. Subsequently however, FN did carve out a niche for itself in this area. In particular, it was associated with the European programme to develop moon robots, although this was too expensive and never came to fruition.

In 1975, one initiative further confirmed the company's long-term interest in space. In those years, the company first worked as a subcontractor to Société Européenne de Propulsion (SEP) for certain components of the Viking engine for the Ariane rocket. This agreement then evolved into the Engines Division's participation in developing components for successive versions of the rocket engine. In 1986, FN became involved in the Ariane 5 programme.

The Ariane space programme was born in 1973. Belgium contributed 5% of the investment needed for the programme and therefore took 5% of the related industrial business. For Europeans, it was all about having the capacity to launch commercial satellites. The first rocket lifted off on 24 December 1979, with valves built at Herstal. FN thereby ensured that it had a key element of its future communications strategy. It was another risky gamble but it paid dividends as FN bought into the capital of Arianespace, along with other big companies, in 1980.

FN worked on different launchers in the Ariane 1 and Ariane 2 programmes through to the Ariane 5 programme, scheduled for flights in 1995. "FN's contribution is as crucial as that of all the other participants because, in space, everything has to be totally reliable."[212]

Through this cooperative work, FN managed to latch on to the European space programmes, as well as to carve out room for itself in high-technology engineering. This was a period when engineering-related requirements began to be incorporated into the new manufacturing processes. The extreme conditions in which FN-made parts had to operate (temperatures, friction, corrosion, etc.) could only lead to a "technological leap forward" and huge progress in the use of superalloys, the development of computing software and modelling, and hardware for test benches, and so on.

A decade drew to a close, one in which FN was transformed. It had of course become a powerful group in Belgium, where it was a true leader in the regional economy, but also and above all internationally, where the strength of its subsidiaries and associates placed it in an excellent position. Thanks to the determination of the heirs of a late 19th century company, the company had managed to transform itself. It had learned new technologies – and even succeeded in them – while changing the way it saw itself and evolving its industrial and especially its commercial practices, undoubtedly helped by the profound changes in the society in which it operated. Over just a few years, this paradigm shift from product to profit enabled (or compelled) FN to embark on a triple evolution: consolidation (horizontal and vertical), diversification and internal restructuring.

FN not only survived, it had prospered amid a difficult economic climate: the end of the convertibility of the dollar in 1971 and above all the famous 'oil crisis' of 1973, with inflation leading to wage demands and global instability. Yet none of these events could divert the company from its objectives. One might even argue that these difficulties accelerated its progress. Were the foundations of this strength solid? Did FN over-extend itself, without sufficient funds? Industry historians are right to ask these questions, yet there is one other thought-provoking question: did FN have any choice? Did it have the choice of growing a little, not too much, of not diversifying its 'core business' too much, or of not joining the economy's financialisation? In response, one should recall the adage already quoted at a Board of Directors meeting in 1964: "not investing is just another way to disappear".

Ariane 3 rocket, 1983
The sixth launch of the Ariane rocket on 16 June 1983 was successful. This confirmed the maturity of the aerospace programme of the Société européenne de Propulsion [European propulsion company], with which FN Engines was collaborating. A year later, a department specialising in space and which had its own design office, was set up within FN Engines to support this fast-growing sector. The Ariane 4 project was then under study.

A time for decisions

A depressed market, more intense competition

There are many reasons why things worsened for FN starting in 1980. However, this was mainly due to chaos in the global economy and trade.

The twelve months in 1980 led to a net income showing a loss of 41 million Belgian francs and of course no dividend was paid out. The following year saw a few signs of hopeful recovery: 1981's net income was in the black and shareholders received a fairly modest dividend. However, the company's financial position was still fragile. Levels of industrial investment continued to decline despite the upturn (812 million Belgian francs in 1981, half of the figure for 1979). Yet financial expenses kept rising, both in absolute terms and as a percentage of turnover. They came to 274 million Belgian francs in 1977, 590 million Belgian francs in 1979 and 1.385 billion Belgian francs in 1981, i.e. respectively 2.6%, 4.1% and 6.4% of turnover. The stock exchange painted a clear picture and the value of FN's shares fell by 56% in 1981, despite what seemed like a fairly good result. Financing expenses were on the rise again, after a slight lull in 1982.

In 1982, the company observed that "a depressed market, more intense competition, excess capacity and serious hesitation with regard to launching new projects were all factors that did not make it any easier [...] for our industry to engage in new activities".[213]

In a text entitled '1984, une année difficile et contrastée'[214] [1984, a difficult and contrasting year], FN chose not to focus on self-criticism. Instead the company offered a lucid and somewhat despondent analysis of the three previous years:

After a decade of unprecedented expansion, diversification and modernisation, for three years now FN has been facing a serious economic situation that is marked in particular by the budgetary difficulties of several client countries and by the crisis in the global aerospace industry. 1984 has been a difficult and contrasting year. Turnover fell by 20% and the fiscal year ended with a negative net result of 159 million Belgian francs.[215]

There were many linked reasons for this and most were totally outside FN's control.

The early 1980s were marked by economic recession globally, which particularly affected the aeronautics sector. This crisis came at the worst time for the FN Group, which then shifted towards civil aeronautics.

THE QUADRUPLE CRISIS OF THE 1980S

The 1980 recession hit the US first, before spreading worldwide, and lasted for a long time. It was the backdrop for global economic activity for many years. Belgium and other countries saw their economic growth shrink and inflation set in, creating commercial and financial uncertainty.

It might have been just one more crisis, one that would only affect FN's business activities at the margins. However, in the context of several other negative issues, this crisis became so severe that it occasionally cast doubt on the company's ability to survive.

GEOPOLITICAL UPHEAVALS

The end of the Cold War and later the collapse of the Eastern bloc – which were among the external causes exacerbating the difficulties of that period – would redraw the boundaries of international politics. The new politics would no longer be defined as the armed opposition of two mutually hostile blocs.

The armaments industry felt the effects of the new political context, through the overall volume of government orders. Over the long term, the industry also perceived how this reduction in orders benefited arms manufacturing groups that could rely on a sizeable national market to compensate them for their export losses. FN was clearly not in this particular situation. In 1980, its exports accounted for 92% of its turnover. The Belgian market was evidently not enough of a counterweight. The bigger and more developed a country is, the more its production can be refocused on its domestic market and the less its external trade will influence its growth. That was very much the case for the US.

OIL CRISIS

Economic activity slowed in the 1980s, leading to weaker demand for oil and eventually a fall in oil prices. Oil-producing countries had fewer resources and the first budget line to be affected by that was often armaments. For several years and in various ways, FN pointed out the "budgetary difficulties of several countries that were customers". In real terms, a barrel of oil imported to Europe cost $18 in 1987, whereas it was sold for nearly $33 in 1980.

'THE DOLLAR IS OUR CURRENCY BUT YOUR PROBLEM'[216]

The US currency had continued to appreciate from 1978 to 1985,[217] and was even overvalued, which helped for instance to boost FN's exports to the US and 'dollar zone' countries.[218] The US trade deficit was growing significantly and protectionism was not considered an acceptable option by US President Ronald Reagan. So the nation sought to turn its trade situation[219] around, by reducing the value of the dollar. The US abandoned its mantra that 'the strength of the dollar is a reflection of a strong America'. As a result, the dollar was deliberately allowed to 'slide' and lost nearly 40% of its value in 1988. US manufacturers were of course helped by this on foreign markets, while it became much harder for European exports into the dollar zone. This severely impacted FN, which had many customers that were in this currency zone. In 1987, the Board of Directors noted in measured terms that "the ongoing decline in the dollar has had an unfavourable impact on exports towards dollar zone countries and has strengthened the position of some of FN's competitors".

CRISIS IN THE AEROSPACE INDUSTRY

One of the sectors most directly and immediately affected globally was the aerospace industry. On 28 October 1980, *Le Monde* ran the headline: "l'année la plus mauvaise pour le transport aérien international [the worst year for international air transport]" before noting "Mais le pire est à venir [But the worst is still to come]".[220] This is how FN assessed the economic downturn in 1982: *The stagnation in air traffic, linked to the economic crisis, as well as the burden of financial costs and fuel costs are weighing heavily on the profitability of airlines; overall, they have recorded significant losses for the third year in a row [...], the Belgian aerospace industry, which is mainly engaged in military activities, may seem less immediately affected by this economic downturn. But it is indirectly affected in terms of research into new activities that were designed to allow it to ensure [...] that jobs are kept in the face of the gradual decline in activity in the F-16 joint production programme.*[221]

This crisis in the aerospace sector came at the worst possible time for the company, which was then moving towards civil aerospace. The Engines Division received its first orders for Pratt & Whitney engine parts for the Airbus A310 and Boeing 767 civil aeroplanes, and it developed the lubrication unit for the CFM56-3 engine for Snecma.

Geopolitics, the dollar, the aerospace industry, and oil: those were some of the difficulties faced by FN Herstal. It had to respond to them as best it could, with clear-thinking and resilience during the worst decade in its history.

CLEAR-THINKING AND RESILIENCE

While the early 1980s had alternately veered between hope and disappointment, difficulties of every kind dominated from 1986, as noted in the General Assembly of

1987: "The 1986 fiscal year ended with the biggest loss that FN has ever known. Only a few balance sheets published in the period between the two wars showed such serious difficulties".

A strike in March 1986 certainly had a negative impact on the company's financial results. However the management drew attention to global causes to explain its poor performance: "The fall in oil prices has drastically reduced the revenue of several governmental customers and the continual fall in the value of the dollar has had an adverse impact on exports to dollar zone countries and strengthened the position of some of FN's competitors". There was acknowledgement that "the situation is worrying [and] creates a climate of concern among staff and among all of FN's partners: it raises numerous questions for those who are concerned about their future".[222]

The following year, when the Group noted a loss of 1.5 billion Belgian francs, an additional reason was identified: "global overcapacity in the light weapons' sector".

In this difficult context, what could a leading company in its sector, region and country do?

FN adopted the same response it had always done, during previous difficult periods for the company – just keep going! Keep on investing, keep on modernising, and keep on inventing and manufacturing. The company had no other choice, because any interruption could sound FN's death knell.

Initially, FN acted defensively by reducing its stocks and cutting the number of jobs. But it went on the offensive in particular by maintaining its research and development activities, improving productivity and increasing investment levels, modernising management methods and adapting its staff to new environments.[223] This at least was the programme set out in the 1980s.

FROM CASH FLOW TO CASH DRAIN

FN's cash flow, which defines a company's 'margin for manoeuvre', dwindled until it was exhausted. In 1986, reference was made to a cash drain, i.e. that the capacity to modernise the company and make it viable was not only zero but negative, thus burdening future results with debt. There were other worrying signs too. Research and development expenditure doubled between 1981 and 1986. This would have been excellent news, if FN's capital had not been sharply declining at the same time.

The importance of this expenditure had already been underlined in 1977: "FN's research and development efforts flow directly from the growth and diversification objectives that the company has set itself for the medium and long term."[224] The aim was to maintain and develop the technological advantages acquired in the company's different divisions.

FN was caught in a vicious circle. It had to invest to survive, but it lacked enough financial resources of its own. So the company had to borrow or resolve to increase its capital, which would be expensive and deprive it of the means to invest in innovation.

THE FUTURE WAS NOT SET IN STONE

Already in 1983, FN was aware of the risks it was running and was quick to announce that in public. It refused to be trapped by a sense of fatalism or decline. In 1984, its Chief Executive, Michel Vandestrick, approached Albert Diehl, a former PDG [chairman and managing director] of Rhône-Poulenc. Working as a consultant, Diehl was tasked with identifying the priority actions to take, in order to improve the way the company worked and its profitability. His observations and analysis would determine the main lines of a restructuring plan. The company knew full well it would take years to implement this plan, which might be effective but could not deliver an industrial miracle.

FN's diversification process, which had begun and been developed in the previous decade, had proved disappointing. There were too many activities and too few synergies; investments were expensive for the Group but insufficient to guarantee significant market positions. Albert Diehl made an insightful observation: the Group should refocus on what it knew best – guns and ammunition – or run the risk of an employment crisis and bitter disappointment.

Conversation between Albert Diehl and Michel Vandestrick, 1987
In 1986, the Board of Administration extended Michel Vandestrick's tenure as Managing Director and entrusted the general management to Albert Diehl. The two men had to successfully implement an ambitious industrial and financial plan, to ensure the company's adaptation to the global economic recession, which had begun in the early 1980s.

The 1986 plan

The main lines of the restructuring plan that the company would put in place were as follows: improve the way the company works and its financial results; prioritise military activities and abandon other activities, with the exception of those that are profitable or complementary to military production; and adapt the workforce to the company's needs. Albert Diehl himself was appointed to carry out this mission as Managing Director.

The first priority was to turn things around financially, with the support of private banks and public authorities. That meant recapitalising, rescheduling debts, taking out loans, as well as the deferral of deadlines by the tax authorities and social security bodies. No stone was left unturned in cleaning up the company's finances, so it could restructure itself.

The next step was recovery at the industrial level. This involved refocusing company activities on its 'core business', i.e. military or civilian weapons and ammunition, as well as improving industrial processes.

In the short term, the plan envisaged a drastic reduction in costs. Hence the reduction of operational costs as much as possible (e.g. by improving the management of stocks, the use of subcontracting, automating and introducing information technology into management and production practices) by ending the expenses linked to haphazard diversification and by reducing staff costs: 12,000 people in 1980, 10,000 in 1985 and 7,000 in 1987.

Browning, a new subsidiary, and takeover of Winchester

The company's own structure was changed. Various branches were empowered, with the implementation of a policy of decentralisation and the creation of subsidiaries. Each entity became autonomous and had the necessary means to define and apply its own policy for production, sales, finance and human resources management; the aim was to simplify operations and improve productivity.

In December 1986, Browning was transformed into a subsidiary and became a public limited company that was independent of FN. It also headed up its own international group. From then on, Browning defined its own commercial, financial, social and industrial policy. The decentralisation process led to the creation in 1986 of Browning SA (plc). At the end of 1987, Browning SA became the parent company of the Browning Group, with the FN Group having contributed to the company from its shares in six companies.[225] Luxury guns continued to be an asset in a market that was still depressed. The business also had to be reinvented in order to survive. In 1987, 16 engravers from FN formed the Société Coopérative "Les Graveurs de Herstal", which could work as a subcontractor with Browning. Lastly, a growing share of the assembly and finishing of civilian weapons had to be transferred to Miroku and to Viana, with the components continuing to be made at Herstal.

H.G. Edwards, Horseman and Bear, Winchester advertising, 1921
By joining forces with the Winchester brand, the FN Group made its entry into a world with deep cultural roots – the Wild West and taming of the natural world.
The Horseman, which appeared regularly in Winchester adverts and would serve as a logo for the brand, was created in 1919 by Philip R. Goodwin, a painter of the American Wild West.

Browning gradually disengaged from all the non-guns products, enabling it to refocus on guns. In 1988, it bought shares in the U.S. Repeating Arms Company (USRAC), which produced and sold the famous Winchester rifles under the licence of the US group Olin. Two months later, a new agreement with Olin allowed Browning to become the exclusive global distributor of Winchester guns produced by USRAC, apart from in the US, Canada and Oceania. Likewise, Browning gave Olin an exclusive licence for the manufacture and distribution of Browning ammunition, apart from certain countries, including those in the Benelux. The integration of the Winchester was part of an effort to strengthen the position of sports guns within Browning, and gave it better coverage of markets and new possibilities for industrialisation. The

company rallied around the idea that the Browning and Winchester ranges were not rivals but were complementary. That position was strengthened in 1988, when the FN Group became the full owner of Browning America. There were strong economic reasons to justify this shareholding. Yet it is hard to ignore the symbolism underlying the union of two brands that had been rivals until then. At times, they had shared a common history, beginning with the key role played by John Moses Browning in each of the two companies.[226] Joining forces with Winchester meant becoming part of the legend of winning the Far West, of Sitting Bull and Buffalo Bill posing with a Winchester rifle in 1895, or John Wayne with the specially designed large loop lever-action Winchester for the film *Stagecoach* in 1939.

WINCHESTER
Lever-Action Rifle
1894 Model

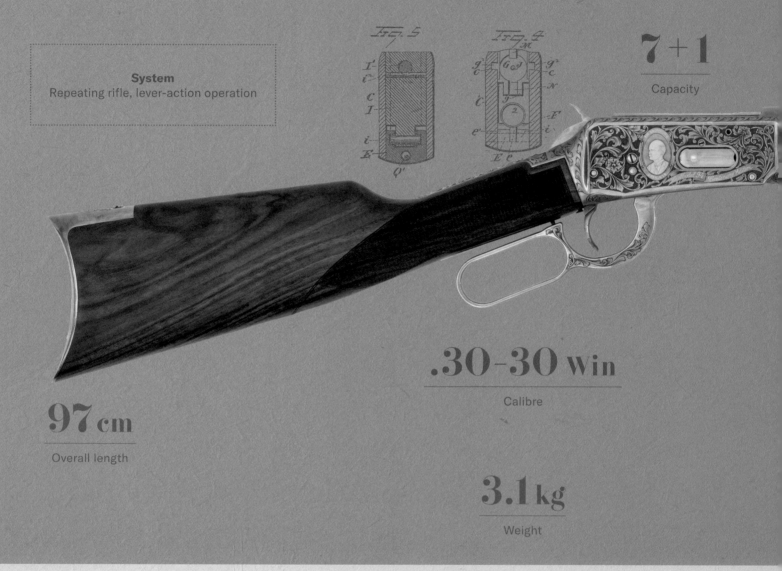

System
Repeating rifle, lever-action operation

7+1
Capacity

.30-30 Win
Calibre

97 cm
Overall length

3.1 kg
Weight

Winchester's brand image has continued to grow stronger over time – be that through the cinema, advertising campaigns or commemorative productions, which have all forged this legend and anchored it in American culture. The story began in 1855 when a factory was founded, in Norwich, Connecticut, to produce 'Volcanic' repeating arms. These featured a tubular magazine and a lever-action loading and locking system, based on an original patent of Walter Hunt and improved by Horace Smith and Daniel B. Wesson. In 1857, the industrial entrepreneur Oliver Fisher Winchester bought up the Volcanic Repeating Arms Company, which had meanwhile set up in New Haven. It was later renamed the Winchester Repeating Arms Company. The factory started producing guns under the Volcanic patent but these were modified by Benjamin Tyler Henry, to shoot new rimfire ammunition. In 1866, the first real Winchester rifles with a patented side loading gate and a brass receiver appeared, leading to its nickname 'Yellow Boy'. Then came the 1873 centre-fire model, considered 'The Gun That Won the West'. A partnership with John Moses Browning began in the late 1870s and early 1880s. He then filed, in 1879, his first patent for a single-shot lever-action rifle, for which the rights were bought by Winchester. John Moses Browning then made his contribution, thanks to the 1886 Model, with a robust locking system and adapted to long and more powerful cartridges. This rifle featured a system with dual vertically sliding locking lugs, replacing Henry's toggle link mechanism. The cartridge loading system was also improved, as was the

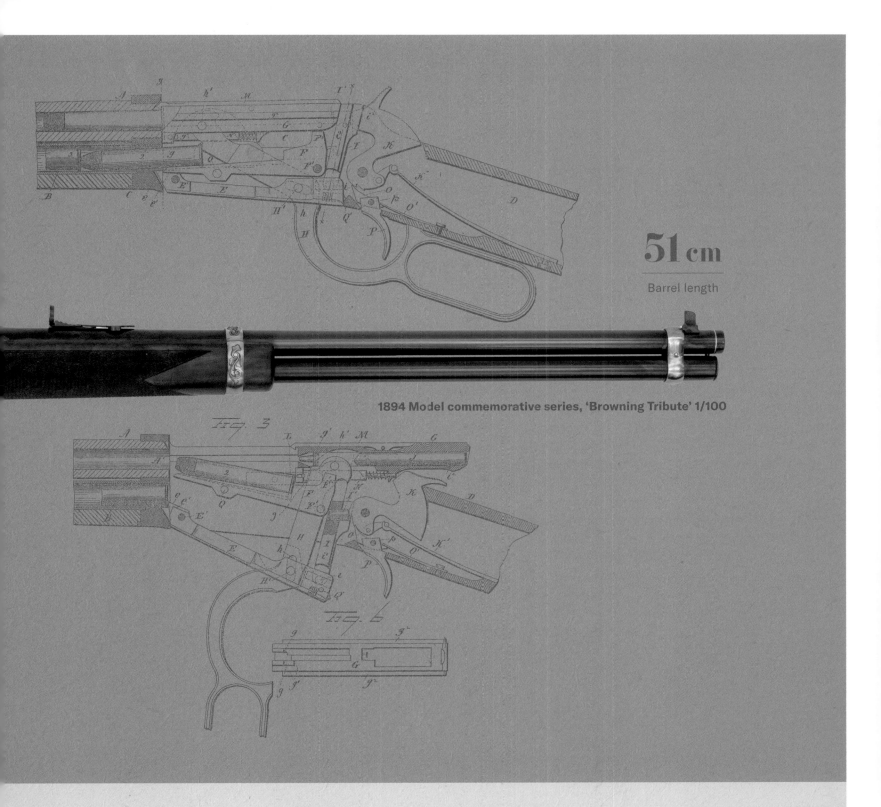

51 cm

Barrel length

1894 Model commemorative series, 'Browning Tribute' 1/100

safety system for the firing pin and the loading via the side port, whose flexible metal part was replaced by a spring-loaded hinge. After the Winchester 1892 came the 1894 Model. This was initially chambered to shoot black powder ammunition (.38-55 and .32-40), and then chambered for .30-30 ammunition with smokeless powder with the introduction of the steel-nickel barrel. This new powder produced more energy and propelled the bullets with greater velocity and a flatter trajectory than with black powder. But it also produced chamber pressures that were far too high for Winchester guns. As a result, John Moses Browning improved the 1886 and 1892 Models and patented a new loading system to accommodate powerful smokeless ammunition. The longitudinally moving breech was retained, but the locking system with

two vertical independent lugs was replaced by a single lug, linked to the receiver plate, pivoting the whole block when the lever was lowered. The **1894 Model** was highly successful and was marketed for over a century, with more than several million of them sold.

BAR MK2 Semi-Automatic Rifle

A500 Semi-Automatic Shotgun

Cover of a Browning catalogue, 1985

Innovation despite the odds

BROWNING, A BREATH OF FRESH AIR

Browning worked to diversify its range of over and under shotguns through the B125, B127, B135 and GTI models. Its marketing touched on motor sport and competition, offering a stock with palm swing, enlarged handguard, a set of three adjustable triggers, ventilated side ribs, plus black and red decoration that was so fashionable at the time. The range of semi-automatic rifles evolved with the BAR MKII, while the range of semi-automatic shotguns expanded with the arrival of the A500 shotgun, which was deliberately similar to its illustrious predecessor, the Auto-5.

The Browning A500 was developed to be the most reliable and fastest semi-automatic recoil-operated shotgun with the lowest recoil. It featured a short recoil-operated barrel linked to a four-lug rotary bolt. Browning engineers succeeded in considerably reducing the usual number of parts found in a semi-automatic. Thanks to its computer-aided design, they were able to test it down to the smallest details before the first prototypes were developed. Satisfying the end-customer, whether a hunter or sports shooter, is essential in a highly competitive market. By calling on increasingly sophisticated technologies, the company came up with the most appropriate designs and it announced those through highly targeted advertisements.

Moreover, Browning started to take an interest in the law enforcement market. This led to the development and marketing of several ranges of product, including the electronic shooting range for training purposes.

FN HERSTAL, MATURITY
AND A BALLISTICS REVOLUTION

The year 1985 saw the mass production of the FNC and the MINIMI. The FAL was still being produced and the Browning HP Double Action (BDA) gradually replaced the HP. The .50 calibre machine gun saw a revival of interest, thanks to the QCB (Quick Change Barrel) system.

For its part, FN Manufacturing Inc. (FNMI) met its productivity goals and supplied the US Army with nearly 10,000 M240 machine guns, replacing the M60 machine gun. At the tail end of the decade, the US Army ordered more than 30,000 MINIMI (made in Columbia, South Carolina, under the name M249) and 26,000 M16A2 rifles to be delivered in five years. FNMI, which had just beaten its competitor Colt to secure the contract to manufacture the famous M16, thus became the US Army's biggest supplier of light weapons.

With its expertise in pyrotechnics, the company studied new calibres and developed future weapons. It offered new products such as a telescopic rifle grenade – based on a new system, 'Bullet Thru' – that could be fired with any kind of infantry ammunition (1987) and notably a submachine gun, the P90, using the brand-new FN 5.7 × 28 mm calibre ammunition. A prototype was also made of a new machine gun, the BRG 15, in 15 × 106 mm calibre, to plug the gap between the 12.7 × 99 mm calibre ammunition used by the M2 and the 20 mm shells. A new bullet was developed to increase its life span and its penetrating power: it had a solid projectile body, without a casing, with a wide plastic belt on the cylindrical part. The industrial production of the BRG 15 machine gun, planned for the end of 1989, was never pursued, for understandable reasons related to the contraction of military markets when the Cold War ended.

Cover of a brochure for 30 mm calibre ammunition for air cannons, 1980s
In the 1980s, Fabrique Nationale Herstal's expertise in ammunition was not limited to small-calibre ammunition. FN also produced full ranges of 20 × 102 mm calibre ammunition, especially adapted to air and anti-air combat and ammunition for 30 mm calibre DEFA air cannons, as well as 40 mm electronic-percussion shells, in partnership with Thomson-CFS, for anti-helicopter defence.

FN M16A2 Assault Rifle
In the mid-1960s, while fighting the Vietnam War, the US Army adopted the XM16E1 Rifle. The ambition was to field a gun that was easier to use and lighter than the 7.62 mm calibre M14. It was based on the innovative design of the Remington .223 calibre Armalite AR 15, which incorporated light materials such as aluminium and plastic. However the gun was introduced with numerous design flaws and required upgrading, which led to the M16A2, put into service in the mid-1980s. This rifle featured several improvements, such as a thicker barrel, to reduce overheating and enhance accuracy and above all to adapt the gun to the new NATO SS109 ammunition. Later, the rifle's automatic mode was replaced with three-round burst firing. The gun also saw ergonomic enhancements, among them a symmetrical two-piece circular handguard, with more solid ridges and capable of greater thermal dissipation plus an integrated spent case deflector, to facilitate firing from the left shoulder.

FN P90®
Submachine Gun

5.7×28 mm NATO

Calibre

50

Capacity

50 cm

Overall length

**Ejection of cases
from underneath**

In May 1986, the Defence and Security Division wanted to complete its range by responding to the Personal Defense Weapon concept defined by NATO. These are guns intended for military personnel – whose main role was not infantry fighting. At the time, they accounted for two thirds of the armed forces, working as drivers, radar personnel, tank teams, signal troops, etc. FN Herstal made good use of its know-how as an ammunition and arms manufacturer. A team was tasked with developing this new gun and producing it on an industrial scale within four years.

This bullpup-type submachine gun, which is made of composite materials, became known as the **FN P90®**. It is compact, portable and can be operated with the left or right hand. Spent cartridges are ejected from below the gun. The magazine, which is positioned in the axis of the barrel and made of transparent plastic, holds 50 cartridges. The new 5.7 × 28 mm ammunition offers considerable stopping and piercing power while avoiding the risk of overpenetration. This is an intermediary gun, which is lighter and quicker than a handgun. It is approximately twice as light as a 9 mm Parabellum gun but is shorter than a 5.56 mm NATO gun. Consequently, the FN P90® can be fitted with a 50-round magazine

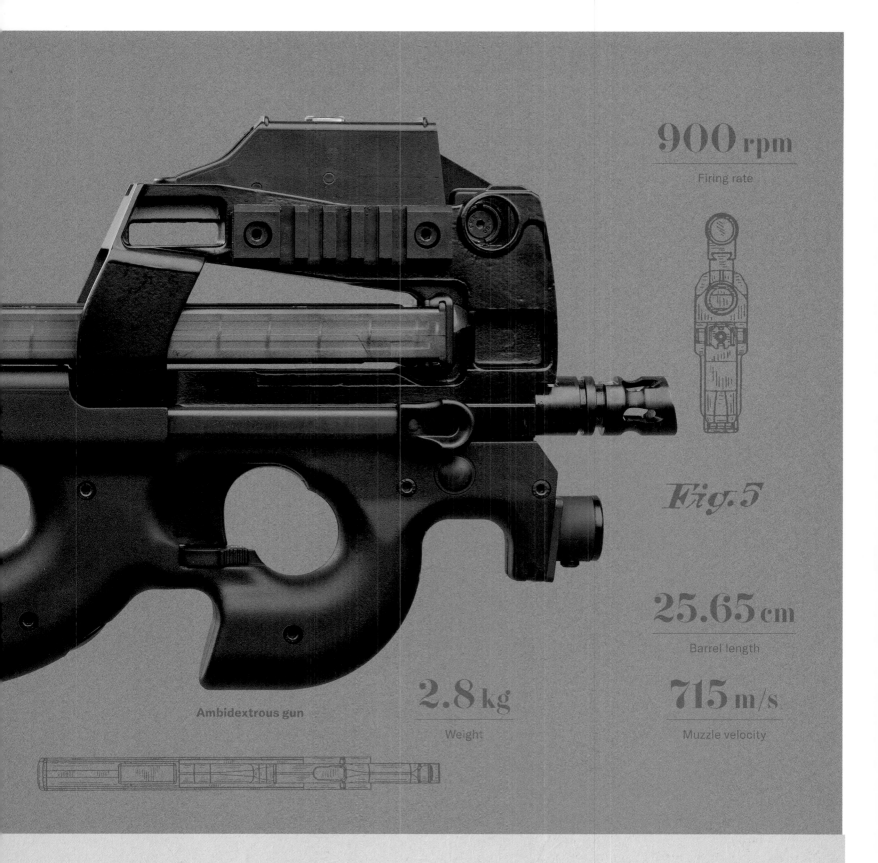

900 rpm
Firing rate

Fig. 5

25.65 cm
Barrel length

2.8 kg
Weight

715 m/s
Muzzle velocity

Ambidextrous gun

rather than the 30-round typically used in submachine guns and assault rifles. It also has less recoil than the 5.56 mm NATO cartridge, enabling quicker and more controllable fire, especially when in automatic mode. However, its kinetic energy is comparable to that of the 9 mm Parabellum but well below that of the 5.56 mm NATO cartridge. It can pierce a layer of titanium 1.6 mm thick reinforced with 20 layers of Kevlar, which corresponded to the personal protection used by the former countries from the Warsaw Pact.

Initially, there was reluctance to take up this gun and its innovative ammunition, because this non-standard calibre had been developed only by FN Herstal. Despite its ballistic particularity, the FN P90® did however slowly and gradually impose itself on all military markets as well as on Law Enforcement markets. This ballistic revolution allowed for the development of the Five-seveN Pistol in 1998.

FN HERSTAL

The rise of FN Engines

In 1983, FN Engines participated in the development of parts for the 850 horsepower TM 333 turbine to provide engines for civilian or military twin-turbine helicopters, such as the new versions of the SA 365 Dauphin, together with the French company Turbomeca. In 1984, the US Air Force finalised a contract for the supply of 32,850 nozzle guide vanes for F100 engines, to be delivered between 1985 and 1987. FN Engines was also tasked with producing four major parts for the General Electric F110 military engine, chosen by the US Air Force to equip its latest generation F-16 and F-15 planes.

The deepest recession in the history of the aerospace industry ended in 1984. Renewal of the civil an fleet began and long-term forecasts were encouraging. In that year, FN Engines increased its involvement in the engine programme for the General Electric-SNECMA CFM56 aircraft engine, which it had been working on as a partner since 1972. This was the CFM56-5 version and FN Aeronautic SA was created to develop it.

Since 1982, the Engines Division had also contributed to the manufacturing of parts for the Pratt & Whitney JT9D-7R4 engine fitted in Boeing 747 and 767 planes, as well as the A300 and A310 Airbuses. In 1984, FN Group and Pratt & Whitney started talks for FN Engines to be associated with the PW4000 programme, a 22-27 tonne class engine set to replace the JT9D-7R4. FN Engines had to pay an admission fee amounting to 3% of the engine's overall cost, in order to jointly produce the PW4000. A finance house, Technofin SA, was set up in 1985, with the participation of the public and private sectors, to raise the necessary funds. This led to three billion Belgian francs being raised. The programme accounted for 15% of FN Engines' turnover and would be spread over 15 years, creating 200 high-level jobs from 1987. In 1986, to meet the growing demand from the aerospace industry, the FN Formétal Division was placed under the control of FN Engines. In September, it became clear that its viability could not be ensured, so the Board of Directors decided to close FN Formétal, while the forge and investment casting were maintained and would be part of FN Engines.

In the space field, FN Engines was still involved in the study and development of elements of liquid strap-on boosters for the Ariane IV rocket. In January 1987, the first FN Herstal employees arrived in Vernon (France), where the liquid propulsion and space division of SEP (the European Propulsion Company) was located. SEP joined the Snecma Group in 1984. They came to work on the test bench created for Ariane rocket engines. The space sector then accounted for 8% of the turnover of FN Engines, which produced around 30 different pieces for the Viking engines of the two first stages of the Ariane IV rocket and for the liquid strap-on boosters. The goal was to become a specialist in valves for the Vulcain engine of the Ariane V rocket.

⬍ New logo of the FN Engines Division, late 1970s
The famous logo with the pedal set, symbolising the first phase of FN's mechanical diversification, gave way to supersonic wings, which evoked the company's more hi-tech products.

◄► Aerial view of the new FM Engines factory in Hauts-Sarts, 1980
FN Engines renovated all of its machine tool, production and control equipment facilities with the inauguration of the Hauts-Sarts factory, which covered an area of 85,000 m². New techniques emerged or became more widespread, such as digital control, electron beam welding, plasma deposition, thermal treatment and vacuum brazing.

CFM56 engine on the Liers test bench, 1985

SPIN-OFFS

In 1987, FN Engines was turned into a subsidiary. The R&D costs, the purchase of licences, the cost of equipment in buildings, machines and new technologies, the wages of highly specialised engineers, and the access fees to take part in large-scale projects... these all became too high for a company that was smaller than the US giants of the sector. The race began to reach a critical size, in order to secure a favourable position for the company among engine manufacturers. On 17 December 1987, the company FN Engines was set up with a capital of two million Belgian francs, which was increased to 4.1 billion Belgian francs in 1988.

In 1989, FN Engines signed its first cooperation agreement, as a plc (public limited company), with the French engine manufacturer Snecma. It joined the programme for the CFM56-5C engine of the Airbus A340, a medium-haul and long-haul plane that entered service in 1992. The contract was for the equivalent of 5% of the value of the engine and included research and development work, the production of various assembled components and modules and lengthy tests. Six months later, FN Group signed a Memorandum of Understanding with Snecma, through which it transferred to Snecma 51% of the shares that it held in FN Engines. Snecma thus became the majority shareholder, with FN and the Walloon Region sharing 49% of the remaining capital.[227] FN Engines became part of an international aerospace industry group, which was the only way of ensuring its future viability and development.

100 years

The 1980s were a difficult period for Herstal, caught between economic downturns and shrinking markets, restructuring plans and the urgency of modernisation, as well as upheavals in financial and oil markets, and so on. Several reasons were cited to explain this disappointing assessment. They included the general decline in armaments companies, higher than expected costs to restructure the company in the mid-1980s, the cost of early retirements and voluntary redundancy, the losses of FN Engines and those due to its sale to Snecma, plus the losses arising from the sale of the foundry, of Rhéo[228] and of FN Industry. Despite achieving good results, FN Industry saw its activities spun off and sold in 1987. The work with emerging countries would be undertaken by a new company combining FN and D'Ieteren – known as FN D'Ieteren Industries S.A.

As FN approached its centenary in 1989, disenchantment was very much in evidence. There were considerable losses, due to an active policy of early retirement and early departures. Expenses due to stopping or ceasing activities were also high, estimated at 4.7 billion Belgian francs for the years 1985 to 1988. The company had to be recapitalised. In late 1988, FN turned to Société Générale de Belgique, which moved from being the company's main shareholder to the majority shareholder.

This news was greeted with cautious optimism, as recapitalisation appeared to enable the company to complete its restructuring. However, the Group lost 2.1 billion Belgian francs in 1989 and its turnover was falling. The prospects were far from bright: "FN Group will continue in 1990 its programme of restructuring, of reducing the number of employees and of limiting the level of its losses".

Yet this did not stop the company from continuing to progress and pursue innovation, by taking advantage of the success achieved with some of its products reaching maturity. So the FN Group's centenary seemed to be a time for renewing and reaffirming the regional role for a group with a global calling.

Luxury B25 shotgun, marking the centenary of Fabrique Nationale, a special series of ten shotguns, 1989
The work of engraving, chasing and inlaying of scrolls as well as references to the history of Fabrique Nationale was completed by Sophie Purgal.

9 The GIAT years

Leclerc tank
In February 1993, the United Arab Emirates placed an order with France for 436 Leclerc tanks for their armed forces. This 'contract of the century' helped Giat Industries to continue serenely doing business in the final years of the 1990s, when the arms sector was very busy restructuring itself.

A period of negotiations

In the spring of 1988, control of Société Générale de Belgique was transferred to the French group Suez and in late 1989, under pressure from its new shareholder, it decided to split off from FN. This decision was undoubtedly taken due to the uncertain future of the arms industry at the time of the Berlin Wall's fall (November 1989), with the emergence of hope for a peaceful world.

Five extraordinary general assemblies were held between June and December 1990. In June, it was noted that the net assets made up less than a quarter of the company's capital, which triggered a legal obligation to deliberate on pursuing the company's activities. The Société Générale de Belgique gave the group a period for reflection, by granting it a subordinated loan of 1.5 billion Belgian francs, thus allowing it to meet different deadlines, until a new General Assembly scheduled for September.

During this period of a few months, the Board of Directors drafted a new restructuring plan but underlined, when presenting it, that the plan's success was conditional on FN securing financial assistance of around 13 billion Belgian francs. At this same General Assembly, it was pointed out that a systematic search for a 'partner' had been organised globally and that, as a back-up, negotiations had started with "a group that had shown its interest in all the arms activities". A decision was taken to pursue these discussions. In October, the Board had received a proposal from the French group GIAT Industries to acquire the firearms manufacturer's business assets.

Ultimately, in November 1990, it emerged that the negotiations between FN Herstal and GIAT Industries had been finalised. From 6 December, part of the assets and liabilities of FN were transferred to a company created by GIAT, Herstal Défense. On 31 December, a notarial act was signed. Through that, FN Herstal sold assets, including the weapons manufacturer's business capital, to Herstal Défense, with the related intangible assets and real estate, the holdings in the direct subsidiaries Browning SA and FN International as well as receivables from certain other subsidiaries. Herstal Défense, which had become FN Nouvelle Herstal (FNNH), also bought up – in successive parts and according to its needs – FN stocks of all kinds. Moreover, 1,300 people were transferred to the Belgian companies of FNNH. The FNNH plc's company capital came to 3.1 billion Belgian francs, 75.81% of which was held by GIAT, 16.13% by DuMex BV and 8.06% by the Walloon Region. The latter took a shareholding, along with the right to veto all key decisions concerning FNNH.

On 1 January 1991, FNNH, in which FN Herstal had no shareholding, took on the industrial and commercial activities previously carried out by FN Herstal in the arms sector. This business included running headquarters located in Belgium and the control that FN Herstal had over some subsidiaries and in particular the Browning Group, FN Manufacturing and FN do Brasil. This excluded the remainder of industrial activities that had not yet been transferred or for which FN Group had some of the capital. Among them were the precision forge, acquired by the JML Company in Herstal, and above all FN Engines. The shares that the Herstal Group had in it, amounting to 19% of the capital, were transferred to Pratt & Whitney, while the Walloon Region increased its shareholding to 30%. Some 30 years later, it is easy now to highlight the wrong turns taken on an industrial journey. However, after the many twists and turns and down-turns, it was clear that FN left Wallonia with powerful tools, most notably the aerospace and space instrument, which became Techspace Aero in 1992 before joining the Safran Group.

Armurerie liégeoise was an ASBL [a non-profit making organisation] whose purpose was to restructure traditional armaments industries in the Liège region. To allow FNNH to step out of its previous activities and related structures, in 1991 it set up the management company Her-Fic, which was tasked with managing the commitments of the former FN Herstal and with identifying partners to take up the activities. These included the FN Engineering subsidiary in Mulhouse, the only French company in the arms sector that had not been acquired by GIAT Industries. Her-Fic was liquidated in 1999.

Finally, on 3 June 1991, the company Herstal SA was created by acquiring FNNH shares and its holdings in Browning and the FN Coordination Center, with capital increased to 4 billion Belgian francs in December. Herstal SA's mission was to set up an autonomous profit-making centre, that included the results of the civilian and military areas of activity; to ensure autonomy in the industrial and commercial management of the two activities; to take responsibility for the political, social and professional relations with the Belgian authorities; to define and ensure the application of a homogenous social policy; and to handle the Belgian group's relations with GIAT Industries and its subsidiary Euro Vecteur. The Chairman of the Board of Directors of Herstal SA would also be the Managing Director of this company and the Chairman of FNNH and of Browning. The Walloon Region's shareholding was transferred into Herstal SA at that time.

GIAT 155 TR Cannons

Reaching critical mass

The industrial and commercial project – which was part and parcel of the takeover of what was now the Herstal Group by GIAT Industries – was seen as an opportunity by economic and political players, and they supported it wholeheartedly. The two parties' production areas were complementary rather than competing (light weapons at Herstal, alongside medium and heavy weapons and vehicles at the other subsidiaries of GIAT Industries). FN was in a position to give GIAT, which was focused on France, an outlet to Europe and the world. GIAT's financial base was impressive. This opened up prospects of attaining the size of a global business through acquisitions in Europe.

There were internal tensions, because of the fear of seeing the group split between civilian and military activities. However, GIAT Industries confirmed that it would

make FN the global hub for its activities in small arms but also for hunting and sports guns. It also confirmed that this hub would enjoy considerable autonomy and even become a focus for closer alliances in Europe.

Recapitalisation and sufficient finance allowed the Group to benefit from the unforeseen geostrategic opportunity provided by the Gulf War in August 1990. That could only lead to a gradual increase in the military capacities of those Western nations involved in the conflict. The financial resources that flowed from that, together with commercial successes (notably the standardisation of models of FN machine guns in some Asian countries), enabled the company to step up its research efforts into the foot soldier of the future (especially by exploiting the new 5.7×28mm ammunition) and to drastically modernise Herstal's industrial facilities.

GIAT *industries*

GIAT

The state-owned company, GIAT Industries, was created at the end of 1990 by the Groupement Industriel des Armements Terrestres. The latter was founded in 1971 and comprised French arsenals under the auspices of the State. The creation of the Groupement Industriel des Armements Terrestres (GIAT) in 1971 was in response to the desire to differentiate State missions from industrial missions. GIAT was a service of the Ministry of Defence. It had no autonomy and no legal status and its management came under the administration's budgeting and accounting rules. This status entailed significant bureaucracy and management complexity. Clearly, this would need to change and that took the form of GIAT Industries. GIAT Industries only emerged from French administration, to become a national company, in the early 1990s. Its aim was eventually to become the leading ground weapons group with five specific business areas: Mobility (tanks and turrets), Arms, Ammunition, Dual-use equipment and Integrated logistics support.

A new factory

The Herstal Group completed extensive renovation work on its Herstal site, so as to bring the different stages of product manufacturing closer together and to organise them in such a way as to accelerate the flow of manufactured products and to reduce their movements. The organisational system was simplified, from the input of raw materials to the packaging and dispatching of finished products. In July 1991, a master plan for this reorganisation was introduced. This began with the renovation of buildings, while maintaining the style and character of the entire factory. The 1889 portico was kept, as were the administrative buildings and the big surrounding wall. The bricks used were chosen for their local style and the window frames were painted blue in homage to GIAT. The ammunition factory for the .50 calibre ammunition was set up at Pré-Madame. The largest projects undertaken were the new barrels manufacturing unit, the machining workshop and heat treatment.

In April 1993, the Delcour barrels manufacturer, which became a 100% subsidiary of Herstal SA the same year, was set up on the site of the former ammunition factory. Machining was redeployed over an area of 7,000 m², while 500 m² were reserved for sharpening, and renamed the 'Tools department'. This department was responsible for the sharpening, initial adjustment, supply, and production of special tools and the management of the

calibres, together with the process planning department. At one point in the past, 10,000 different tools had been in service. Now there were no more than 3,000, but to achieve full efficiency, that number had to be cut in half again. Each tool was categorised and accessible, thanks to a location barcode managed by an IT system.

The new building for surface treatment, where the old equipment had become obsolete, was now located between machining and assembly; it was also close to the refurbished effluent treatment plant. New operating methods were introduced, in small series, on a just-in-time basis, while ensuring the traceability of processes and by using new technologies to reduce the cost price and to increase the quality of operations. The equipment was also made to comply with future Belgian and EU legislation on liquid and gaseous waste. Respect for the environment was evident in the surface treatment and machining processes, with centralised collection for metal shavings and the recycling of cutting fluids.

The industrial surface area was reduced by two-thirds, from 218,000 m² to 70,000 m². The whole programme cost one billion Belgian francs.

The new machining hall, featuring the latest Heller digital control machines

Prosperous military business

Large military orders from the United States and the confirmation of traditional streams of business (especially with Taiwan and Saudi Arabia) guaranteed, for a while at least, the prosperity of the Group's military branch and allowed FNMI to break even finally.[229] FNMI got more involved in arms development programmes launched by the US. In 1993, the Board approved the creation of the subsidiary FN Special Projects, which was tasked with following all the US Defense Department projects requiring Security Clearance.[230] In 1995, FNMI received a new order from the US government for the provision of 16,000 M16A2 assault rifles, which had to be produced in eight months.

The MAG and the MINIMI continued to be safe bets. In 1994, the US Marine Corps adopted the M240G whereas the US Army equipped itself with the M240B the following year. The M240B was in response to the Medium Machine Gun Upgrade Kit Program, which was launched to replace the M60 SACO 7.62 mm machine gun.[231] From 1995, the company was subjected to global turbulence that negatively impacted its profitability. It saw a contraction in defence spending in many countries, the end of conscription in Europe, a lack of financial resources in non-Western countries that were receiving second-hand weapons from the former Eastern Bloc, and the West's hesitancy to intervene in war zones.

The Delcour barrels manufacturer

The barrels manufacturer was established in the early 20th century in the Vesdre Valley. It focused on producing smoothbore barrels, assembling pairs of tubes and reforging. After the Second World War, it diversified and started the production of rifled bore barrels. In the 1970s, FN Herstal, which at the time purchased its .50 barrels from army surplus, saw its stocks run out and placed orders with the Delcour company. In the early 1980s, Browning also called on Delcour for its A500 barrels. The barrels manufacturer adapted to strict demands, thanks to its excellent drilling, honing and hammering machines. But the company's facilities had become dilapidated and redeployment was essential. They were revamped under the auspices of Herstal SA, which had taken ownership of the company. The new barrels manufacturer was inaugurated in 1993 and had to be capable of producing 150,000 barrels per year. As with Browning and FNNH, the Delcour barrels manufacturer was finally attached to the Military Division in 1997.

FN F2000®
Assault Rifle

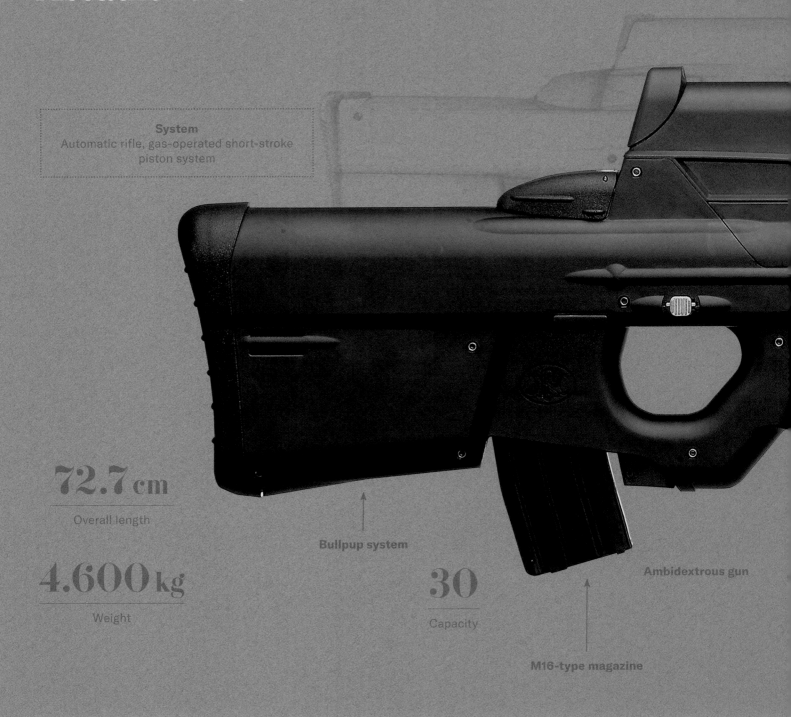

System
Automatic rifle, gas-operated short-stroke piston system

72.7 cm
Overall length

Bullpup system

4.600 kg
Weight

30
Capacity

Ambidextrous gun

M16-type magazine

In 1995, the United Industrial Corporation's subsidiary, the US research company AAI Corporation, joined FN Herstal in a consortium tasked with defining the gun of the future, better known as the Objective Individual Combat Weapon (OICW). It was a 5.56 mm calibre rifle that was configured to combine the roles of an assault rifle and a grenade launcher for a weight of 5 kg in one firearm. The **FN F2000®** 5.56 × 45 mm calibre, which was unveiled in 1996, was developed against this backdrop.

With its futuristic design, this bullpup-type rifle is ambidextrous. This was in response to a common flaw of this type of rifle, namely a fixed ejection at the cheek rest, which can be dangerous and make it difficult to use on the left shoulder. The FN F2000® has a unique solution for the forward ejection of cases far from the shooter's face. Finally, the rifle is modular, notably allowing the fitting of a 40 mm calibre grenade launcher or a sighting system.

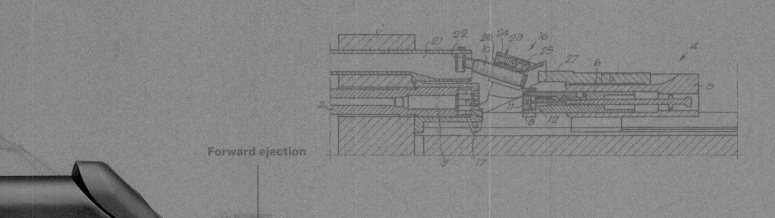

Forward ejection

5.56×45 mm NATO

Calibre

40 cm

Barrel length

900 m/s

Muzzle velocity

850 rpm

Firing rate

GL-1 pump-action grenade launcher and 40 mm rotating bolt

FN HERSTAL

Since 1964, Winchester has produced a limited series of rifles and carbines based on the Model 1894. Their decorations have celebrated characters, events, places and animals that are associated with the history of the US and Canada. In 1994, Browning SA, which had just bought the New Haven factory, relaunched a commemorative series to celebrate the centenary of the Model 1894. These firearms were a huge success, once again highlighting traditional demand.

Difficulties in the civilian business

The civilian business was also experiencing a difficult period. Browning and Winchester were hit by the recession and by the unfavourable transfer of some assets, such as fishing equipment to the US company Zebco, but also by various industrial decisions.

In 1991, the U.S. Repeating Arms Company, which was mainly producing Winchester rifles and shotguns licensed by Olin, was fully bought up by the Herstal Group and put under Browning America's responsibility. The Winchester production centres had to be modernised and this investment seriously hampered Browning's capacity to develop new products. The new Winchester factory was inaugurated in New Haven (Connecticut) in 1994. GIAT benefited from the support of Olin, the state of Connecticut, the town of New Haven and even Yale University in building and equipping this factory. Covering an area of 21,000 m², the factory had 550 employees. They used 321 machine tools, including numerous digitally controlled machines, structured around three assembly lines. There was one for the 9422 and 94 Model rifles, another for the Model 70 rifle and a third for the Model 1300 pump-action shotgun. A heat treatment unit and a rifle range were added to that. The first results of the Winchester range were encouraging. Sales of all kinds of rifle rose from 225,000 in 1992 to 247,000 in 1993 and reached 270,000 in 1994, while the total number of orders was close to 400,000 guns. The management's objective was to produce 500,000 rifles per year by 2000.

In 1992, the acquisition of the Anagni ammunition factory, which produced and distributed Winchester brand ammunition in Europe, under licence from Olin, appeared to make sense. However, there was a significant price war going on in the ammunition industry. In May 1996, the Board of Directors reacted and approved the restructuring of the Anagni factory. At the same time, a decision was taken to reorganise Browning Europe's distribution network. The Browning France company was transferred to Saint-Étienne and Browning Sport Italia was recapitalised and reoriented towards mainly production activities. Innovation continued, despite the difficulties. In 1997, the semi-automatic Gold rifle was presented. It sold particularly well in the US. The Gold represented a new generation of semi-automatic shotguns, fitted with a revolutionary gas extraction system. As a result, in the same configuration, it could fire all the ammunition on the market and in any order. This achievement resulted from team work between departments in Browning Europe and Browning USA.

Advert for the Browning Gold Semi-Automatic Shotgun
This gun has a longer receiver and a barrel chambered to fire the powerful magnum calibres.

The A-Bolt II rifle and its Ballistic Optimizing Shooting System (BOSS) also stood out. The A-Bolt system was based on a breech with a three-lugs locking mechanism and a counter-pin system. It superseded the BBR Lightning, which was manufactured by Miroku from 1979 to 1984.

In the pistols range, the Browning Double Mode (BDM), which was made from 1991 onwards, was fitted with a selector switch through which its operating mode could be chosen. In 'double action pistol' mode (P mode), the BDM operated like all 'double action' pistols of this kind. In R mode, the BDM operated like a revolver and each shot was similar to the first shot fired. The advantage of this latter mode was to maintain a constant weight from the first to the last shot, which allowed the shooter to be more precise.

The Walloon Region, from veto rights to stop-gap support

The GIAT years laid the foundations for Herstal's industrial modernisation, as well as for a new company culture to match the expectations of an economy that was undergoing a major transformation. Although not turning its back on its century-old roots, the company initiated an in-depth reflection process, helped by the external view provided by GIAT Industries.

In the first quarter of 1995, however, staff were worried about the future of the company and their jobs. The trade unions feared that the Group would be dismantled and sold back to the civilian sector in order to boost GIAT's cash reserves. There was also criticism about maintaining the Famas assault rifle, which was made in Saint-Étienne and was a competitor of the FNC, not respecting the agreement by which GIAT Industries had to transfer all its small arms business to Herstal in exchange for transferring the Belgian production of ammunition (except for the .50 range) to the factory in Le Mans. A strike was called and lasted for a month. The conflict ended with an agreement, thanks to mediation by the Walloon Region. Albert Diehl's role was confirmed. He was tasked with carrying out his mission until his successor took over. The civilian and military divisions were kept as a whole and it was officially noted that important subjects would need to remain the exclusive preserve of Herstal SA and its employment-related bodies. The financial management, together with the human resources and communication management, were brought under Herstal SA.

In October 1995, GIAT Industries replaced Albert Diehl with Claude Elsen as the head of the company. Elsen failed to draw up a credible recovery plan. The business climate deteriorated and confidence was lost.

GIAT decided to throw in the towel and went looking for a buyer. A recovery plan was drawn up and accepted at the end of 1996. This plan was key to attracting possible buyers, but its employment-related section envisaged a new and drastic reduction in staff numbers. In 1997, earnings fell further and some investors/buyers withdrew. There was only one American group left in the running,

Colt had very strict demands. For example, it wanted the Walloon Region to finance a new employment-related plan at Herstal, and it wanted to remove the Walloon Region's right of veto over important company decisions. Wielding its right of veto, the Walloon Region refused this suggestion. Thanks to a bold move by the minister Robert Collignon and his head of cabinet Pierre Sonveaux, the Region offered to single-handedly take over the whole of GIAT Industries' shareholding in the Herstal Group.

This decision was obviously crucial for the company and more generally for the whole of Wallonia's industrial sector. The Walloon Region had no intention of allowing this sector to be dismantled. At the outset this was really only a stop-gap operation, limited to the time needed for a serious buyer to make itself known. The 'Herstal 2000' industrial plan was put in place in 1997. Its effects would be felt from 1998 to 2000. After an audit completed by the Société Régionale d'Investissement de Wallonie [Walloon regional investment authority], an employment-related package accepted by the staff went hand in hand with further efforts to restructure the Group.

On 25 June 1998, after reducing the Group's debt by eight billion Belgian francs, GIAT Industries transferred its shareholding to the Walloon Region for a symbolic amount. The Walloon Region increased the capital to 2.5 billion Belgian francs, thus providing the means to pursue the 'Herstal 2000' plan.

**The general management's buildings now displayed
the Walloon Region's colours**
The 1988-1989 revision of the Belgian Constitution foresaw new transfers
of competences – especially in industrial and economic areas – towards the
Regions and the Communities. In 1993, a new reform officialised the state's
federal structure, with federated bodies gaining all their competences.
The Walloon Region's executive body would now be in charge of Walloon
arms companies. In 2003, competences for the import, export and transit of
dual-use firearms and goods were also handed over to the Regions.

Philippe Tenneson

A former naval school student, controller of the armed forces,
director general of the French Ministry of Defence's general
administration, and then deputy director of the French Minister of Defence's civil and military cabinet, Philippe Tenneson
later became the head of GIAT in 1996. Appointed the Chief
Executive Officer and Chairman of the Herstal Group, he was
tasked with a major new mission: implementing the Herstal
2000 plan and restoring the company's fortunes.

10 A new century

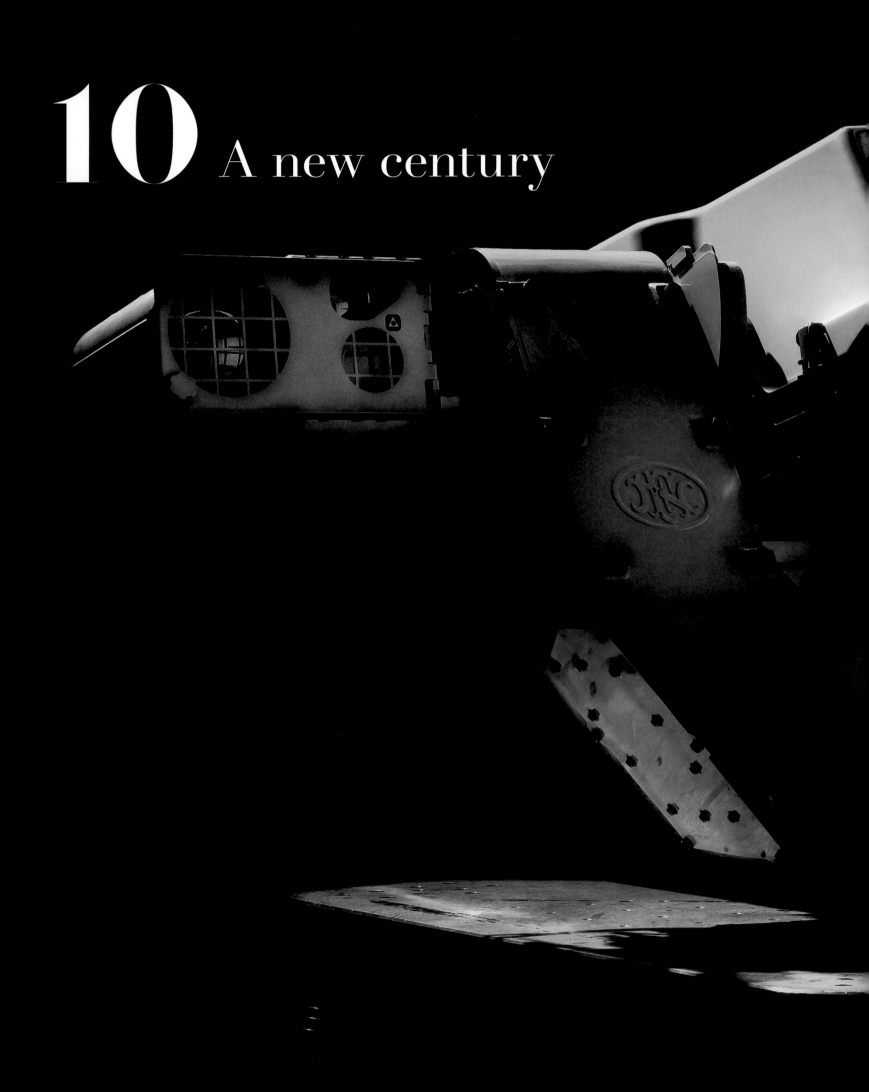

FN Sea deFNder® remote weapon station on 6DOF stabilisation platform, 2023
In the 2000s, the Herstal Group expanded its business as a systems integrator, which had started in the aeronautical field in the late 1970s. By launching the production of land and sea remote weapon stations, the company continues to operate in its core businesses, integrating new cutting-edge technologies. This expanding activity will enable it to diversify its strategic partnerships and strengthen its leadership position in the field of small arms.

Recover or disappear

THE HERSTAL 2000 PLAN

"In 1996 and 1997, the Herstal Group lost nearly 1.8 billion Belgian francs each year [...]. Either the group recovers or it fades away."[232] With these words, Philippe Tenneson, the Chief Executive Officer and Chairman of the Herstal Group, highlighted the need for a real recovery plan to overhaul the company to a level where it could face 21st century challenges. Implementation of the Herstal 2000 plan enabled it to reach this goal. The plan was based on three pillars: adapting its industrial facilities to a difficult global marketplace; having the capacity to design, develop and manufacture innovative, competitive and high-quality products; and setting up top sales teams to win back market share.

Trade globalisation bolstered by the ongoing digital revolution, plus ongoing economic and financial globalisation, were becoming increasingly entrenched and deep-seated. These trends determined the conditions for the company's necessary adaptation. The Herstal 2000 plan called for the Group to strengthen its global approach to markets, in order to reach an optimal size. This would ensure the Group achieved a rapid return to profitability and secured long-term financing for research and industrial investment. A new global structure, that was better coordinated and solely focused on its traditional core business, should be set up. The Group's range of products would need refreshing. Innovations and costs would also have to be managed better, through better synergies between the different entities in the Group. Cost reductions should be achieved by rationalising and modernising the production facilities across the Group, and adapted to reasonable growth prospects for turnover. Structural costs as well as administrative and commercial costs, particularly expenditure on staff, needed to evolve and the Group's staffing levels fell from 2,900 to 2,500 employees. The Walloon Region, which had been the company's only shareholder since 1998, played a key role in implementing the Herstal 2000 plan. As a public sector shareholder, it provided stability plus long-term economic and social support within Belgium and abroad. Initially, this was only a temporary umbrella type operation for the Region, but the synergies grew stronger with the passing of time. This single shareholder, which was not an industrial player, entrusted the management of its shares to the Walloon company SOGEPA and then to its operational arm, the Société régionale d'Investissement de Wallonie (SRIW) [Wallonia's regional investment agency]. This linked the Herstal Group in effect to the aerospace, space and defence area of business – because SRIW also had a holding in Sonaca and the Herstal Group's former Engines Division, Techspace Aero. This ecosystem became a major area of competitiveness, supporting Wallonia's reorganisation.

Defence and Security, in tune with a changing world

A DIVERSE SET OF THREATS

Ten years after the fall of the Berlin Wall, the bipolar world in which the Soviet bloc and the West stood on opposite sides was superseded by a multipolar world dominated by the United States. Conflicts based on ethnic or religious differences grew in number worldwide and reached Europe's doorstep. In parallel, new transversal dangers – among them terrorism and cross-border crime – became more serious threats for international stability. In 1998, the Kosovo War triggered the first major military action led by NATO since its establishment. From then onwards, the Atlantic Alliance entered a complex process, redefining itself as a whole, its pillars and its out-of-area peacekeeping missions covered by the North Atlantic Treaty.[233]

The terrorist attacks of 11 September 2001 marked the emergence of a new geostrategic era. Terrorism, previously limited to being a domestic concern for individual states and rarely perceived as an international challenge, became a global phenomenon. On 12 September, for the first time in its history, NATO invoked Article 5 of the North Atlantic Treaty, which foresees the possibility for all of its members to intervene if one of them is attacked.[234] For the first time, NATO carried out land-based military operations. Foot soldiers were deployed. Supported by modern technologies, they faced a diverse set of threats and asymmetric fighting. The soldiers were also assisted by implementation of tactical concepts on controlled risks, mobility and special forces. These would be seen in the front line of all 21st century operations and came with sophisticated weaponry.

FN HERSTAL: RESTRUCTURING AND REPOSITIONING

Military expenditure, which had been considerably reduced up until 1996, was unstable at the beginning of the 21st century. Europe's adoption of the euro, its new single currency, and convergence plans led to delays in arms programmes. So the recovery came mostly from other regions of the world,[235] especially from the US after the 9/11 terrorist attacks.

FN Herstal needed all its dynamic strength, in both sales and engineering, to get through its restructuring phase. This strength was also crucial for the company to adapt to the changing marketplace and to make its mark in the arms industry of the 21st century. The company's Defence

US Marine Corps soldier, with the M249 (FN MINIMI®) Machine Gun, based in Helmand province, Afghanistan, 2008
The early 2000s were notable for the United States' intervention in Afghanistan and Iraq. The Herstal Group responded to the US Army's growing needs for firearms through its FNMI subsidiary.

business repositioned itself by ploughing money back into pistols. It also worked toward that goal by offering an assault rifle adapted to the Future Soldier concept, by establishing a new arm focused on the security forces, and above all by maintaining its dominance in the global market for traditional machine guns (FN M2HB-QCB, FN MAG and FN MINIMI). The Group therefore consolidated its business activities in the US via FN Manufacturing (FNMI), so as to maintain its position as the main supplier of light weapons for the US Armed Forces.

THE FUTURE SOLDIER

Since the early 1990s, most modern armies have sought ways to optimise the performance of their fighting forces while affording them maximum protection. These studies would drive the 21st century defence industry via the concept of the Future Soldier – who would now be redeployed on the ground and connected to chains of command. These soldiers' skills in combat and protection should be enhanced by technology. US research centres were well ahead in this area and were investing heavily in it.[236] For the arms industry, the goal is to meet the requirements set out in new programmes, such as the

Objective Individual Combat Weapon (OICW), which aims to equip the US Army with new generations of compact weapons together with integrated technology.

FN Herstal was quick to join this new technological concept by developing a firearm with a modern design, the FN F2000. The gun's design phase began in the 1990s and it was presented in 2001. It met the demand for a modular gun, enabling a user on the ground to change all its components according to specific needs of the current operation. The grenade launcher's fire control, backed by an integrated laser rangefinder, was developed with the Finnish company Noptel in 1998. This increased the system's effectiveness by range finding the target and calculating the angle of fire. With its futuristic design, the FN F2000, which was a true engineering masterpiece, marked a new departure in the design of assault rifles. For FN Herstal, the gun was an opportunity to test separately the market for Bullpups. This mechanical configuration was designed to reduce the gun's length, but it was ultimately abandoned, because it no longer suited the new operational theatres.

A NEW SECTOR: LAW ENFORCEMENT

Alongside efforts to adapt its range of defence guns to the new requirements of military strategy, the Herstal Group reacted to growing demand from law enforcement bodies for the appropriate means to respond to growing insecurity within civil society.

The Herstal Group had already embarked on projects to equip police forces, in the late 1970s. In 1998, this sector became a business activity in its own right within the Group, effectively a third branch, between the civilian and military activities. The aim was to enter a booming market, especially in the US, where the terrorist threat and organised crime had sparked the creation of special units calling on standardised methods and sufficiently equipped to deal with major incidents such as hostage-taking. In 1998, the Herstal Group established FNH USA in McLean, near Washington, to break into this new market. By locating this structure in the US, the Herstal Group set itself a major challenge, in terms of setting up a logistical, commercial and industrial process to match the complex organisational structure of the US police forces. Initially, FNH USA was tasked with developing a marketing strategy linked to the needs of the US military market, before expanding to include the federal police forces. The plan was for FNH USA to gradually move into civilian commercial operations at the local level. This strategy was based on coordination of its distribution networks, making the structure more stable.

One of FNH USA's business objectives was to offer a wide range of products adapted to this new niche – and to reinvest in the pistol market as a priority – while optimising production processes. Key to this was adapting military and civilian products to the law enforcement market, in order to build a robust Defence and Security business. It would also be essential to strengthen this business's profitability by increasing export growth potential as well as sharing research and development costs. FN Herstal diligently developed a new pistol, the FN Forty-Nine, and offered its range of 5.7×28 mm calibre products. These comprised the FN P90 submachine gun and the FN Five-seveN pistol, which entered production in the 1990s but were now adapted to the new market. They came in a Tactical version, with a laser pointer and light. A second version boasted an internal hammer integrated into the linings of the breech, to increase the speed with which the gun could be used during operations. The gun was small and light, and its 5.7×28 mm calibre ammunition has low recoil. This enabled quick and precise shooting for single shots and bursts of fire. This ammunition also reduced the ricochet effect and avoided any risk of over-penetration, making it a suitable weapon for police forces. In 2002, the FNH USA's efforts were rewarded when the Secret Service – which provided protection for the President of the US – adopted the FN P90.

THE SMART GUN

From the late 20th century, FN Herstal had focused on developing a 'Smart Gun', a security-enhanced firearm that could only be used by its owner. To prevent accidents, such as a shot being accidentally fired when the gun was replaced in its holder, the technology that was developed allowed the shooter to be uniquely identified. It also offered an assessment of the light and included a 'black box' function to record the time and date of shots.[237]

In 1998, FN Herstal presented its first prototypes of such a gun. This included a patented system based on ultrasound transmission between a transmitter bracelet worn by the owner and the firearm fitted with a receiver. The US National Institute of Justice quickly expressed interest, because this system would make it impossible for someone to use a gun stolen from a police officer and would prevent a child from firing it.[238] A grant was then given to FNMI to launch the first phase in the development of a handgun, which would be for use by law enforcement forces and would only allow its designated user to fire it. The first tests were carried out on the FN Five-seveN and FN Forty-Nine pistols, thanks to the use of ASIC (Application-Specific Integrated Circuit) and RFID (Radio-Frequency Identification).

The Smart Gun technology would not be used directly, because it was too advanced for its time and because there was no specific market for it. However, in addition to the laser rangefinder for the assault rifle, it did allow FN Herstal to take its first steps in enhanced weaponry, a technology with a promising future.

FN Five-seveN® MK2
Semi-Automatic Pistol

20.8 cm
Overall length

20
Capacity

12.2 cm
Barrel length

Blank · L191 tracer · FR199 frangible · SB193 subsonic · SS190 standard · SS192 Soft

5.7 × 28 mm NATO
Calibre

0.645 kg
Weight

System
Semi-automatic pistol, delayed blowback operating system with double-action only or single-action for the Tactical

The **FN Five-seveN®** is a light and low recoil semi-automatic pistol, with no external hammer. Its first version only came with an obligatory double-action only trigger. The lack of a manual safety is compensated for by the high base weight. All the safety mechanisms are immediately reactivated after each shot. A tactical version, which was initially intended for special forces, has become the standard version and so the pistol can be used with a single action. That enables reduced trigger resistance and travel, important factors for rapid and precise firing. The FN Five-seveN® Tactical includes a manual safety that blocks the trigger mechanism and has a manual slide stop catch.

FN HERSTAL

Headquarters of Browning North America in Morgan, Utah, 2021
In 1964, the headquarters of Browning Arms Company was moved from Ogden to Morgan. This was only a short distance and helped to preserve the spirit of the very place where, some 90 years earlier, the Browning brothers had opened their first store. Today, the facilities serve as the business management as well as research and development centre of Browning North America, a subsidiary of the Herstal Group.

Hunting and Sports Shooting Division is reorganised

REORGANISATION FIRST

In 1998, implementation of the Herstal 2000 plan proved to be a radical change in the commercial and organisational approach pursued by Herstal Group's Hunting and Sports Shooting Division. The aim to create more cohesion between the brands and the production locations led to the establishment of a common global management structure: Browning SA. This management holding company, a subsidiary of the parent company Herstal SA, directly controlled the main entities of the Browning Group: firstly Browning International, which was created in 1999 on the Hauts-Sarts site and which directly managed the company's European branch in Belgium, France, Germany, the UK, Italy and Portugal; and then Browning North America, headquartered in Morgan in Utah, which managed the US and Canadian branches. Browning SA's mission was to define a global strategy for the Browning Group and to coordinate its implementation. The North American and European branches, which had always worked in two separate markets, would join forces in a new global organisation with a view to optimising the use of all resources.[239] To support its sales policy and to increase the product renewal cycle,

Browning SA could count on production centres located in Herstal, in Viana do Castelo in Portugal (Browning Viana), in Anagni in Lazio (Browning Italia), and in New Haven in Connecticut (USRAC), not to mention its strategic partner Miroku (Japan), an independent entity in which the Herstal Group held a minority shareholding. The reorganisation helped FN Herstal to handle the decline in the civilian gun sector, a decline that started in 2001 and followed on the heels of considerable growth in the US since 1995. The priority was to bring Browning International back into profitability and to face up to the new competition from low-cost producers, by showing the company was agile, responsive and able to stand out from competitors. The business plan of the Hunting and Sports Shooting Division would be based on exploiting and strengthening the identity and image of its brands, and would rely on the quality of the products, know-how and customer services.

THEN CENTRALISATION

In a second phase, the European distribution network was reorganised to achieve economies of scale. Since the creation of distribution subsidiaries, communication

Headquarters of Browning International in Hauts-Sarts, Herstal, 2022
Since 1999, Browning International has been the commercial and logistics management centre of Browning and Winchester for Europe. The company is located in the industrial zone of Hauts-Sarts, a key Walloon economic and strategic hub, bringing together 350 companies and around 10,000 employees.

mechanisms had evolved and information systems were more developed. The decentralised organisation became expensive, with a number of identical tasks carried out in different places. A decision was therefore made in 2001 to shut down the European subsidiaries for distribution and to reorganise Browning International's business approach. The process was strengthened until 2007.[240] The administration for sales (dispatching, billing and accounting) was centralised at the sites of Browning International in Hauts-Sarts in Belgium as well as in Morgan and Arnold, within the Distribution and Service Center, in the US. Information technology was used to bolster the management of warehouses and distribution, with the introduction of the Warehouse Shipping System.[241] On the ground, teams now comprised a network of gunsmiths and freelance agents representing the two brands – Browning and Winchester.

In Europe, centralisation also impacted the ammunition side, where Browning International had two business activities: shotshell, produced in the Anagni factory in Italy and distributed under the Winchester brand name (and partially Browning and Légia) and metallic ammunition, for which Browning was the distributor for the Olin

Group. Difficulties with the ammunition market prompted the Herstal Group to give up Anagni, an uncompetitive factory with obsolete equipment. In the future, the cartridges would be sold under licence by the Olin Group by specialised partners. This was what happened with metallic ammunition, whose distribution was organised in Europe from Hauts-Sarts.

Finally, in 2001, the Custom Shop was relocated to the Herstal factory, because the luxury market was getting tougher. The artisanal workshop, which was now tailored to receive customers, benefited from FN Herstal's industrial experience in automating operations where the artisan's work did not add value. Wood was sanded down on numerically controlled machines, an operation carried out through mechanisation. Barrels were assembled in the barrels manufacturing unit and the engraving was done with laser technology. However, the hammer and the chisel continued to be the most used tools, during the 100 to 500 hours of work required to complete the engraving of a B25.

A BREATH OF FRESH AIR

In the early 2000s, the Hunting and Sports Shooting Division adapted its structure to remain an upmarket producer. It also aligned itself with new practices in hunting and civil society's aspirations for sustainable development, within an increasingly strict legal setting for the possession of firearms. For example, in North America, a market of 25 million hunters,[242] Browning and Winchester joined networks of associations that defend wildlife and nature: Ducks Unlimited, the National Wild Turkey Federation, the Delta Waterfowl Foundation and the Mzuri Wildlife Foundation.

The strategy for renewing product ranges was a response to the decline in the populations of small game, as well as the sharp and simultaneous increase in populations of wild boar and big game. New technologies and new designs emerged, which required more R&D activity, and focused on reducing production costs. Communication was improved between the design offices of Herstal and Morgan; and the production of components was optimised in the most appropriate production sites. Browning's policy was enhanced by production of the Browning Gold and Winchester SX2 semi-automatic shotguns, chosen as 'Best Shotgun of the Year', in 1998 and 1999 respectively, by the Shooting Industry Academy of Excellence. These shotguns were designed by a team from Morgan, developed and industrialised by teams from Liège, with components made in Liège and in New Haven. They were then assembled in Viana do Castelo and finally distributed from Arnold and Hauts-Sarts.

In 1999, for over and under shotguns, Browning marketed a new 28-gauge shotgun – the B425 Elite. Three years later, it launched the B525, the fifth generation of over and under shotguns. The shotgun features a wide flat bolt, which makes it very strong, and a pin with a large contact surface, ensuring durability. The gun has a mechanical adjustable trigger and hammer ejectors for a clean ejection. The barrel assembly was updated to include Invector technology for firing steel shot cartridges. The B525 is available in three families: Sporter, Hunter and Heritage, representing over 30 models, with three calibres and four different barrel lengths plus a choice between an action frame in steel or alloy. At the start of the new millennium, the B525 was destined to become the traditional platform for Browning over and under shotguns. For Winchester, the 12-gauge Supreme arrived in 1999. Its low-profile action frame was available in a hunting and sporting version. The sporting version had more technical finishes, with a back-bored barrel: Back-Bored technology results in better performance from the shotgun blast with less recoil and higher initial speed. In 2002, the shotgun was launched in the Super Sporting version and was followed a year later by the Energy Trap and Sporting versions.

Finally, for handguns, Browning refocused on its traditional production of the Browning HP pistol, which could be enhanced by the engraving workshop and was aimed at collector customers. Above all, the company worked on the range of .22 calibre Buck Mark target shooting pistols. Manufactured by Arms Technology Incorporated (Salt Lake City), the Buck Mark was derived from the Challenger and Medalist pistols, but had a simplified form of production and a more competitive sales price.

FN-Browning and .22 calibre

The oldest calibre still in use nowadays, the .22, because of its widespread and easy use – as much for hunting or sports as for shooting training – quickly caught the eye of the FN. Early in the 20th century, the company started producing the FN Carabine, with bolt action and rimfire, and as a 'school training model', followed by semi-automatic and pump-action rifles designed by John Moses Browning shortly before the First World War. In the 1960s, FN revived interest in this calibre with the Browning T-BOLT straight-pull bolt action rifle, as well as by launching the production of a semi-automatic .22 calibre pistol for competition shooters. Based on the Woodsman Model, designed by John Moses Browning in 1915, it was available in many variants: Nomad, Challenger, Medalist, 150, and International. In the mid-1980s, it was named Buck Mark. These pistols are fitted with an ergonomic grip and counterweight, ensuring full control of the gun.

Adjustment of a B525 Over and Under Shotgun, Miroku factory, in Nankoku, Japan, 2007
Careful adjustment after machining of the Browning new-generation over and under shotguns. This manual work, based on experience and the sharing of knowledge, is a source of pride for the Japanese company and of confidence for the Herstal Group, which ensures that it plays a growing role in the strategy of its civilian division.

A factory for the 21ˢᵗ century

MANAGING A GLOBAL INDUSTRIAL FACILITY

In 1998, a corporate entity governing industrial policy was created for the Herstal Group.[243] It aimed to ensure that the various different factories coordinated their purchases and positioning of industrial equipment and investments, while improving productivity and responsiveness. At the system's centre was Liège-Zutendaal, with two sites in Herstal and Zutendaal, which brought together the workshops and functions needed to produce military products sold by FN Herstal plus some components of civilian guns for other units of the Group. In addition, the Herstal site brought together the functions relating to quality, logistics, purchasing, and use of the IT network, general services, management supervision, human resources, plus the research and industrialisation offices of FN Herstal and Browning International – which were now grouped together. The cartridge production business – for the machining of metallic elements for 12.7 and 5.7 mm calibre ammunition – was also set up there. The Zutendaal site concentrated on pyrotechnics, powder and fuses and the shooting range for trials. Besides these structures, Herstal Group's industrial facilities extended to Columbia, with the FNMI and New Haven with the USRAC factory; covered Portugal, with the Viana do Castelo factory; and covered Japan thanks to the industrial facility of its partner Miroku.

To manage this industrial ecosystem, IT was vital for improved communication and working conditions within the Group. Although IT was not a new field, the internet's development and accessibility made it possible to create historically unparalleled platforms for exchanges, and the dissemination of information and communication. Computer technology was also extended to the pool of numerically controlled machines, which were connected to the network, electronically identified and managed by the Flexible Manufacturing System. This was a key element for production according to the new machining process.[244] Parts were now checked using the Statistical Process Control system and the plans were now available in electronic format in the workshops.

The digital revolution, which sped up the globalisation of trade, reorganised the whole technical system over a few decades. This created a worldwide network of information, which has generated a real culture based around computer screens.[245]

Aerial view of the FN Herstal factory and headquarters of the Herstal Group's General Management, 2022
Located in the urban landscape of Herstal town, the factory of FN Herstal today covers 13 hectares. Reflecting its industrial strength, this space has been a source of pride for the company since its history started, as shown by the aerial photographs included in its catalogues since the early 20ᵗʰ century.

End of the assembly line for ammunition of calibre 5.7×28mm, Zutendaal factory, in Limburg, 2022
In the early 2000s, ammunition production focused on calibres 9mm, 12.7mm and above all the new-generation calibre invented by FN Herstal and standardised in 2021 by NATO 5.7×28mm.

MODERNISING AND EVOLVING

The process of modernising industrial facilities that began in the 1990s continued at the turn of the century. Investments were made in the barrels manufacturing unit from 1998, through the acquisition of new machines, automation of production lines and strict supervision of expenditure and productivity processes. The modernisation concerned both the numerically controlled turning and deep drilling machines for smoothbore barrels, together with the core forging machines for cold forging of rifled barrels. The Herstal Group also invested in cartridge production, for which the activity had been refocused on the production of 12.7 and 5.7mm calibre ammunition during the GIAT period. A new production and maintenance facility was inaugurated at Herstal. At the same time, the production line capacities, especially in tracer bullets, were increased at Zutendaal. The 7.62 and 5.56 calibre ammunition would now be produced through a partnership, but with a follow-up, final inspection and guarantee of quality provided by FN Herstal. The company kept its know-how and its capacity to satisfy the demand for ammunition, for all NATO calibres, including by proposing the setting up of industrial production facilities in countries that asked for this.

The industrial plan also evolved by taking into account the technological input that firearms of the future required, i.e. that they must be lightweight, strong and reliable at the same time. In the 1980s, FN Herstal already had a small plastic injection workshop to produce the forearm for the FAL and the FNC.[246] But it did

not produce sufficient volumes to be profitable. Ten years later, steel working continued to decline while composite materials and polymers increased. This meant the company had to master new production techniques. For example, in the FN F2000, the grenade launcher, the receiver and the sight system are made of polymers and other composites, with properties similar to aluminium. The overall technologies to mould parts using heat or cold make it possible to produce thousands of (based on injected powder) complex and high-precision parts. These parts are shock resistant and resistant to ageing, heat and compression at a competitive cost and in short time spans. In the late 1990s, FN Herstal invested to expand injection processes at an industrial scale. In that period, the Metal Injection Moulding process, which was

harnessed to produce small-scale parts, used extremely fine metal powder mixed with an organic binding agent. The mixture was then injected into a mould to obtain the shape of the part. This minimised any loss of metal and made it easier to adjust the material's composition and thus to produce complex small parts at low cost. This evolution of techniques and materials, designed to make products lighter – and the addition of optical, electronic and digital systems – has significantly changed industry's way of working.

Healthy accounts again

The Herstal Group returned to profitability in the middle of this century's first decade. Credit should be given to the restructuring and adjustments to the business model, as well as investment in industrial facilities and their modernisation and reorganisation. The Group's resurgence led to the manufacture of more products and more kinds of products, particularly for the Defence and Security line of business.[247] This period also saw a turnaround in the business cycle. The Hunting and Sporting sector had accounted for 60% of the Herstal Group's turnover since 1994. In 2001, that figure had even climbed to 68% before finally reaching the same level as the Defence and Security sector in 2006.

This newly improved financial balance quickly allowed the Herstal Group to envisage fresh investments and to cope with the contraction in markets following the 2008 crisis. For the Group, the crisis, which came after an exceptionally bright spell, had an impact on the US civilian market especially.

The rise of Defence and Security

Faced with new international threats, NATO members adapted their weapons to the context of counter-insurgency and expeditionary warfare. A period of renewal and modernisation of arsenals followed the period of austerity. At the same time, other countries across the world, whether in South America or Asia, increased their military investments to reach the levels of NATO members. The Herstal Group responded to demand by strengthening its leadership position in machine guns, by developing a new assault rifle, by pursuing its considerable progress in Law Enforcement, by getting started on the concept of less lethal weapons, and by expanding its Integrated Weapon Systems business.

MACHINE GUNS AS 'BATTLE WINNERS'

In the 2000s, FN Herstal strengthened its foothold in the US, Europe and in the rest of the world, where several armies were standardising their pool of heavy, medium and lightweight machine guns by adopting the FN M2HB-QCB .50 BMG calibre, the 7.62 mm calibre FN MAG (M240 in the US), and the 5.56 mm calibre FN MINIMI (M249 in the US). These guns still meet the prerequisites of different operational theatres, thanks to continual adjustments.

The FN MAG or M240B (L7A2 in the Commonwealth countries) proved itself essential in both Afghanistan and Iraq. This gun's firing capacity and high range made it the main gun for suppressive fire within an infantry company. The machine gun was modernised over the years, notably with a polymer version fitted with an optical rail, with a cold-forged barrel and hard chrome-lined bore, making it more durable and enhancing the precision.

Contracts for MINIMI were secured worldwide, in each case meeting the requirements of customers. For instance, the British Army ordered a short-barrel FN MINIMI, known as the L110A2/A3. The gun quickly became popular among soldiers deployed in Afghanistan, thanks to the flexibility when used.[248] However, the fighting in operational theatres revealed several problems that Western armies were not expecting. Until then, the typical combat range for light weapons was estimated at 300 to 400 m, for which 5.56 mm calibre weapons were perfectly suited. Yet in the open spaces of Afghanistan, allied forces frequently found themselves facing an enemy armed with the powerful 7.62 × 54 mm cartridge, forcing them to engage targets at distances of over 600 m. In 2001, the United States Special Operations Command (USSOCOM) launched a call for tenders for a 7.62 × 51 mm calibre lightweight machine gun. FN Herstal responded by submitting a revised version of the FN MINIMI with this calibre, the MK48. Developed since 2004 and produced on an industrial scale in 2007, the gun has the same ergonomics as its 5.56 mm version, but is lighter and shares the same calibre as the FN MAG. This gives the MK48 an advantage, because its users are already familiar with the product.

Assembly workshop for FN MAG® Machine Guns, FN Herstal factory, 2022
In this early part of the 21st century, FN Herstal remains the leader for sales of light and medium machine guns, which the company exports worldwide.

FN SCAR®-L Assault Rifle with folding stock and technical drawing of the receiver

THE FN SCAR ASSAULT RIFLE

Starting in the late 1980s, the US had launched and increased the number of its arms programmes (including the Objective Individual Combat Weapon) designed to replace the M16 rifle, whose first version dated back to the 1960s, and then the M4 carbines. The final choice of a replacement firearm was postponed from one decade to the next, due to a very strict set of specifications and the high budget required. So the M4, M203 and M16A4 (the fourth generation of the M16) were still produced, notably by the FN Manufacturing factory. In late 2003, a Joint Operational Requirements Document was again published in the US, calling for bids by the arms industry for projects linked to the adoption of an assault rifle. In 2004, FN Herstal was chosen unanimously, by a panel made up of USSOCOM operators, for the product that it submitted: the FN SCAR.

Since 2001 and the beginning of the war on terrorism, USSOCOM's operational engagements have emphasised the quality of their firearms and the reasoning for their choices. To meet the demands of front-line combat, special forces soldiers have called for a range of advanced modular firearms that can be adapted to different calibres, the aim being to optimise the number of

common parts and functions of the FN SCAR 5.56 mm and 7.62 mm versions, as well as to be able to add accessories that are specific to a given mission.

By proposing a platform adapted to two calibres, FN SCAR fulfilled the criteria of USSOCOM, with which FN Herstal worked closely in the development process. The 5.56 × 45 mm version was called FN SCAR-L (*Light*) or MK16 in the US, while the 7.62 × 51 mm version was called FN SCAR-H (*Heavy*) or MK17 in the US. The gun has an open architecture thanks to its Picatinny rail, which enables lights, lasers and other targeting devices to be added. It can also be fitted with the grenade-launcher module, the FN40GL (MK 13 EGLM in the US). In August 2007, the initial assessment phase was carried out by various units of the US military community, especially the Rangers, the Navy SEALs and the Green Berets. These tests were conducted until 2008 in a variety of operational theatres – including urban warfare, maritime, jungle and mountain environments. In 2010, FN SCAR and FN40GL finally received Milestone C, i.e. authorisation to be put into service. After three years of successful in-theatre combat evaluations, the firearms moved into the acquisition phase.

Combat swimmer equipped with the FN SCAR®, 2015
Developed in collaboration with the United States Special Operations
Command or USSOCOM, the mechanism of the FN SCAR® had to cope
with the most adverse conditions, including amphibious use. USSOCOM
was officially established in 1987. Each of the US Army's four main
branches – Army, Navy, Marine Corps, and Air Force – has a Special
Operations Command (SOC) governing its units and activities, as well as
Joint Special Operations Command (JSOC) providing, by its own definition,
"a joint command charged to study special operations requirements and
techniques; ensure interoperability and equipment standardisation; plan and
conduct joint special operations exercises and training; and develop joint
special operations tactics".[249]

Deployment of law enforcement forces, 2016
FN Herstal has responded to the changing nature of law enforcement with its new FN 303® Less Lethal Launcher.

FN 303® and M16A2 Assault Rifle fitted with the FN 303®
This semi-automatic compressed-air launcher has 15 projectiles of 8.5 gram. When fired at an assailant with an energy of 33 joules, they have a neutralising effect like that of a long-distance punch. The projectiles feature an aerodynamically stabilised polystyrene body and a non-toxic bismuth pellet for more accuracy and greater effective range. The stock of the FN 303® can be detached in order to mount the launcher beneath the barrel of an assault rifle. This gives soldiers a less lethal system, enabling them to deal with any type of military operation in an urban environment.

LAW ENFORCEMENT

Herstal Group's security-related business, which had gained a foothold in the US in 1998, was able in the early 2000s to offer an initial range of pistols with different calibres, to meet the requirements of police forces. This business was developed thanks to the development of polymer frame pistols – the FN FNP, with the FN Forty-Nine, produced within the FNMI, in .40 and 9 mm calibres, then followed in 2003 by the FNP-9 and in 2006 by the FNP-45, chambered in a .45 ACP, 9 mm and .40 S&W. The appeal of the 5.7 mm calibre ammunition, as used by the FN Five-seveN pistol and the FN P90, was also boosted by the establishment of additional production facilities in Italy and extra distribution channels in the US.

The adaptations of military technologies to the civilian market also had to comply with the many regulations and types of police unit. On US territory, for example, most of the units had to comply with the regulations of the BATFE (Bureau of Alcohol, Tobacco, Firearms and Explosives), which stipulated that these guns adapted from military technologies had to be longer and only fire semi-automatically. In 2006, FN Herstal responded to these demands by developing the FN PS90, a version with a long barrel and with a magazine of 30 rounds (instead of 50) or 10 rounds in those US States where the law imposed limited-capacity magazines. Similarly, the Tactical FN FS2000 became the civilian-legal version of the military FN F2000. By optimising the use of its know-how in military firearms, FN Herstal could better recoup the costs of mass production and boost its competitiveness.

LESS LETHAL WEAPONS

FN Herstal also made a name for itself in another area, that of less lethal weapons. This concept emerged in the US, after the establishment of the Non-Lethal Warfare Study Group in 1991. The recommendations of this working group made little impact, until international and domestic circumstances revived the debate in the late 1990s.[250] In today's strategic context, armed forces are increasingly being called on to intervene in missions such as crisis resolution, the protection of sensitive places, maintaining peace and humanitarian assistance. These missions often play out in the presence of civilians, in a situation where the forces that are intervening have to restore order through proportionate means. The security forces face situations such as domestic violence, disturbances during sports events or riots and protests, setting them against crowds, organised gangs or isolated individuals. The goal is to incapacitate people or equipment, without risk of death or serious injury and with a minimum of collateral damage.[251]

The arms industry therefore had to develop a weapon positioned somewhere between a truncheon and a pistol. In 2003, FN Herstal unveiled the FN 303, a less lethal weapon that uses the kinetic energy of a projectile propelled by compressed air, in order to slow down and stop a target 50 metres away. With the development of the FN 303, Herstal Group found itself in a new field of expertise, in terms of engineering and society. To strengthen its mastery of the product and its engagement in society, the Herstal Group initially contacted Liège University in January 2004. This led in 2005 to the creation of a centre of excellence on less lethal weaponry. The centre was tasked, through sociological and technological studies, with developing a range of products in line with users' needs, while taking into account issues of a medical, legal and socio-environmental nature. The programme, which was subsidised by the Wallonia Public Authority's Department for Technological Research and Development, aimed to become a European reference in this area.

INTEGRATED WEAPON SYSTEMS

Airborne systems

FN Herstal's Integrated Weapon Systems business emerged in the late 1970s, with the development of the first pod concepts, the 7.62 mm calibre Twin Mag Pod and then the 12.7 mm calibre Heavy Machine Gun Pod. Through contact with helicopter and aircraft manufacturers, FN Herstal strengthened its partnerships and its know-how, underlining its international credentials as a producer of systems and as a lead contractor. This helped the company to secure different contracts, such as fitting pods to the German version of the Eurocopter Tiger helicopter or Beechcraft T-6 subsonic aircraft flown by Greece, which demanded intense development work on the connecting interface.

In the 2000s, the Integrated Weapon Systems business consisted of Air Axial, a systems of pods controlled by the flight deck, and Air Sabord, a system controlled by a crew member. For Air Sabord, FN Herstal proposed a range of mounts for helicopters, with the integration of a system for the FN MAG machine gun on Eurocopter's Cougar in 2006 for example, or systems for the .50 calibre FN M3M machine gun. This machine gun was initially developed during the Second World War by the US Army, which was searching for a gun with a high rate of fire for its combat aircraft. The machine gun was then totally redesigned and optimised by FN Herstal to be adapted for long-distance firing from helicopters, subsonic aircraft, land vehicles and ships. Over a range of nearly 1,850 metres, it provided fire power of 600 rounds in a continuous burst without overheating, compared to 150 rounds for the original version.[252] In 2004, negotiations were concluded with NAVAIR (Naval Air Systems) for a contract to supply M3M machine guns and to mount them on a door or ramp. The system was adopted by the US Air Force, the US Navy and the US Marines with the standardised name of GAU-21/4 and fitted to CH-46, CH-53 and H-60 helicopters.

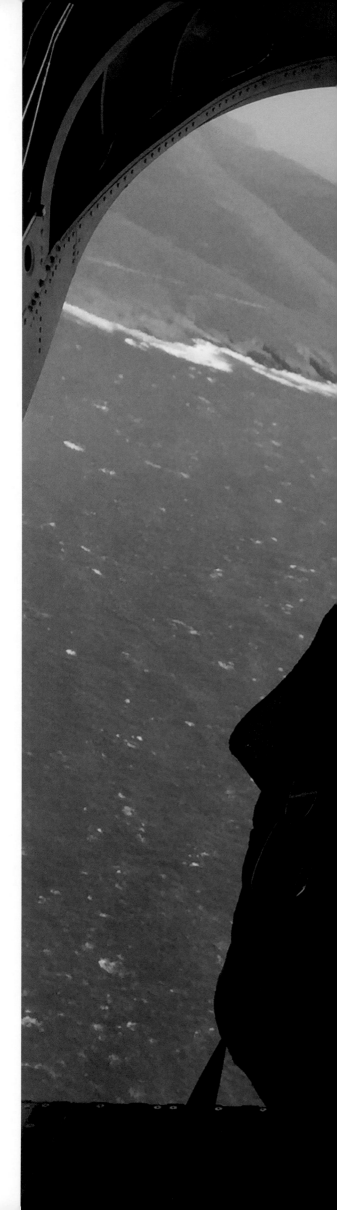

FN® M3M GAU21 machine gun system mounted on a CH-47 helicopter, 2015
In the mid-2000s, FN Herstal won a sole-source contract with the US Navy and US Marine Corps to supply their rotary wing aircraft with .50 calibre FN® M3M machine guns, known as 'Gun, Aircraft, Unit-21' (GAU-21).

Integrated Weapon Systems workshop, mounting of FN® HMP pods, 2018
Since the first pods were produced in the late 1970s, FN Herstal has continually improved their ergonomics in order to equip the latest subsonic aircraft and helicopters. FN Herstal has gradually become recognised worldwide as an integrator. This led to the creation, in 2017, of a new assembly workshop to handle the expanding business.

Land systems

The defence industry has experienced an upswing of US research and development projects since the first Gulf War in 1991. Partners had to join these programmes, which launched the overall concept of 'Revolution in Military Affairs'. This concept optimises the transformation of military strategy resulting from the use of new capacities introduced by computers, networks, spatial systems and precision-guided missiles.[253] One of the notions that flowed from this was 'Network-Centric Warfare', which involves networking of all the available sensors, processing data and information in real time, transforming them into knowledge and their transmission to the fire units for precision combat. The development of artificial intelligence was destined to shake up the way that military interventions were approached. The war in Kosovo had accelerated research into detecting threats, through use of various cameras and sensors, and ways to protect the operator inside vehicles during urban operations. After the attacks of 11 September 2001, these concepts were given fresh impetus. The war on terrorism, conducted almost exclusively with heavy military means, was seen to be inadequate when facing an extremely mobile and invisible enemy that moves around within civilian populations. NATO now wanted to have forces that were capable of moving fast, to cope with multi-faceted threats, in any kind of climate, within coalitions and using high-tech weapons.

To respond to this new market, FN Herstal drew on its skills in the company's core businesses while also relying on its original Airborne systems business to develop their application on land. The first land systems were produced in 2002, in partnership with the company Oerlikon Contraves Canada, today known as Rheinmetall Canada. This was a remotely operated weapon station called ARROWS (Advanced Reconnaissance & Remotely Operated Weapon System) fitted with 12.7 mm (MRWS) machine guns or 7.62 mm (LRWS) machine guns as well as a stabilisation system, a laser range-finding system plus CCD cameras for use during the day and infrared cameras for night use. The two companies linked up to respond to the GIAT call for bids to equip French VBCI vehicles from 2006. Winning that first order allowed FN Herstal to enter the market and to win other contracts, such as, in 2005, within the MPPV (Multi-Purpose Protected Vehicle) and AIV (Armored Infantry Vehicle) programmes, those equipping Belgium with mobile armoured vehicles to replace heavy Leopard tanks. For this purpose, FN Herstal linked up with the Munich-based vehicle manufacturer Krauss-Maffei Wegmann (KMW) to produce remotely operated stations.

ARROWS remote weapon station
The ARROWS remote weapon station offers the operator complete protection, thanks to its remote control system via a display unit and an operating handle located inside the vehicle. An actuator allows the mount to move through 360° and to elevate between +55° and -20°. Target acquisition is possible both during the day, with a CCD camera system, and at night, with a thermal vision system. The station can be equipped with machine guns of 7.62 and .50 calibre or a 40 mm grenade launcher, with a capacity of 300 to 500 rounds depending on calibre.

Hunting and Sports Shooting: pursuing change

In the 21ˢᵗ century, consumer behaviour has been changing considerably. Emotions, pleasure and the authenticity of the experience – these are all key criteria that companies have had to take into account when differentiating themselves from their competitors. In addition, since 2003, the development of Web 2.0 has triggered the participatory internet age, which is very driven by social media. There has also been an information explosion, creating a more anti-institutional mindset among the public, who now have great power to sanction. Companies need to be on the internet, while adapting to the image-based culture initiated by social media. Potential customers must be constantly updated about changes in brands and new products. Companies also have to satisfy consumers who are increasingly well-informed.

To meet consumers' new expectations, Browning and Winchester set about boosting their brand image. They highlighted the real passion that inspires the sports shooter and hunter, by underlining the values of personal achievement and returning to one's roots. FN's Hunting and Sports Shooting Division therefore adapted to environmental ethics by offering a complete range of ammunition, featuring a new alloy of steel balls to replace lead (with its harmful effects); it also paid great attention to its products' packaging. In 2009, Browning created its own official team, which one year later did well at the European Sporting Championship and the Compak Sporting Championship. Browning also embarked on creative marketing, with a campaign targeting partner gun manufacturers. The aim was to strengthen links with customers and to create a real community around the brand. For example, this campaign worked to improve merchandising displays and to develop a range of hunting and shooting accessories (camping, fishing, discovery, and hiking), as well as gun security safes.

QUALITY AS A DIFFERENTIATOR

Browning International and Browning North America were faced with the arrival of low-cost competitors and the consolidation of traditional rivals, whose presence was growing all the time and who were becoming more assertive in the traditional markets of over and under shotguns and the semi-automatic shotgun. In 1998, they responded by engaging in a project on 'total quality' and eliminating waste. This led to the creation in 2005 of quality assurance units at the global level, which enabled certification of all Browning and Winchester products, with a guarantee that they were designed and produced according to specific standards. The aim was to improve the launch of new products – from the marketing stage through to production – as well as to standardise the checks in each form of production, establishing common indicators of quality and updating the unique procedure

Cynergy and B525 Over and Under Shotguns
With the Cynergy and B525, Browning has designed the over and under shotguns of the future, both for sports shooting and hunting – catering for fans of tradition with the wood version, or innovation with the polymer version.

manuals for BACO (Browning Arms Company) and WACO (Winchester Arms Company) in all the factories. The level of requirements became identical to those for the military sector. Guns were rigorously checked and subjected to various very strict tests of their operation and endurance.

In 2002, a production cost-cutting programme was launched, by forming study groups around the semi-automatic BAR rifle, as well as the Browning Gold and Winchester SX2 semi-automatic shotguns.[254] These projects helped to underline the need to use new technologies when producing the receiver pre-case, to simplify gas cylinders, to reduce the number of components and to standardise all the barrel profiles while reducing polishing time. Intensive use of a rapid prototyping system was key when producing parts made of resin or plastic for new generations of the BAR, called Light and Evolve. Moreover, the Evolve received a nitriding treatment and nickel plating on its aluminium body, which is a difficult task and requires special baths.

Modifications to the semi-automatic carbine benefited from a cutting-edge gas-operated system and they were mainly focused on its ergonomics and modularity. The year 2003 was a new turning point for its aesthetics and technical aspects, with the ShortTrac and LongTrac ranges. The BAR ShortTrac, which was shorter and easier to handle, had a gas-operated system that was revised

to meet the development by Browning-Winchester of short magnum calibres. The Winchester Short Magnum ammunition, which was voted ammunition of the year by the Shooting Industry Academy of Excellence in 2001, was shorter but equalled or exceeded the performance of traditional magnum calibres. In 2003, the Winchester Super Short Magnum (WSSM) in .243 and .223 calibres was launched for small game hunting.

In 2004, Browning also stood out from competitors thanks to its range of over and under shotguns, notably with the new Cynergy. Composite and wood versions were also offered in 2004. Featuring a 12-gauge chamber, this shotgun was revolutionary because of its low action frame, which refined the gun's general lines and allowed more effective shooting. Its Reverse Striker Trigger mechanism offers rapid firing. The innovative recoil pad system helps to reduce the gun's recoil. Unlike the B25, the Cynergy was perhaps too modern and was not the success with customers that the company had expected, as they were too attached to the traditional nature of hunting guns. The Citori 625, a new model of over and under shotgun with a refreshed look, was launched in the US. From 2006, Browning worked more with design professionals to make each product unique and with a view to continually updating the look of guns.

Staff member of the USRAC factory, New Haven, Connecticut, c. 2000
Despite substantial investments, the Herstal Group was obliged regretfully to close the New Haven factory in 2006. The factory had been located in the industrial landscape of Connecticut since the 1860s. The Winchester brand, still under licence to Olin Corporation, was successfully redeployed in the other production sites of the Herstal Group.

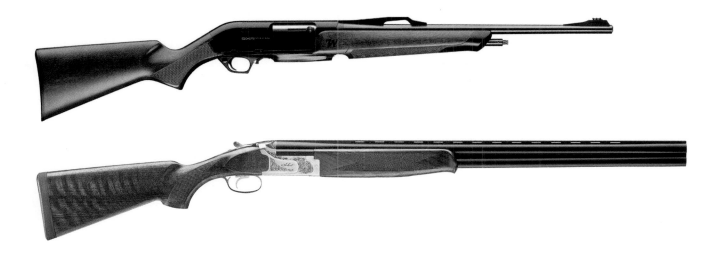

Winchester SXR Vulcan Semi-Automatic Rifle
Half-sister of the Browning BAR, the SXR has the same gas-operated loading system, but integrated into an aluminium receiver.

Winchester Select 2 Sporting Over and Under Shotgun
Winchester has benefited from Browning's research and development to diversify and attract a broader customer base by investing in mid-range over and under shotguns.

WINCHESTER, THE TOUGH DILEMMA

At the start of the 21st century, FN Herstal mass-produced the metallic components (barrels, receivers and breeches) for Browning while the Custom Shop produced high-end handcrafted guns. The Viana do Castelo factory focused on the assembly of guns and their finishing, such as polishing and surface treatment: oxidation, anodising, phosphating, varnishing, chrome or nickel plating. Across the Atlantic, the New Haven factory, which was then producing the Model 70 and 94 Winchester rifles as well as the Model 1300 pump action shotguns, had a production overcapacity. This situation was exacerbated because these products were legendary but becoming obsolete. In 2002, the wood, polishing and finishing activities for Winchester guns were transferred from New Haven to Viana do Castelo, and the barrels manufacturing was moved in 2005 to the FNMI factory in Columbia, South Carolina.

There was renewed interest in the Winchester brand in 1998. This was thanks to the production of new ranges of rifles, semi-automatic and over and under shotguns, produced in Viana do Castelo, a factory that was already working on the same families of firearms for Browning. In 2006, the Super X Rifle Vulcan was introduced into the entry-level market. It took its name from the Volcanic rifle, which underpinned the success of the Winchester and its lever-action mechanisms. A half-sister of the BAR, the SXR used the same gas-operated loading system but was integrated into an aluminium frame. The Winchester Select over and under shotgun was marketed the same year. Its Light version had an aluminium action receiver, which was the lowest on the market at the time.

However New Haven, which henceforth concentrated exclusively on the assembly of traditional products, could not cope with the decline in orders, the price reductions and the excessively high operational costs, which represented a major cash drain for the Herstal Group. The Group's slow return to stability could not be achieved without making a clean break with the past. In 2006, a decision was made to shut down New Haven's operations. Yet despite the closure of this factory, the Herstal Group had no intention of abandoning the production licence of Winchester guns, for which it held the design authority. With the Olin Group, the owner of the brand, Browning SA renewed a long-term licensing agreement for the production and distribution of Winchester rifles and shotguns. The brand was redeployed across other production sites, with an overall redefinition of design. The goal now was to ensure that each product came from a production unit capable of providing the best quality at the best price, in addition to meeting deadlines.

Thanks to the reorganisation efforts, the Browning and Winchester brands were able to cope with a market that was becoming increasingly difficult and competitive. Yet another key factor had to be taken into account: the way that sensitivity to exchange rates between the euro, dollar and yen affected the profitability of the Civilian Division.[255] In Europe, Browning International's financial situation stabilised in 2005. It saw a marked increase in its business in 2007, characterised by a demand for products with high added value. The Custom Shop was clearly attractive again, as could be seen by its good results in 2005, and an order book that was constantly growing.[256]

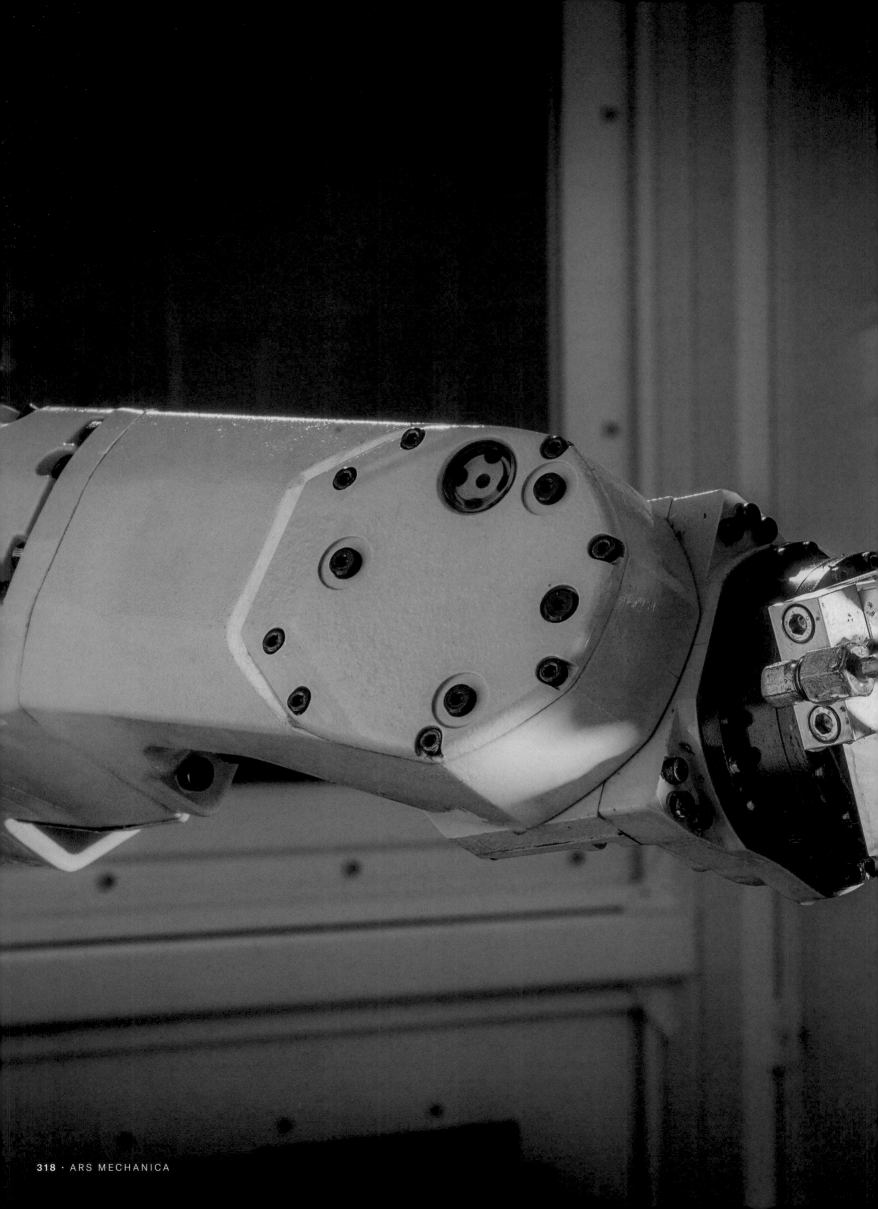

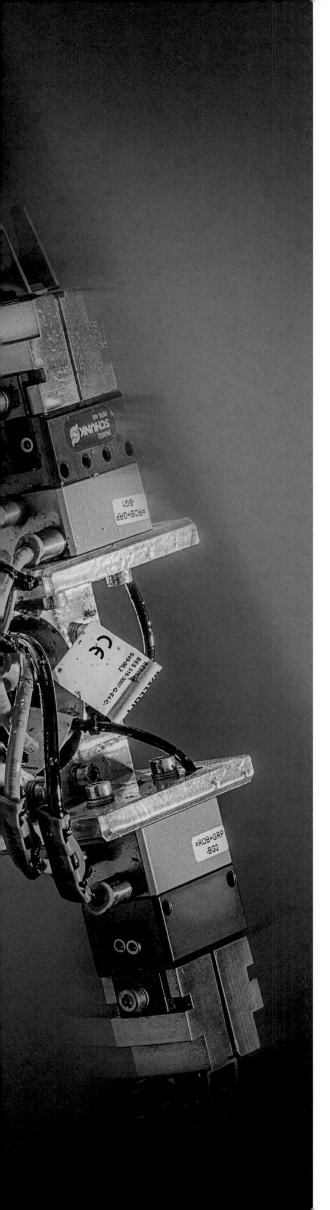

Modernising industrial facilities, a constant target

Returning the Herstal Group to stability could only have been achieved by reconfiguring its industrial facilities. The aim was to guarantee an increasingly diversified type of production and to meet profitability objectives, which had become tougher due to fiercer competition. In 2004, FN Manufacturing was equipped with one of the most modern barrels manufacturing facilities in the US. Investments were also made in heat treatment, machining and assembly in order to support the growth in its activities, which were linked to the needs of the US Army – itself engaged in conflicts in the Middle East. In 2005, new investments by Herstal ensured that the ammunition factory had systems for automatic handling, cutting to length and case cutting. In Zutendaal, work was more targeted around the furnace and the line connecting to the coating. In 2007, the machining workshop focused its investments on increasing the production capacities for high added value military and Law Enforcement components. In 2008, new machines with five axes were installed, as a vertical machining centre for small parts.

The quality of production was essential for ensuring the strength of the internal chrome coatings. The work on these was carried out in the surface treatment workshop. The overall modernisation project on this unit had begun when GIAT was in control, with the construction of a new building in 1994 and the installation of the first equipment the following year – with lines for phosphating, painting, chroming and sanding. In 2004, the new anodising line achieved productivity around 30% higher than the previous line had, and it could process more than 100 different aluminium parts.[257] This figure was constantly rising, to meet the needs of the lightest materials, with the aim of reducing the weight of guns. Between 2006 and 2009, the surface treatment was supplemented with new chroming layers to increase the longevity of the barrels. In the same period, Viana do Castelo took advantage of investments to set up a new surface treatment line and new tooling.[258] The factory's organisation and production were reviewed, so as to handle the increase in new products sold by the Hunting and Sports Shooting Division, while remaining focused on assembly and finishing.

In the mid-2000s, FN Herstal's modernisation plan continued. Robotics were progressively integrated into workshops, such as for the barrel system, to facilitate handling and enable operators to concentrate on high value-added tasks.

Simulation run at the Test and Evaluation Centre, FN Herstal, 2018
Investment in this new structure, inaugurated in 2008, was part of a major strategic programme to organise the Herstal site and its activities, as well as to modernise it, complete the necessary technological and environmental changes, and to secure its sites. For example, new simulation techniques enable the company to assess the lifespan of firearms and systems parts without the need for live firing.

In parallel, the plastics branch was undergoing a profound evolution, thanks to research into the thixomoulding process. This technology calls on magnesium alloys instead of plastic alloys, which gives the part better mechanical properties, a low porosity and a better quality of surface and tighter tolerances. The mechanical characteristics were further improved by the thixoforming process, which results in a technological evolution of the foundry through the processing of thixotropic aluminium. Alloying is increasingly used in new generations of rifles, such as the FN SCAR.

In late 2003, the Herstal factory was equipped with a new laser sintering machine. This enabled the production of moulds by superimposing layers obtained by the fusion of metallic powder[259] – in just one day and with reduced costs from 25% to 50%.

MANAGING THE TECHNOLOGY

The contribution of new production technologies meant there was a need to ensure the interface between subcontracting, internal production, acquisition of the knowledge needed to produce tooling and the final quality control. In 2003, research and development capacities were strengthened by forming a Process and Technologies unit within the factory. It is a real centre of technological skills within the Herstal Group. The goal was to create a hub between the R&D teams and all the services producing components for finished products and tooling, in order to adopt the new technologies. Afterwards, an electronic and optoelectronic skills centre was opened, with a goal to increase expertise in embedded IT and to ensure strategic monitoring in these areas. This constant willingness to keep updating the business was evident again in 2004.

The Herstal Group and Liège University signed a

Anodising line in the surface treatment workshop, FN Herstal, 2022
The anodising technique is based on a series of baths, which give the aluminium parts greater resistance to wear, corrosion and heat. The anodising itself comprises an acid bath, through which an electric current is passed. The aluminium dissolved by electrolytic reaction is used to produce a new layer of aluminium oxide, which has different properties from natural alumina, such as hardness, melting temperature and expansion coefficient. The bath parameters and temperatures are set by the Process Planning Office, based on the material's required properties.

Kinematic test performed at the Test and Evaluation Centre, FN Herstal, 2018
Using CCD or high-frequency cameras, the kinematic test can determine the path of the moving parts inside the gun.

cooperation agreement for the development of smart guns and microelectronics, as well as to develop complex armoury procedures with mixed materials.

This spirit of openness to new technologies notably allowed the Group to be a pioneer in Additive Manufacturing or rapid prototyping, which helps to validate the geometry of a part quickly. Furthermore, in 2005, increasingly powerful software was introduced into the IT, databases and mechanical simulation, so as to guarantee a leading industrial process. Since the early 1990s, ideas translated into 3D form using CAD software have enabled visualisation of the parts, although without any indication of how the mechanisms would behave. Mechanical simulation software was then installed in all the R&D groups, to facilitate the building of virtual prototypes, which can be used to create mathematical models of the behaviour of weapons and mechanisms.

Movements can be visualised and analysed in 3D. This enables 3D visualisation and analysis of movements, kinematic surveys of model components and faster error detection before mass production.[260]

Finally, in 2007, the new testing and evaluation centre was officially opened. This infrastructure centralised the testing equipment – simulation room, ballistics lab, instrumentation lab, operational firing, precision firing, firing weapons in adverse conditions for both civilian and military products and firing tunnels – for civilian and military products and shooting ranges. It enables testers to host clients and to conduct tests in a modern and safe environment. The building would eventually become home to the research, development and industrialisation arm of the Integrated Weapon Systems business. This facilitated the grouping of the Testing section's previously dispersed activities on the factory premises.

Expansion

Ten years after the Walloon Region's acquisition of the Herstal Group, the recovery plan continued to be resilient and a financial balance was gradually restored. The emphasis was on maintaining the momentum, by making a success of the structural changes. Above all, the aim was also to expand the Group against the backdrop of the 2008 financial crisis. This was a crisis which, given its scale and the effects on global trade mechanisms, dragged countries into policies of austerity, devaluation as well as economic and political downturns. The crisis led to a weakening of the globalisation system, or at least its economic aspects, which had begun in the 1970s and heralded a change in the balance of international relations.

The global economic crisis and geopolitical changes proved to be fresh challenges for the Herstal Group. The upswing gave way to a decline in the market, coupled with an exchange rate crisis, which resulted in an economic slump of the Civilian market (particularly for top-of-the-range products) and on the Defence and Security market. Tumbling defence budgets led to greater competition and increased protectionism, which benefited national companies. Customers were becoming increasingly demanding about quality, delivery times and levels of service. The tricky task of recreating growth in a pessimistic climate was entrusted to Philippe Claessens, then Director General of FN Herstal, and appointed Chief Executive Officer of the Herstal Group in 2009, while Jean-Sébastien Belle became Chairman of the Board of Directors.

The team's mission was to reinforce the leadership of the Group in its core businesses and to maintain the dynamism that it had shown in the civil-military pairing for over a century. An ambitious programme was set up throughout the Herstal Group. It sought to consolidate efforts undertaken over the previous ten years, to strengthen its global strategic base, to reposition itself on the market, and to create real momentum in the Group's expansion. As before, this programme was driven by the goal to extend the Group's portfolio of activities. To create profitable growth, the Group embarked on a process to diversify and modernise its products and services, so as to reoccupy all its traditional niche areas. To survive in the market, the Group also had to continue innovating and standing out from the competition. That goal was achieved thanks mainly to increased investment in R&D, as well as through changes in technologies and business lines. There was pressure to continually update industrial tools, in order to keep pace with evolving

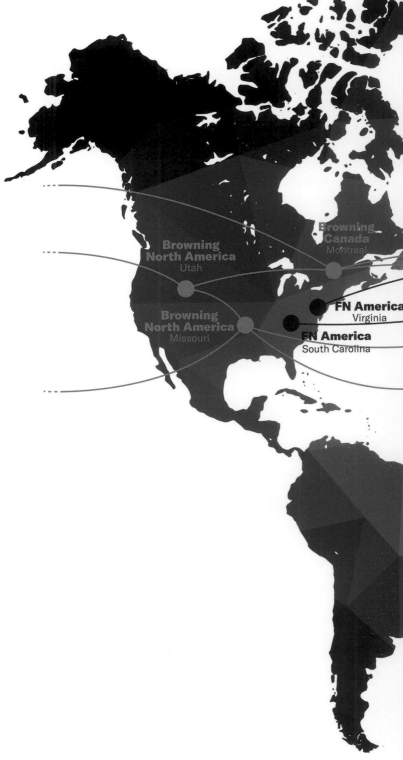

market needs plus the diversification of suppliers and customers. These expanded to include police forces and producers of aerospace, naval and vehicles. This new strategic approach called for adaptations to the organisation and operational process. It also required continual updating of key skills, by embracing the regular changes in the legal framework and compliance rules. That was the Group's direction of travel for the next decade.

Defence and Security

In the early 2010s, the Herstal Group's strategy in its Defence Division was to create growth in a flat market. With the 2008 crisis, numerous governments were compelled to reassess their budgetary priorities, to ensure

Location of the parent company, subsidiaries and partners of the Herstal Group
In the 2010s, the Herstal Group reinforced its global strategic footprint by consolidating its subsidiaries and acquiring new entities related to its core businesses

their public debt did not skyrocket.[261] Yet improvements could be glimpsed by mid-decade, except in the US, which was in the process of military disengagement from the Middle East, beginning in Afghanistan in 2011 and then in Iraq two years later.

For the Herstal Group, the heart of its Defence and Security activities was located in North America and in Europe, whether with NATO members or within programmes developed by the EU's defence policy. Elsewhere in the world, the Group's aim was to remain a point of reference and a strategic partner. This could be done by strengthening market share in traditional niche areas, ensuring that new systems of small-calibre weapons had a long-term future, and chiefly by expanding the Group's range of technological activities.

Philippe Claessens

Managing Director of FN Herstal since 1998, Philippe Claessens is appointed Chief Executive Officer of the Herstal Group (FN Herstal – Browning) in 2009.

When he takes over his function, the world and the markets are deeply impacted by the financial crisis of 2008.

Philippe Claessens has to address three major challenges for the Group in order to ensure its long-term viability: reinforce its financial stability, pursue its repositioning among the worldwide leaders in its two fields of activity, and conduct a major redeployment policy.

These main axes will be the pillars of an ambitious expansion strategy to consolidate the leadership of the Herstal Group.

Assembly bench for M240 (FN MAG®) Machine Guns, FN America, South Carolina, 2022

Assembly bench for FNX-45 Pistols, FN America, South Carolina, 2022

EUROPE/US, TRANSATLANTIC CONSOLIDATION

The Group could only operate on the US market by focusing for years on adaptation and by being tenacious. It also worked hard to foster mutual confidence in relations – based on the reliability and innovative nature of Herstal Group's products. The Group consolidated its special links with the US Army, for whom it continues to be one of the main suppliers of small-calibre firearms, by extending its product range across all the niche areas of the Defence Division. Meanwhile, the Group's Law Enforcement business continued to penetrate the US market, by constantly expanding its product ranges and its sales network.

Originally operating as a production facility with FN Manufacturing in Columbia (created in the 1980s),

followed by constant modernisation, FN Herstal in the US expanded into R&D while acquiring more industrial autonomy. It then moved into Law Enforcement with FNH USA in the Washington D.C. area. This complemented the overall set-up with a sales force targeting all market segments. A structured platform took shape, prior to FNH USA and FN Manufacturing merging in 2014 to create FN America. The aim was to rationalise and optimise the way in which FN Herstal's American structure was run and to respond to a market that had become tougher and more competitive.

In that same year, the Herstal Group acquired the Manroy Engineering factory, located in Slade Green, England. This English small business, which was initially based in East Sussex, was founded in 1975 to supply the British Army with spare parts for light weapons. It

Assembly bench for FN® M2HB-QCB Machine Guns, FNH UK factory, Slade Green, 2019
In 2014, the Herstal Group acquired the Manroy Engineering factory, specialised in assembling medium and heavy machine guns; the company was renamed FNH UK in 2017.

later expanded into assembling M2-HB machine guns followed by the GPMG, plus the machining and design of certain components for mounts. Acquisition of the Manroy Engineering factory was part of the harmonisation of product plans and the broad reflection on sales approaches. FN Herstal created a forward base on the UK market, with additional production capacity for machine guns while maintaining its leadership in the production of .50 calibre machine guns. In 2017, Manroy Engineering became FNH UK, an entity of approximately 100 people within the Herstal Group.

By consolidating its sales and production capacities on either side of the Atlantic, the Herstal Group forged a solid ecosystem, giving it sufficient agility to manage growth, control its diversification and expand its markets. The global expansion of its subsidiaries and partners was part of an overall vision, which revolved around mastering all the production phases, from research to management of the sales network. This cohesive approach sustained the company's strength and reputation.

Assembly bench for FN® FCU fire control units, Noptel, Oulu, Finland, 2022
In 2011, FN Herstal strengthened its technology position by acquiring Noptel, a company specialised in laser telemetry.

TECHNOLOGY AS A SPEARHEAD

In the 21st century, connectivity, programming and new technologies have become the new paradigms of industrial activity, which has now become part of a global knowledge economy.[262] In the arms industry, this cognitive revolution has led to major economic, social and cultural transformations. These have enabled implementation of new activities, new products, new businesses and new roles, with the aim of maintaining a technological lead and achieving economies of scale.[263] To safeguard its competitiveness, the Herstal Group has therefore had to continually adapt to the new knowledge-based values. It has done this by focusing on the quality of training of its staff, the efficiency of the company's organisation, the capacity for continuous improvement of the production processes, increasing investments in research and development, as well as the quality and integration of the products in the company's marketing strategies.

FIRST STEPS IN OPTOELECTRONICS

FN Herstal was very active in the innovation race. It developed the 'Smart Gun', followed by optoelectronics, with the first versions of the FN F2000 and the remotely operated ARROWS station. These all signalled the early development of weapons with augmented capacities, thanks to input from electronics. New impetus followed with the production, jointly with Noptel, of the fire control of the FN SCAR rifle – called the FN Fire Control

Unit (FCU). This system maximised the chances of a grenade hitting its target, with the support of an automatic adjustment of the reticle. Its overall design and the ballistic calculations were now internally managed, to keep full control of the process of integrating the system with the grenade launcher. This partnership highlighted the reversal of a trend, which began in the 1990s, whereby a major share of military applications was derived from civilian innovations, and especially procedures linked to sensors, precision and information processing. A range of products known as 'Armatronics' (a contraction of armaments and electronics) was created and based on all the electronic applications developed like this.

DIGITALISATION OF SYSTEMS

In parallel, FN Herstal set about modernising its Integrated Weapon Systems business and started on the digitalisation of pods. To position itself as a major player in this area, it also decided to develop, fully autonomously – after securing its first contracts with Oerlikon – its own and a differentiated range of land systems. This range is called deFNder. The Group had to set up a large-scale industrial project to consolidate this booming activity, to respond to the complex demands of customers, and above all to meet the needs for new technologies. This work involved numerous departments working together plus numerous resources for research and development. Gradually, this business was reorganised, by adopting a new direction and by focusing and adapting resources to the expertise required by this new business. In 2008, the department comprised around 30 people, before growing to 200 in 2020.[264]

NOPTEL, A TECHNOLOGICAL EXTENSION

To underpin its technological knowledge, FN Herstal acquired the Finnish company Noptel in 2011. The takeover allowed the Group to enhance its expertise in all aspects of arms production, both mechanical and electronic, and to retain full intellectual property rights to the knowledge required for the Future Soldier specifications. Created in 1982, Noptel had evolved from a spin-off of the Electronics Faculty of Oulu University. The company then developed two activities. Firstly, an NTS (Noptel Training System) branch, a shooting training system based on laser technology and reflectors, which was targeted more specifically at the military and police forces. Secondly, the NMS (Noptel Measurement System) activity, which was based on laser range-finding technology, especially for the civilian market, with stakes in projects such as the alignment of railway lines or speed control radar systems.

Family of laser telemetry modules developed by Noptel, 2019
These small modules employ a special method to measure the flight time of laser pulses, with an advanced signal processing technique that travels several kilometres and is eye-safe.

FN E-NOVATION

Together with Noptel, FN Herstal developed a new range of products, renamed FN e-novation, in 2014. The aim was to harness digitalisation, so as to optimise the use of weapons by integrating a predictive maintenance system; increase the shooter's capacities thanks to the application of decision-making support; and propose training solutions by monitoring handling and shooting performance.

FN Herstal developed the FN SmartCore system to optimise operational availability and to reduce a weapon's maintenance costs throughout its lifecycle. A kind of black box, the system features an integrated chip to analyse and record the conditions in which a weapon is used according to the type of ammunition used. Thanks to its communication system with the soldier, FN SmartCore could be integrated into the Battle Management System. It would then report back, in real time, on the fire contacts on a digital battlefield system and keep the top commanders informed at all times. In addition, its FN RFID tag integrated chip was extended to other applications. These included FN SAM, an equipment inventory system, and more recently the FN Smart Armory, which focuses on managing arsenals of small-calibre guns. This digital application covers the full management of a pool of guns and accessories, thanks to the input provided by the FN SmartCore shot counter to the FN SAM software.

This application helps to register the entry and exit of guns and accessories, to know in real time the operational status of the guns, and to manage and plan maintenance operations. Herstal Group's technological contribution made it suitable for managing a country's military assets, army logistics and staff training.

The shooter's capacities were increased, thanks to the updated FN Fire Control Unit. Its MK3 version, released in 2017, can store and calculate the ballistics for around 50 different types of grenade. FN Herstal also developed a new ballistic calculator for precision rifles, snipers and support weapons, the FN Elity. This calculator, which is installed above the sight, offers the operator a wide range of corrections based on standard data transmitted, such as the wind's strength and direction, the distance and angle of attack when shooting at moving targets. When developing the IT system, FN Herstal partnered with ApexO, a Canadian company specialised in ballistics.

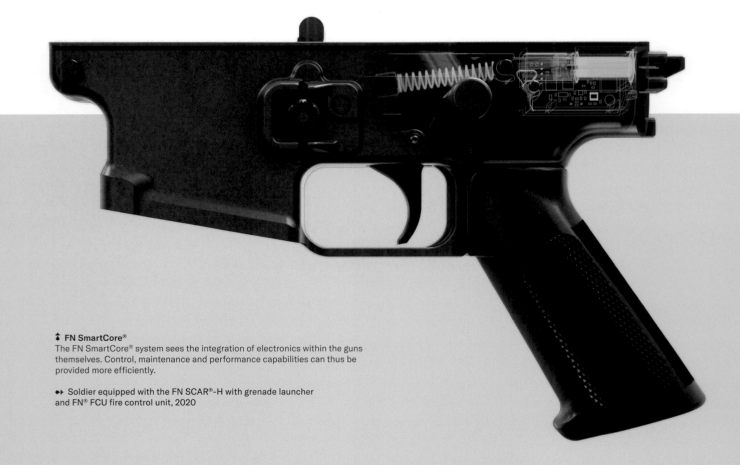

↕ **FN SmartCore®**
The FN SmartCore® system sees the integration of electronics within the guns themselves. Control, maintenance and performance capabilities can thus be provided more efficiently.

↦ Soldier equipped with the FN SCAR®-H with grenade launcher and FN® FCU fire control unit, 2020

↕ FN 15® fitted with the FN Elity® sighting system, 2021

↕ FN® Expert training system

Lastly, FN Expert enabled indoor and outdoor digital training. This marksmanship training system records how the gun is handled and the number of shots. The module analyses breathing, weapon grip, aiming and trigger finger action to highlight the shooter's weak points. It also helps to reduce costs for transport, maintenance of stands and supply of ammunition. In 2021, FN VictoR was launched. This new product significantly increased FN Expert's capacity to identify targets. Lighter and more compact, it now enables tactical monitoring of shots on the move and prepares the shooter for the use of integrated electronics in combat theatres, which will be reinforced in future.

New assembly line for electronic cards, 2022

In 2018, the Herstal Group decided to equip itself with an electronic card assembly line to consolidate its mastery of digital information technology integration. With its capacity to produce 4,000 cards per year, the SMD (surface mounted device) line takes FN Herstal one step further in developing high technology and diversifying its activities. The Group's goal is to make integrated electronics, for weapons or within systems, one of its main differentiating factors. This would for example allow it to position itself as a major player in large-scale programmes. One example is the European Defence Industrial Development Programme, a precursor of the European Defence Fund, which aims to streamline certain European industrial capacities to strengthen Europe's competitiveness.

FN SCAR®-L MK2
Assault Rifle

7.62 × 51 mm NATO (SCAR®–H)
5.56 × 45 mm NATO (SCAR®–L)

Calibre

System
Automatic rifle, gas-operated short-stroke piston system

6 5 4 3 2

Ambidextrous

Modular and foldable stock

90.3 cm
to 96.9 cm

Overall length

By developing the **FN SCAR®**, FN Herstal has created a unique product. This is also a genuine family of firearms that can, starting from a common base, be configured for the user's personal preferences to meet their mission objective. Throughout the 2010s, more than 200 different variants were delivered. These included the precision rifle to the subcompact carbine, with a possible choice between different calibres, barrel lengths, and types of stock, colours, receivers, Picatinny or Keymod rails, in semi-automatic or in semi-automatic/automatic. The FN SCAR® is now the rifle of choice for the most demanding units, thanks to digitalisation (through the integration of different systems from the FN® e-novation range) complementing the advantages of mechanical modularity. The FN SCAR® is partly made of aluminium and polymer, which makes it light without reducing its mechanical resistance to corrosion. The firing and ejection mechanism operates with the short-stroke gas piston system. The advantage of this – compared to the direct gas-operated system – is that the heat, carbon and fouling from the combustion of the propellant are not carried into the breech or chamber, thus enhancing the reliability

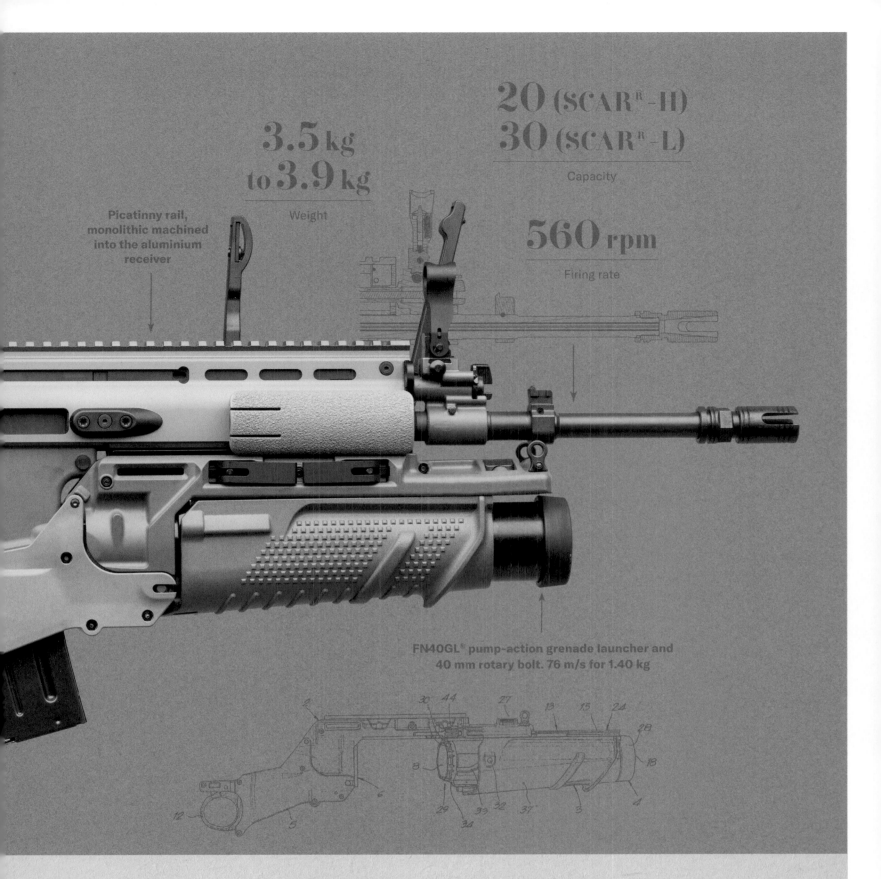

3.5 kg to 3.9 kg
Weight

Picatinny rail, monolithic machined into the aluminium receiver

20 (SCAR^R -H)
30 (SCAR^R -L)
Capacity

560 rpm
Firing rate

FN40GL® pump-action grenade launcher and 40 mm rotary bolt. 76 m/s for 1.40 kg

of these critical areas. A setting on the gas regulator also ensures that there is no increase in the cyclic firing rate when using a sound suppressor. The rifle features an open architecture thanks to its multiple Picatinny rails, which allow the addition of lights, lasers and other targeting devices. Finally, the FN40GL® grenade launcher, or the MK13 EGLM in the US, which fires all types of standard 40×46 mm low-speed grenades, can be mounted under the rifle or can be configured as a standalone launcher (it also has an open architecture). The FN SCAR® brings together open architecture, modularity, a

robust firing system and a highly resistant barrel to face the most adverse conditions.

FN HERSTAL

Belgian Special Forces vehicle fitted with FN® M2HB-QCB and FN MAG® Machine Guns, 2020

A NEW GENERATION OF MACHINE GUNS

In the 2010s, FN Herstal strengthened its status as a leader in machine guns. It achieved this by constantly adapting its products and developing a new generation of lighter weapons.

The aspiration to reduce the machine guns' weight gave rise, on the US market, to the M240L. It shed two kilograms, thanks to the use of new materials, particularly titanium and polymers to replace some of the steel operating parts.[265] Operational in 2010, the gun – which came with a shorter barrel, a foldable buttstock and a kit with tactical rails – helped to increase mobility in combat units. In Europe too, FN Herstal continued to call on its expertise in the field of support machine guns, with

France, Norway and the Netherlands notably adopting the FN MAG. Over 70 countries have purchased this machine gun, whose first model was launched over 60 years earlier.

The new 7.62 calibre FN MINIMI also notched up several successes. It was for example adopted by Australia in 2012 (under the name Maximi), by the UK, Greece and more recently by Norway and the Czech Republic.[266] The 5.56 mm calibre version was, in parallel, updated at FN Manufacturing: this resulted in the MK46, which included improvements to portability, accessorisation and weight, being reduced by around one kilogram. In 2013, after a decade or more of operational experience, FN Herstal unveiled the FN MINIMI MK 3 series, whose

Navy Seal with the M249 (FN MINIMI®MK3) Light Machine Gun training in wintery conditions. 2014
After four decades of production, the FN MINIMI®, in its various forms and developments, is extensively used by armies worldwide,
thus meeting the critical need for firepower within small infantry units.

adapted ergonomic versions were incorporated into both its 5.56 mm and its 7.62 mm variants. More than 40 years after its launch, the updated FN MINIMI continues to be a mainstay in the armed forces, and most recently in Portugal and France.

Nevertheless, modern intervention scenarios are increasingly focused on portability. New international research programmes, based on Lightweight Small Arms Technologies, a programme initiated in 2004, have focused on reducing the weight of ammunition and therefore of weapons, by 20%. Another target is including a sound suppressor and reducing muzzle fire while maintaining shooting accuracy.[267] In 2010, FN Herstal responded to the demand through the concept of the

Infantry Automatic Rifle (IAR), at the request of the US Marine Corps. Based on the FN SCAR platform, the FN HAMR (Heat Adaptive Modular Rifle) fulfilled the functions of an individual assault rifle and of a light machine gun in a light and compact receiver but equipped with a heavy barrel. It also used the mechanism necessary to ensure the transition between the open and closed bolt modes, based on the temperature of the chamber, and thus to avoid overheating.

FN EVOLYS®
Light Machine Gun

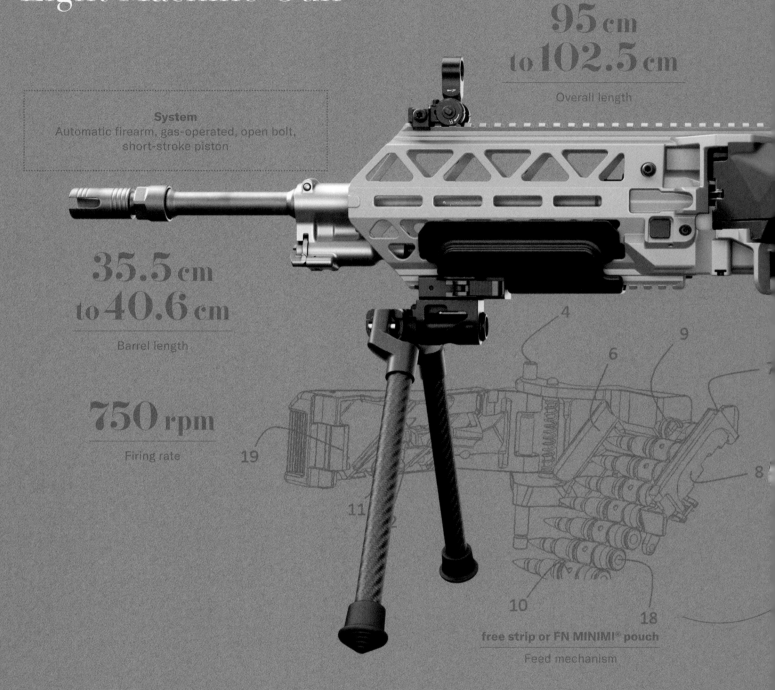

95 cm to 102.5 cm
Overall length

System
Automatic firearm, gas-operated, open bolt, short-stroke piston

35.5 cm to 40.6 cm
Barrel length

750 rpm
Firing rate

19

11

4

6

9

8

10

18

free strip or FN MINIMI® pouch
Feed mechanism

In the 2010s, FN Herstal actively worked on developing a new generation of light machine guns and created a splash in 2021 with the launch of the **FN EVOLYS®**. Weighing between 5.5 and 6.2 kg depending on the calibre, the gun's portability is close to that of an assault rifle but it boasts a machine gun's firing capacity. It also has low recoil, thanks to its hydraulic buffer integrated into the mobile parts. Available in two calibres (5.56 mm and 7.62 mm), the receiver of this ultralight machine gun is made of a single piece of aluminium, produced by moulding and then machined and equipped with a single-piece long top rail for the addition of accessories. The gun's precision is enhanced by including a free-floating barrel and a single shot mode. The effective range is between 800 and 1,000 metres, depending on the calibre. The lateral feed mechanism is a patented innovation and guides the ammunition almost automatically by closing the cover, considerably reducing the risks of jamming on

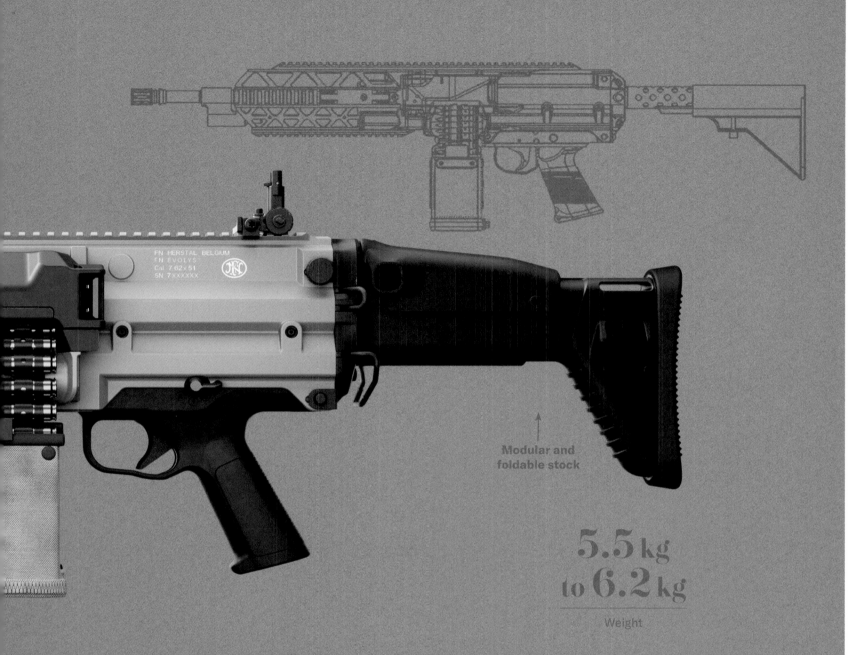

Modular and
foldable stock

5.5 kg
to 6.2 kg

Weight

5.56 × 45 mm NATO
7.62 × 51 mm NATO

Calibre

first use. Reloading, as with the firing selection, is ambidextrous and is done via an FN MINIMI® type free belt
or pouch.

FN HERSTAL

EVOLUTION OF INTEGRATED WEAPON SYSTEMS

Air

In the 2010s, FN Herstal extended its expertise to integrating FN D-HMP400, FN D-HMP250 and FN D-RMP pods into subsonic planes or hybrid Integrated Weapon Systems for helicopters. In early 2015, the helicopter maker Sikorsky selected FN Herstal to fit out HH-60W helicopters, as part of the US Air Force's CRH (Combat Rescue Helicopter) contract. It launched the first integration of non-standard FN Herstal guns in the Integrated Weapon Systems business, because the specifications not only combined the side-mounted fire and axial fire functions but also provided for full compatibility for three guns – the two FN .50 machine guns and the 7.62 mm Minigun.

In the Air Axial field, the Group achieved a technological leap forward in the pods business. In 2018, instead of analogue systems, it developed a range of digital products called FN Airborne Digital Suite. These were integrated into the latest generation of planes and helicopters. The suite enables real-time communication between all the Integrated Weapon Systems, and facilitates the maintenance and integration of a ballistic calculator, for greater effectiveness.

Airbus H145 helicopter fitted with FN® HMP pods
Through collaboration with companies such as Airbus, FN Herstal is part of major European consortia developing tomorrow's military helicopters.

FN Sea deFNder® remote operated station with the Belgian marine section, 2020

Land and sea

While FN Herstal strengthened its position in Air Axial and Air Sabord systems, it was above all the increase in land and naval activities that posed a new challenge for the company. In response, the company had to redesign a new range to meet the needs of the market. It also had to rethink a new value chain to facilitate the integration of systems, while establishing new sales approaches between customers and vehicle manufacturers or OEM (Original Equipment Manufacturers) for the sale of customised development services. The aim was to follow the product throughout its lifetime, guaranteeing to the purchaser its operation, installation, maintenance, emergency repair, repair, renovation and upgrading.

In 2010, FN Herstal started developing its own land systems, known as deFNder. These systems notched up their first major successes in the standardisation programmes for light-armoured vehicles in Europe and the Middle East. Building on these initial commercial successes, the product underwent continuous development, culminating in 2019 in the production of the deFNder MK2. This new generation of ground systems achieved a strategic success, in 2020, when it was included in the Franco-Belgian JLTV (Joint Light Tactical Vehicle) and CaMo programmes to equip the Griffon multi-role armoured vehicles (VBMR) and Jaguar reconnaissance vehicles, of French design.

The system has been fitted into the latest generation

of Leclerc tanks. The deFNder MK2 enables overhead and supertraverse firing, thanks to better coordination and interdependence between the gun and the camera. Backed by gyroscopic stabilisation, the tracking function improves multi-target identification and follow-up of multi-targets. Finally, a camera breakthrough was achieved by moving from analogue to digital and fibre optics. This increases the sharpness, range and flow of images. The module is better connected and fitted to interact with the Battle Management System, especially for identifying friendly forces as well as the secure networking of different stations operated by a single operator.

Technological updates were essential in the naval field, because ships require the installation of several coordinated target tracking stations, while taking into account swell movements and compensating for lower frequencies. The Sea deFNder system, developed to respond to these needs, was fitted to the Belgian frigates Castor and Pollux, plus minesweepers in the Belgian-Dutch programme MCMV (Mine Countermeasure Vessels) in partnership with Naval Group and Kership.

THeMIS Remotely controlled tracked robot equipped with a FN deFNder® Medium remote operated station at the Eurosatory fair, Paris, 2018
Participating in international defence and security trade fairs is a key mission for the FN Herstal marketing and sales department. This activity enables the company to promote its products and solutions, as well as to understand the market expectations and developments. In 2017, FN Herstal revealed THeMIS, a further step in remotely controlled operation by taking part in the design of a UGV (Unmanned Ground Vehicle), in association with the Estonian firm Milrem Robotics.

EUROPEAN UNION DEFENCE POLICY

The Kosovo crisis had helped speed up the European Union's common defence policy. To catch up with the US, several programmes were put in place and, in 2004, the European Defence Agency was set up to define common needs for armaments. After the 2008 crisis, the EU called on directives and action plans to try to harmonise the legislative and regulatory provisions of its Member States. The aim was to promote industrial cooperation and the competitiveness of the European defence sector. During the NATO summit in Wales in 2014, the fear of losing a large part of their technological and industrial know-how prompted the Allies to reconfirm their commitment to spend 2% of their GDP on defence.[268]

The EU waited until the end of 2020 to launch a European Defence Fund, to support its Member States' research and development activities in military matters.[269] In parallel, a political agreement was reached for the new European Peace Facility (EPF), which is designed to finance military activities and to support the capacities of their EU partners.[270] By pooling some of their investments, Europeans are committed not only to improving their military capabilities but also to achieving economies of scale and streamlining an overly fragmented sector. Common development projects include protecting soldiers on operations, and initiatives promoting the interoperability of the various European weapon systems and their certification. These are areas where the Herstal Group excels. The CaMo (Motorized Capability) programme, equipping multi-role armoured vehicles in which FN Herstal was involved, was a trailblazer. The same can be said of its FN e-novation range, with multiple applications, from increasing the shooting skills of soldiers to the management and maintenance of military weapons.

A French armoured personnel carrier (VAB) equipped with a FN deFNder® Medium remote weapon station, 2020
After equipping the VAB, FN Herstal is set to further strengthen collaboration between European defence industries by providing remote weapon stations for the new-generation of multi-role armoured vehicles (VBMR) developed by France.

FN deFNder® MK2
Remote Weapon Station

63 cm
Height

**Gyroscopic stabilisation
and target tracking**

200 kg
Weight

FN® M3R machine gun

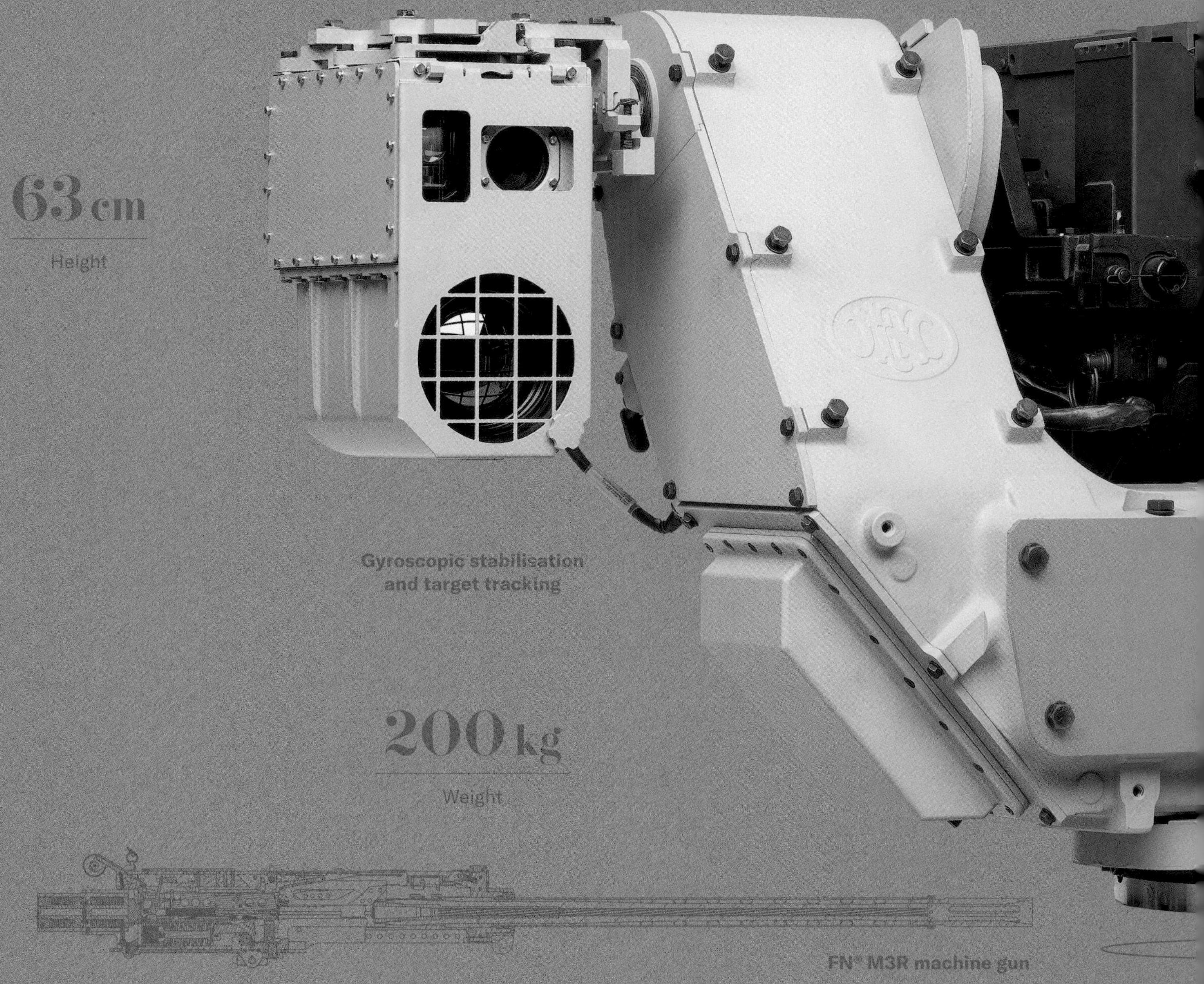

The first version of the **FN deFNder® MK2 Remote Weapon Station** was notable for its high-resolution digital optical path, which optimises the transmission of power and electrical signals from the fixed to the rotating structure. Other key features are an extreme traverse and extreme elevation system, i.e. two degrees of freedom for the camera, a capacity for identification and tracking by artificial intelligence (in processing, not in decision-making), as well as a new security architecture, a networking capacity for collaborative applications such as the defence of ships, perimeter defence, and finally steering capacity through a wireless network, addressing the problem of latency.

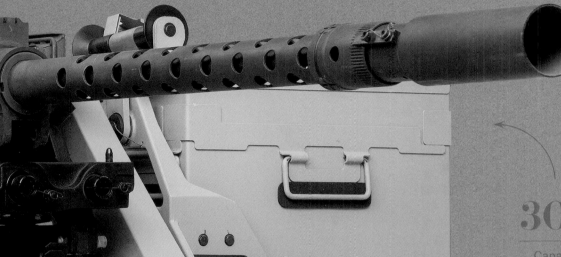

-40° to +70°

Angle of elevation

1000 rpm

Firing rate

300

Capacity

12.7 × 99 mm NATO

Calibre

N × 360°

Traverse angle

Protected in the vehicle's control station, the operator controls
the remotely operated station via a console and control handle,
which can be a one-handed joystick or a gamepad

FN HERSTAL

‡ Sharpshooter in the Liège anti-banditry platoon (PAB) carrying the FN SCAR®-H PR, 2020

↔ Assembly of the FN P90® in the FN Herstal factory, 2022
Featuring a bullpup design, integrated optical sight, ambidextrous control and its various different evolutions over almost 30 years, the FN P90® has been successfully redeployed in the Law Enforcement sector and today it is a major part of the assembly workshop's workload.

EXPANDING LAW ENFORCEMENT AND COMMERCIAL OPERATIONS

In the 2010s, the Law Enforcement sector's development revolved around the continuous improvement and enlargement of product lines. Chief among these were the company's pistols, which were expanded with the FNX series in 2009, and followed by the FNS and their Pre-loaded Striker system, a hammerless-system like Glocks. The FN 509, based on the architecture of the FNS, was conceived and developed in 2015 to comply with the American MHS (Modular Handgun System) programme. In 2022, this pistol became a big success, being selected to equip Los Angeles police, the second largest police force in the United States.

Also in 2022, FN Herstal unveiled its latest 9 mm calibre model, the FN HiPer. It takes its name by referencing the High Power, a weapon launched some 90 years earlier and one that enjoyed global success. The new semi-automatic pistol benefits from several patented innovations, in terms of ergonomics, reliability and safety. One of the differentiating elements is an ambidextrous, rotary operating magazine catch, which allows a quick magazine change without shifting the grip or pointing away from the target. The magazine is made entirely of high-strength polymer and the handle design allows for optimal control of the gun, with a well-proven angle, similar to that of the HP 35. Lastly, the FN HiPer comes with an innovative architecture, with a two-spring firing mechanism, which enables the firing of both civilian and military ammunition. Because although the pistol is intended for the Security market, it is also designed for Defence and offers the same power as its illustrious ancestor.

The company's sniper rifles are based on the design and system of the Winchester Model 70 rifle, with the FN Manufacturing factory producing the FN A1 model and, from 2009, the FN A3. The latter was selected by the FBI as a standard tactical sniper rifle. Other sniper rifles marketed include the FNAR, directly descended from the BAR and its gas-operated reloading system, chambered in 7.62 × 51 mm, plus the FN SCAR sniper variant, as well as the FN SCAR-H PR, in 2013. Its American equivalent, the FN SCAR 20S has been chambered since 2020 for the 6.5 Creedmoor long-range cartridge.

⬍ **Five-seveN® MK3 and FN HiPer® Pistols**
Over the years, FN Herstal has established a lasting presence in the pistols market by developing a diverse range of products covering the calibres most commonly used by police forces: .45 ACP, 9 mm, 5.7×28 mm. The Five-seveN® MK3 and FN HiPer® Pistols highlight FN Herstal's capacity to constantly improve its models, alongside offering new high-performance systems.

↦ **Deputy Sheriff in Richland County, South Carolina, with the FN® 509, 2018**
With the FN® 509, FN Herstal provides law enforcement with a handgun that offers great accuracy. The pistol features a tapered firing pin, secure flat trigger, 17-round capacity magazine and a low-profile optical mounting system for rapid target acquisition.

The 2010s also saw the launch of the FN SCAR range in two calibres, for the Security market, and in the United States the FN 15 Tactical II came to market. It is the latest evolution of the FN 15 family of carbines. The upgrade to the existing platform included the addition of the Rail System, an enhanced lower receiver and a floating barrel, which is chrome-lined and cold hammer-forged to provide increased durability and performance.

In the mid-2010s, the Paris and Brussels terrorist attacks led to new and specific requirements for police forces. In 2017, FN Herstal responded by developing a subcompact variant of the FN SCAR. Initially available in 5.56 × 45 mm calibre, and then in 7.62 × 35 mm/.300 BLK from 2018, this gun combines a compact architecture with firepower to meet the ever-increasing demands of better armed and trained attackers. To ensure they are prepared for any eventuality, police forces are increasingly aware of the need to improve their firearms training. FN Herstal has satisfied this need by developing 'training' guns similar to a service gun in their sizes and shapes, although their internal mechanism does not allow the use of lethal ammunition. FN Herstal also offers digital systems developed for the military, such as the FN Expert module and, more recently, the FN VictoR, which is more adapted for integration on the pistol.

Police officer in the Liège anti-banditry platoon (PAB) carrying the FN 303® MK2, 2021
With the FN 303®, police forces are offered a choice of five different types of standard projectile:
Clear Impact (training), Washable Paint, Indelible Paint, PAVA/OC Powder, and Inert Powder.
Officers can quickly change from inert, marking, impact or irritant projectiles, with a clear view
of the projectile type and remaining count in the magazine.

LESS LETHAL

'Less Lethal' has emerged as a key field among the new challenges facing the security sector. In 2010, a compact variant of handgun was produced, known as the FN 303-P Less Lethal Pistol. In 2020, the FN 303 MK2 Less Lethal Launcher came to market: this is a lightweight, rugged polymer variant offering increased accuracy, together with new adjustable flip-up metal sights and a top rail for integration of optics or red dot sights. In 2022, this system took a technological leap forward with the release of the FN 306, which still uses kinetic energy, but its optoelectronics enhance accident prevention during use.

◄◄ Macroscopic view of a FN 303® round
This unique .68 calibre, 8.5 gram projectile features
a polystyrene body and a non-toxic bismuth pellet.
This technology was developed under FN Herstal's
research to meet the growing demand for less lethal
systems.

►► FN Smart ProtectoR®
This system includes the FN 306™ launcher, the FN
VictoR®-SP image recognition system and a red dot
sight. Fitted with a 5-tube magazine, the FN 306™
launcher fires 12.55 FN SP calibre cartridges featuring
an elastomer projectile, designed to drastically
reduce the risk of bodily harm. The FN VictoR®-SP
captures, and analyses in real time, local images of
an environment to detect the shape of a human body.
It also alerts the shooter of parts of the body that
must not be engaged: for instance, if the head was
targeted, firing of the launcher would be blocked.

Soldiers on operations and waterfowl hunting scene
Two different environments, but a single purpose expressed by the Herstal Group: to inspire user confidence by providing them with quality, reliable and innovative products, in both its Defence and Security Division and its Hunting and Sports Shooting Division.

Browning B725 Hunter Over and Under Shotgun, 2019
Almost a century after the B25 was designed, this famous over and under shotgun and its various successors are still a guarantee of reliability and durability. This genuine legacy is passed on from generation to generation.

Hunting and Sports Shooting Division

BROWNING, "THE BEST THERE IS"

In the United States and Europe, the Hunting and Sports Shooting and the Defence and Security sectors both suffered enormously due to the financial impacts of the 2008 crisis. Yet in the following decade, the American market – which accounted for 75% of Browning's sales – experienced a gradual growth, with especially large increases in demand from 2020. Browning North America and Browning International were not simply satisfied with one-off peaks in demand growth to return to a high cycle, but relied on Browning's ability to update its traditional ranges, in order to create short cycles and offer diversified products while remaining at the top of the segment and preserving the strength of the brand. Concurrently, competitiveness had to be improved through delivery performance, by constantly reducing the time between order placement and shipment. At the same time, the concept of design-to-cost was emphasised. In other words, costs are no longer determined by technical constraints, but rather the technical solution is based on market constraints. Here, technological support has been essential, such as the use of metal injection moulding for the production of small parts, thereby reducing costly manual operations.

WINCHESTER: A STRATEGIC SHIFT AND GROWTH

Cost-cutting efforts had become essential, so as to attract an ever-growing base of customers who were seeking more affordable prices. This has applied notably to Winchester, which today has to focus on mid-range products. Like Browning, it offers products that meet the needs of each section of the civilian market. Revitalising Winchester was a new growth area, yet one that ultimately proved to be a real strategic success. However, this required a fundamental change in the products' DNA and a design overhaul, while maintaining the brand's legendary status. This new vision entailed an in-depth transformation of the company's industrial system and commercial approach. Closure of the New Haven, Connecticut factory in 2006 was therefore a challenge, since it required the rebuilding of a strategy

Winchester XPR Rifle, 2019
Even though the Wild West days are long gone, Winchester upholds its image of freedom, adventure, authenticity and simplicity.

for production and marketing on a wholly new basis. After some provisional repositioning, the Winchester product line was finally shared between Miroku and Viana do Castelo. Miroku made the lever-action carbines, including the famous Model 94s, with a small production of top-quality replicas that were well matched to Japanese know-how. The Japanese company would, over time, centralise the production of Browning's over and under shotguns as well as some bolt-action rifles. As for the Viana do Castelo factory in Portugal, it mainly focused on the bolt-action and semi-automatic rifles from both the Winchester and Browning brands, in addition to the straight-pull action rifles. Viana would receive substantial investments, aimed at boosting its original field of operations and moving beyond assembly to the production of components. Browning rifled and smoothbore guns continued to be manufactured in Herstal. Moreover, Silah, a new partner in Turkey, helped the company establish itself in the entry-level market. Since 2008, this has involved producing the Winchester SXP pump-action shotgun. This gun has inherited the four-lugs bolt system of the Model 1300, for optimal locking and fast, clean ejection, but at a reduced production cost. Silah would also produce Winchester .22 calibre rifles, enabling the company to compete through a low-price policy.

A NEW ECOSYSTEM

Calling on the R&D departments in Herstal and Morgan, the production of the Hunting and Sports Shooting Division therefore relies – without ever losing control of product quality – on a robust network of subsidiaries and partners worldwide, together with subcontractors in Europe, Turkey and Japan. This strategy is supported by a leading distribution network. This is managed from Hauts-Sarts for Europe and from Morgan for the US, with an operational arm in Arnold, which is the storage, distribution and service centre for Browning, which also supports FN America's commercial work. This centre has seen a major modernisation of its warehouse systems, through improved IT management and automation, in order to keep pace with the growth in orders. Thanks to this new ecosystem, the Herstal Group continues to uphold the legend and reputation of these two legendary names in the arms industry: John Moses Browning and Oliver Fisher Winchester.

STAY TRUE

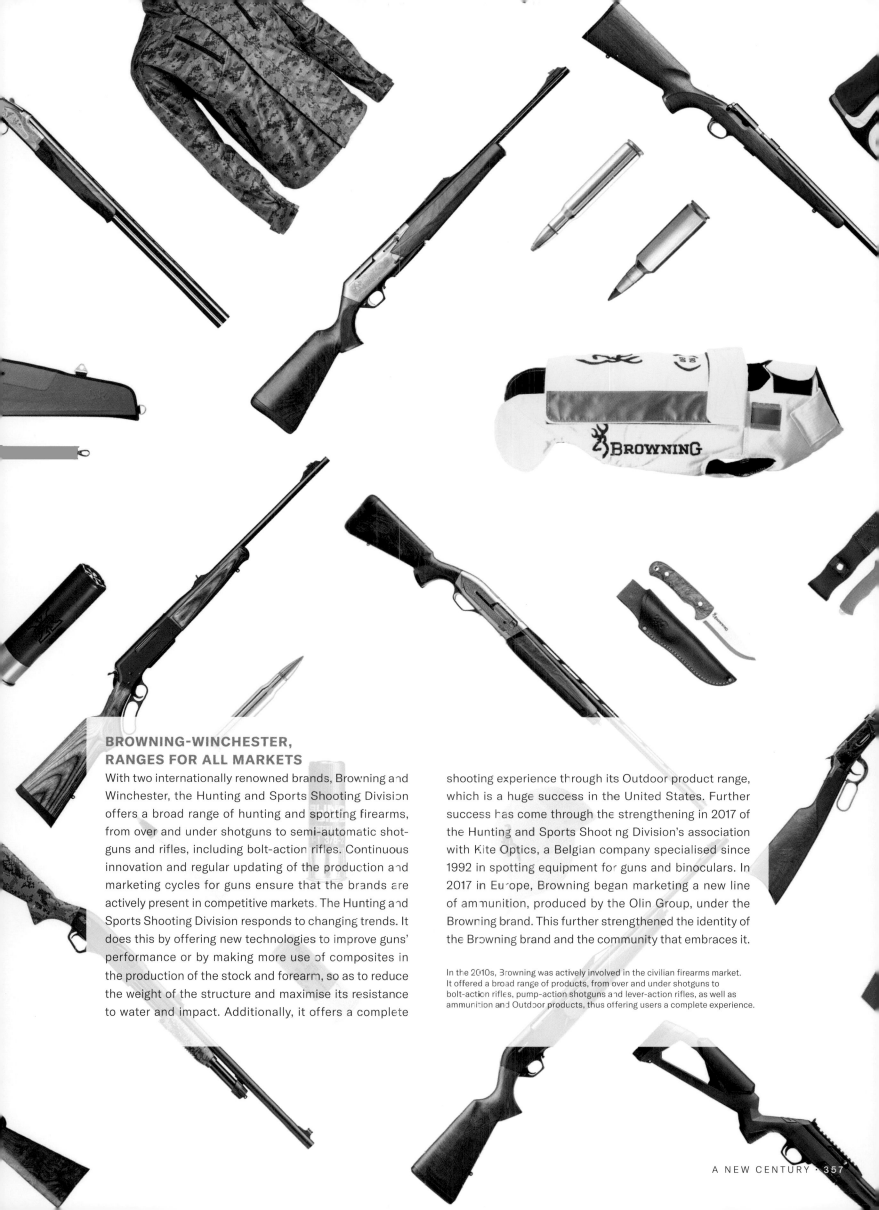

BROWNING-WINCHESTER, RANGES FOR ALL MARKETS

With two internationally renowned brands, Browning and Winchester, the Hunting and Sports Shooting Division offers a broad range of hunting and sporting firearms, from over and under shotguns to semi-automatic shotguns and rifles, including bolt-action rifles. Continuous innovation and regular updating of the production and marketing cycles for guns ensure that the brands are actively present in competitive markets. The Hunting and Sports Shooting Division responds to changing trends. It does this by offering new technologies to improve guns' performance or by making more use of composites in the production of the stock and forearm, so as to reduce the weight of the structure and maximise its resistance to water and impact. Additionally, it offers a complete

shooting experience through its Outdoor product range, which is a huge success in the United States. Further success has come through the strengthening in 2017 of the Hunting and Sports Shooting Division's association with Kite Optics, a Belgian company specialised since 1992 in spotting equipment for guns and binoculars. In 2017 in Europe, Browning began marketing a new line of ammunition, produced by the Olin Group, under the Browning brand. This further strengthened the identity of the Browning brand and the community that embraces it.

In the 2010s, Browning was actively involved in the civilian firearms market. It offered a broad range of products, from over and under shotguns to bolt-action rifles, pump-action shotguns and lever-action rifles, as well as ammunition and Outdoor products, thus offering users a complete experience.

Browning Maxus 2 Semi-Automatic Shotgun, 2022
A decade after the launch of the first generation of the Maxus Semi-Automatic
Shotgun, Browning has identified users' latest expectations in terms of handling,
comfort, versatility and ergonomics: this resulted in the development of the Maxus 2.

Winchester SX4 Silver Performance Semi-Automatic Shotgun
Browning Maxus 2 Camo Semi-Automatic Shotgun
Browning A5 Ultimate Partridges Semi-Automatic Shotgun

Semi-automatic shotguns

In 2002, the launch of Gold Fusion helped to reposition the Browning semi-automatic shotgun on the market. Using the Gold as a foundation, the Silver and Phoenix began production in 2005. The semi-automatic shotgun platform saw a complete redesign between 2009 and 2012 with the Maxus – a new generation of semi-automatic shotguns. Development of the Maxus resulted from an international collaboration: the prototype was designed, manufactured and tested in Morgan. The industrial plans and tolerance tests were in Herstal, and the tests and adjustments were in Viana do Castelo. The study corrected the Gold's shortcomings, thanks to new loading and unloading systems, a magazine round limiter, rapid assembly and disassembly of the gun, mechanical retraction of the firing pin and a gas-tight system that guarantees operation with low-pressure cartridges. Performance had been improved, with reduced recoil, muzzle rise and a faster firing time. The gun also benefited from new technologies in terms of recoil, barrels and interchangeable choke tubes, and it is equipped with a fast-locking handguard. For purists, these technological contributions have been complemented by the availability of an engraving grade, to recall traditional gunsmithing. In 2020, the launch of the Maxus 2 offered ergonomic improvements such as better fitting of the stock to the body, as well as over-moulded grips on the fore-end and on the pistol grip, ensuring a better grip. The bolt handle and over-sized trigger guard also allow for easy handling. The fast-reloading system, which is Browning's own, also saves the hunter valuable time by feeding the first round from the magazine directly into the chamber.

In 2013, the launch of the new A5 revived interest in the original Auto-5, whose production had ceased in 1999, with a total of five million units produced over 96 years. Although the lines of the new A5 recall the illustrious shotgun invented by John Moses Browning, its design has been completely changed. It has an aerospace-quality aluminium alloy receiver with a humpback integrated into the rib, to lengthen the line of sight and facilitate faster targeting. The shotgun uses kinetic energy with the barrel's short recoil system, which uses the bolt's inertia and converts it into mechanical movement for actuating parts. The gun's reliability has been proven through a series of severe tolerance tests, under all extreme conditions of weather, temperature, moisture or fouling. The gun also comes with a warranty of 100,000 rounds or five years, which is a first for a civilian semi-automatic.

At Winchester, the semi-auto shotgun had been modernised in 2006, under the name SX3, by the addition of an excess gas regulation system that facilitates the use of a wide range of loads. The SX3 was then considered the world's fastest shotgun, thanks to a record achieved in 2012 by Raniero Testa, who managed to shoot twelve clay pigeons with this gun. He beat his own record in 2017, with the SX4 shotgun, this time shooting 13 clay pigeons in 1.6 seconds.

Winchester SX4 Waterfowl Semi-Automatic Shotgun, 2017
By pairing the world's fastest shotgun with innovative ammunition, Winchester offers unmatched shooting performance to both hunters and sports shooters.

Straightening of a smoothbore barrel in the FN Herstal factory, surface treatment and assembly of semi-automatic shotguns in the factory of Browning Viana, Portugal, 2022
Before reaching the hands of customers, semi-automatic shotguns undergo an industrial process based on a logical distribution of tasks between the Herstal Group's different entities: the barrel system is completed in the FN Herstal factory; the surface treatment, assembly and adjustment in the factory of Browning Viana; and shipping is from the Browning International logistics centre in Hauts-Sarts.

Winchester Model 70 Extreme Weather Bolt-Action Rifle
Browning X-Bolt Pro McMillan Bolt-Action Rifle
Winchester XPR Sporter Bolt-Action Rifle

◆◆ Browning X-Bolt Rifle and Browning Outdoor equipment, 2019
The values conveyed by Browning focus on the natural environment and the pass on driving hunters, by emphasising the values of personal achievement and a return to one's roots.

Bolt-action rifles

In the early 2000s, Browning bolt-action rifles were continued with the A-Bolt II, followed in 2008 with the reissuing of the T-Bolt, for .22 LR rimfire enthusiasts. It was not until 2009 that a new generation of rifles was marketed, with the X-Bolt, which was to bolt-action rifles what the Cynergy was to over and under shotguns.

The rifle has modern, aesthetic and unconventional lines. It is distinguished by the quality of its trigger, its composite rotary magazine and its new safety system. Some 15 years after the launch of the X-Bolt, the dozens of models produced satisfy all the needs of users: these models offer different finishes, colours or materials used, as well as the extension of the chamberings available, plus the addition of options and features. The surface treatment ranges from blue with gold highlights, with the Medallion Safari model, to Cerakote finish (polymer-ceramic) with the Pro Carbon model. A free-floating barrel, perfect trigger action, plus straight and direct shooting have also made the X-Bolt a popular product for long-range target shooters in the United States. Here too, Browning developed the models to meet the requirement: 10-round box magazine, new muzzle brake, adjustable comb, aluminium stock frame, fluted Varmint medium-weight barrel and tactical rails.

At Winchester, the closure of the New Haven factory had little impact on the legendary Model 70 rifle, which returned in 2008 in a modernised version, with a trigger improved by Browning. In 2016, Winchester launched the XPR (Expert), with its medium-weight barrel, threaded for fitting a muzzle brake to reduce recoil, which enhances accuracy when shooting large game. The XPR was the first bolt-action rifle developed entirely by Browning's research and development team for Winchester. Moreover, Winchester reinvested in .22 calibre by marketing the Wildcat in 2021, for 'pest control'. Nor should we forget the Model 94, the celebrated lever-action rifle, which is still very popular in the United States with collectors and 'cowboy action shooters'.

Browning X-Bolt
Bolt-Action Rifle

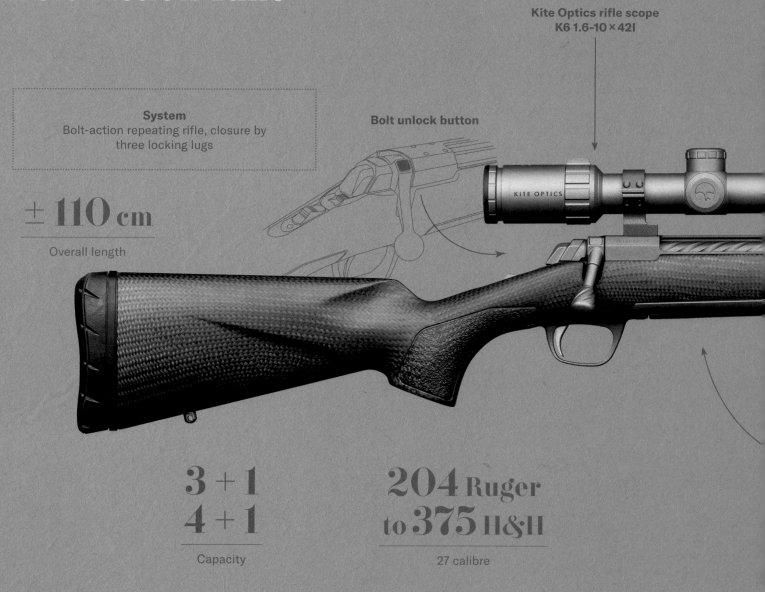

System
Bolt-action repeating rifle, closure by three locking lugs

Bolt unlock button

Kite Optics rifle scope
K6 1.6-10 × 42I

± **110 cm**

Overall length

3 + 1
4 + 1

Capacity

204 Ruger
to 375 H&H

27 calibre

From the mid-1980s to 2014, Browning sold the A-Bolt rifle. This firearm acquired a good reputation among its users, evolving according to the trends, in particular those introduced by stainless steel and polymers. In the late 1990s, Browning no longer only wanted to improve the rifle but also to create a new generation of rifles for the next century. There was now a need to offer a new product, one that would stand out thanks to its style, precision, light weight and functionalities – focused on its magazine, trigger and the safety system.

This wish to refine a new rifle and make it lighter initially led to a reduction in the diameter of the A-Bolt's receiver and the modification of some mechanical parts in order to optimise the action. The three locking lugs were kept, but the positioning was modified to avoid steel-brass friction between the bolt and the ammunition. The 60 degree rotation of the bolt handle, which had made the A-Bolt a success, was also kept in order to ensure quicker reloading while preventing the cocking bolt handle knob from interfering with the optics.

A new magazine made of solid polymer was then developed. The proposed rotary system maximised the magazine's capacity. It also enabled a reduction in the height of the gun and protected the bullet tip, as the cartridge was held by the shoulder, preventing it from hitting the magazine wall after each shot. The system also features a central loading design, with the cartridge positioned directly in line with the chamber, which avoids feeding problems. Finally, the magazine release latch is moulded into the magazine for ease of use.

2.7 kg

Average weight with scope

**53 cm
to 66 cm**

Barrel length

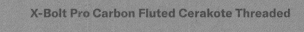

X-Bolt Pro Carbon Fluted Cerakote Threaded

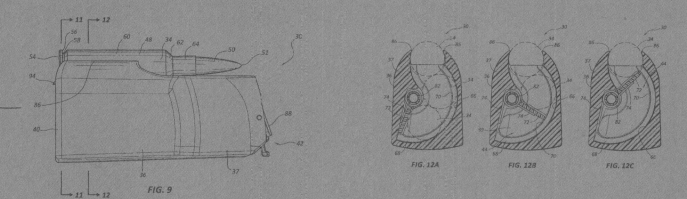

FIG. 9

FIG. 12A   FIG. 12B   FIG. 12C

The trigger assembly, whose quality is essential to ensure accuracy, has also been redesigned: it has a 'feather' trigger system, which minimises play and overtravel. Additionally, a near vertical interface between the sear and the firing pin has been developed, instead of the usual 45 degrees, which ensures the right angle for a clean hit.

On the first prototypes, the safety of the gun was first based on the Winchester's three-position system, now positioned on the tang and no longer on the side. But a more intuitive solution was finally found. In addition to a two-position safety tang and a cocking indicator, a bolt unlock button connected to the trigger has been integrated into the bolt handle. When the gun is in a position where it has been made safe, this button allows the

cocking bolt handle to be unlocked and for the chambered cartridge to be unloaded.

With the **X-Bolt**, Browning has therefore designed a lightweight rifle that users are likely to appreciate as much for its performance as design. The new models incorporate carbon fibre, which has proven to be more rigid, lighter and more durable and resistant to poor weather than wood or polymer. They feature a slim breech and fluted barrel, to improve reloading fluidity, ensure better heat dissipation and to reduce weight.

Browning Maral Straight-Pull Action Rifle and Kite Optics scope
Browning BAR MK3 Reflex Semi-Automatic Rifle
Winchester SXR2 Tracker Semi-Automatic Rifle

◄◆ Browning BAR MK3 Eclipse Semi-Automatic Rifle, 2019
With the BAR MK3 and Maral rifles, Browning follows trends and offers universal platforms adapted to both driven and stalking hunting: the latter is a style increasingly popular with new generations of hunters. Whether in the production of its guns or its Outdoor range, Browning retains a subtle blend of tradition and modernity, yet always with safety in mind.

Semi-automatic and straight-pull rifles

In 2009, the release of the BAR Zenith (Europe) aimed for the top of the range with its new refined line and its sets of customisable plates and handguards. The third generation of BAR arrived on the market in 2015, with the MK3. Winchester's SXR II was launched in a composite version in 2021.

The semi-automatic rifle has a significant place in the European market, where the evolution of hunting practices has also led to growing interest in linear reloading systems. In the late 1990s, Browning launched the Acera to enter this market. The gun was based on the system that had been proven on the BAR, calling on the lugs, locking and barrel assembly. Cartridge loading through gas-operation had been replaced by a lever operation, with a bolt sliding longitudinally inside the receiver. The bolt was blocked directly in the barrel by a forced guide, by means of a rotary head with seven slightly oblique lugs. In 2012, this system was revised on the Maral (Europe), a more advanced product, featuring a patented constant-force spring system plus the inclusion of patented ergonomic technologies offered by Browning. Composite and left-handed versions would be produced in 2018.

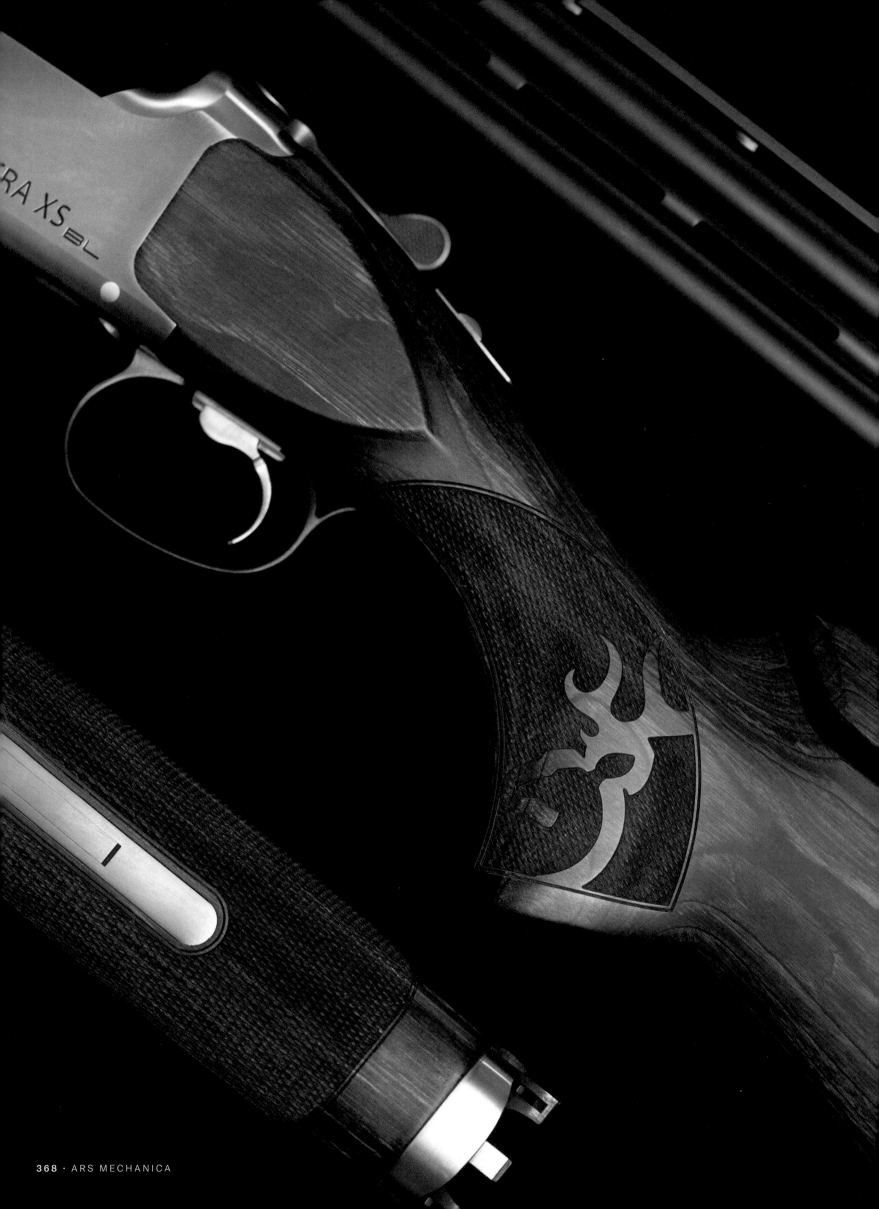

Browning Cynergy Wicked Wing Over and Under Shotgun
Browning B525 Heritage Over and Under Shotgun
Winchester Select Sporting Over and Under Shotgun

◀◆ Browning Ultra XS Over and Under Shotgun, 2022
Whether for hunting or sports shooting, the passion conveyed by Browning s reflected in a taste for work well done, knowledge of the craft
and the performance of its products. This reputation for excellence is promoted by Sam Green, a sponsored shooter equipped with the Ultra XS
and the 2019 World FITASC Sporting Champion.

Over and under shotguns: modernity meets tradition

Today the B525 is the classic entry-level over and under shotgun. It has seen several updates, such as the launch of the B525 Liberty Light in 2018. This gun is adapted for smaller shooters, with its adjustable stock offering a better fit for different body types. In 2019, the B525 Sporter Laminated offered a laminated/glued birch wood stock, a technique normally only found in rifles, a high-density wood that is more resistant to abrasion and humidity and is free of environmentally harmful substances.

Although well integrated into sporting clays, Browning also develops specific ranges for the shooting sports, such as the Ultra XTR over and under shotgun in 2009. However a revolutionary model arrived in 2012 with the B725. It has a contemporary look, and is the fruit of collaboration between Browning North America, Browning International and Miroku. The mechanism and receiver system are almost identical to the B525, but modification of the hammer stroke has resulted in a shorter striking time and a faster start. Aiming is also improved thanks to a lower act on receiver. The shotgun also benefits from three patented technological contributions for the butt plate, the barrel and the brass seal on the choke tube, the latter of which reduces fouling due to the gas penetrating between choke and barrel.

Clay target shooting with Browning B725 Pro Master Over and Under Shotgun
Browning meets the exacting requirements of competitors by offering reliable and high-performance guns adapted to every body type.

Production of over and under shotguns in the Miroku factory in Nankoku, Japan
Miroku is today a strong partner of the Herstal Group, in charge of producing all the new-generation over and under shotguns together with other products in the Browning and Winchester ranges. Almost 300 employees proudly work to produce precise and reliable arms, with a close-to-zero clearance allowance between parts thanks to adjustment by qualified workers. The guns are also good-looking, produced by a team of engravers who concentrate on the ornamentation – whether by hand, roller, etching or laser.

Browning B725
Over and Under Shotgun

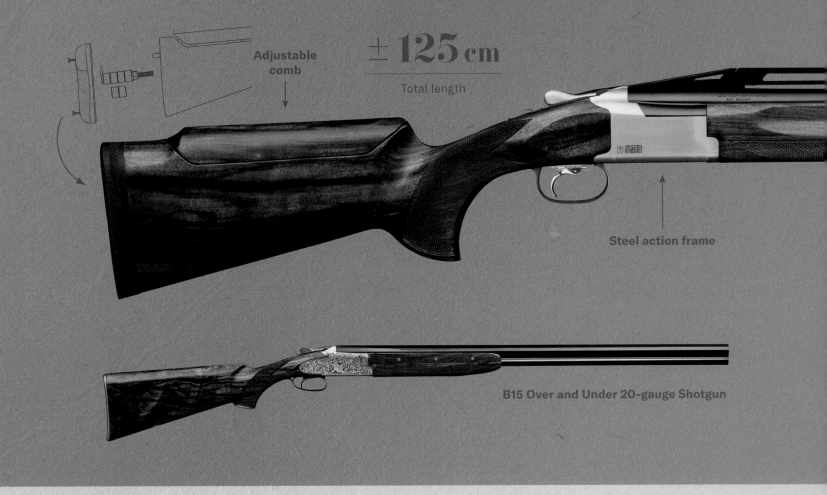

Adjustable comb

± **125 cm**
Total length

Steel action frame

B15 Over and Under 20-gauge Shotgun

Almost 100 years after the design of the famous B25, the Browning over and under Shotgun has gained a solid reputation for reliability and robustness, among both hunters and sports shooters.

In the mid-1970s, Browning made the decision to upgrade the B25. Ever since then, it has been offered in numerous models: Sporting Hunter, Lightning, Skeet, Trap, Liège, GTI, B27, and Cynergy, etc. This was a time when the company's over and under Shotguns adopted the generic name 'Citori' in North America. To mark the different periods and to assign its over and under Shotguns to new and successive generations, Browning regularly high-lighted the major upgrades with an incremental number preceding the design date of the first model: B125, B325, B425, B525 and B625.

Consequently, the **B725** is the latest version of the B25. It is based on the B525's platform, but is more innovative than its traditional predecessor. The B725 has a lower action frame profile than the B525, which speeds up aim-ing and target acquisition, and results in more instinctive

shooting. The low action frame also optimises in-line recoil, which limits the upward motion of the barrel and provides improved shooting comfort, thus enhancing effectiveness when shooting.

The B725 has a mechanical trigger system. This is pro-vided by a part called a 'disconnector': unlike an inertial trigger system, this guarantees a second shot can be fired whatever the circumstance. The mechanical trig-ger has resulted in shorter, faster and lighter shooting: it limits the movement of the parts and optimises their geometry, thus reducing the distance travelled by the sear before it comes into contact with the sear as the shot is taken. The revised architecture between sear and hammer allows the two parts to slide over one another and to disengage very quickly. Finally, a clean shot is achieved by limiting the distance travelled by the trigger when the hammer has fallen, by means of a simple stop. The shotgun features the latest Browning technolo-gies: new Back-bored Vector Pro shaped barrels and Invector DS chokes, resulting in ergonomic use of the

Back-Bored Vector Pro shaped barrels

66 to 81.3 cm

Barrel length

B725 Pro Master Over and Under Shotgun

3.8 kg

Weight

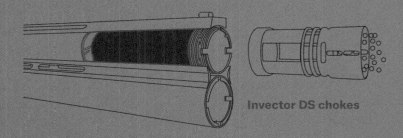

Invector DS chokes

12-76 / 20-76 28-70 / .410

Gauge

B725 in all shooting conditions, including with the use of steel shot. The principle of the Back-Bored Vector Pro system is to increase the diameter of the barrel bore and to extend the forcing cone. This results in increased speed and improved pellet penetration by reducing friction, improved pellet grouping by reducing any pellet deformation, and reduced recoil.

The Invector DS chokes provide unrivalled performance thanks to their optimal length, their exclusive internal shape and their patented gas seal segment, which limits the penetration of combustion gases between the choke and the barrel. Lastly, the addition of an ultra-flexible Inflex II recoil pad enables better recoil absorption while moving the stock away from the shooter's cheek.

The B725 offers numerous innovations, yet maintains the major design elements of Browning over and under Shotguns. It also symbolises what the brand represents today: an international group. Because although the B725 is produced – just like all the contemporary over and under shotguns – at Miroku, Browning's traditional partner, its more luxurious side-plated boxlock is produced at Herstal under the B15 name. Featuring an entirely handmade finish, the B15 was introduced in 2015 and fits into the range among the 'industrial' shotguns, such as the B725, and the B25. The B15's assembly, woodworking, shaping of the rounded-body action frame, and engraving in four different grades: these are carried out by the master artisans of the John M. Browning Collection, using the same processes and materials used for the B25. Based on the mechanics of the B725, the B15 is not only elegant, it is also very sturdy and reliable, a genuine marriage of tradition and modernity.

Engraving of an action frame, using the chisel hammer technique, carried out in the John Moses Browning Collection, Herstal, 2015

Legacy of five centuries of arms manufacturing in Liège

Browning maintains the subtle balance between modernity and tradition, with its Custom Shop. In 2015, this became the John Moses Browning Collection, in honour of the legendary inventor, whose name has been perpetuated by the Herstal Group for more than 120 years. The workshop still offers customisation of the famous B25, in various engraving grades, not to mention the B15, the CCS25 rifle and the HP Renaissance. The workshop is a unique tool, whose reputation is sustained by highlighting its expertise. This is underlined by its use of the finest materials as well as its precise craftsmanship, which bear witness to the ancestral knowledge passed down by generations of gunsmiths.

Browning B25 Renaissance Over and Under 12-gauge Shotgun

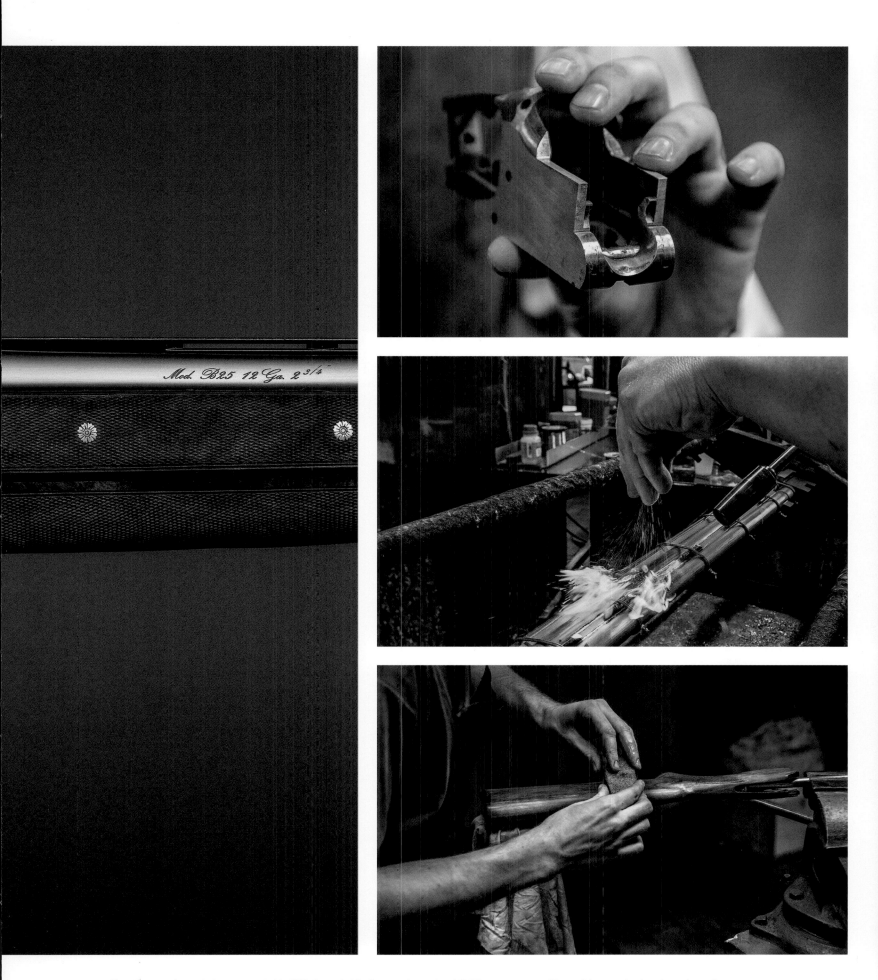

Since the engraving workshop was created in 1926, the market for fine guns has matured. Yet the very essence of the craft has remained unchanged over the decades: working with the best steel and walnut wood to produce efficient, reliable, sturdy and elegant guns. Assembly, finishing, ornamentation, chasing, inlaying of metal parts and wood mounting – these tasks are entrusted to different master craftsmen, reflecting the age-old division of work. All the craftsmen contribute their know-how, experience and sensitivity, thus making each gun a unique product.

Transition to Industry 4.0

After 130 years of operation, the Herstal Group's global industrial park embarked on a major transformation. It has incorporated the technological changes generated by digital developments and has adapted to the increasing need for innovative production methods. Industrial sites have become global and connected systems, dedicated to quality and productivity. The industrial process has gained in flexibility and cross-disciplinarity, ensuring the delivery of more and more sophisticated product ranges. To fund the research and development policy, which would double over the decade, a sufficient level of profitability and cash flow had to be maintained. The company's efforts were focused on continuously reviewing product positioning, commercial strategies, cost prices, margins, overheads and inventories.

In 2012, an operation to reduce production time was launched. It aimed to improve the flow of activities, by identifying any obstacles to efficiency and performance. The operation involved intensive modernisation investments in all sectors, which took place in 2014. The company also introduced a new cross-functional operating mode, centred on the customer: the Supply Chain, which is a complete process for managing information linked to an order, from sales management through to fulfilment. The goal is to better communicate, share information between sectors, anticipate needs, and verify the right decisions are made, as well as to manage inventory and equipment flows. The principles of integrated planning between workshops, heat treatment, machining and assembly have to be combined with dynamic subcontracting. Because these are crucial for managing the coordination of in-house and outsourced production, as well as boosting the arrival of components at assembly.

The new Gröb machining centre, machining workshop, FN Herstal factory, 2022
In an increasingly competitive world, the Supply Chain enables costs and lead times to be controlled for the deve lopment, launch, production and marketing of products. This just-in-time production is ensured by a technological leap in the range of numerically controlled machines, with the support of robotics.

Assembly line for 5.7×28 mm NATO calibre ammunition, Zutendaal factory, 2022

Three-dimensional control of a remote-controlled station cradle, machining workshop, FN Herstal factory, 2016
Since the 2000s, coordinate metrology, using coordinate measuring machines, has become increasingly precise. This has raised the standards of accuracy, speed and error reduction in the manufacture of parts that must be totally reliable.

In 2017, the company launched a new transformation plan for industrial tools, with a view to continuously adapting to the new diversity of products and technologies. In 2019, the assembly shop's physical flow was further improved by computerisation and better visualisation of administrative tasks. The Production, Supply Chain and Sales departments now had real-time access to the status of orders. The objective was also to adapt to limited production runs and to the increasingly specific needs of customers, while controlling production costs. Pilot projects were launched around the FN SCAR assault rifle. Because of its modular design, this weapon has the most variants, with no fewer than 230 different configurations

in 2020. The goal was to manage backlogs by creating multidisciplinary and flexible teams.[271] Industrial thinking will be revolutionised by the digitalisation and the fluidity of information and processes.

Alongside the new management system, industrial facilities were modernised through continuous investments. They were also re-scaled to meet the planned production volumes. Old buildings were torn down, administrative buildings were renovated. There was also a review of the ergonomics, working environment, and safety regulations. The firearms Assembly and Packaging operations were relocated in the same (renovated) building, so as to simplify the order administration process. The Integrated

Drilling bench for .50 calibre rifled barrels, barrels manufacturing, FN Herstal factory, 2022
Among the numerous operations completed in the industrial manufacture of firearms, barrels manufacturing is one of the key areas of the trade. This assembly has unique machines to carry out the olive-cutting – the drilling of a metal bar on a horizontal lathe – and the hammering, i.e. the pressing of a barrel on a mandrel.

Weapon Systems operations – which continue to grow within the company – were also relocated here, before acquiring their own infrastructure in 2017. In 2016 and 2017, the Herstal site's phosphating line was replaced and a new sandblasting plant was installed. Further modernisation is in progress for the anodising, chromium plating and painting lines.

Between 2016 and 2018, renovation work was carried out at Herstal's ammunition factory, where the latest equipment was installed to produce 12.7 mm calibre tracer bullets. The Zutendaal Factory is introducing new APEI bullet rolling machine: this will upgrade the operation and increase its production capacity. Moreover, the site is continuing to modernise its administrative infrastructure and its test centre, with the setting up of a new 100-metre shooting range and a tactical course. Lastly, automated shrink-wrapping machines were installed in the factory. In 2020, a new and secure chemistry laboratory was opened. It is designed for the synthesis of new molecules, formulation, testing and analysis of energy compounds.

The modernisation of barrels manufacturing operations has mainly resulted in an increase in milling capacity for military barrels, plus nozzles for civilian smoothbore and rifled barrels in 2015. In 2018, the barrels manufacturing unit was equipped with state-of-the-art .50 calibre

Assembly bench for FN® M2HB-QCB Machine Guns, FN Herstal factory, 2022

Assembly bench for FN SCAR® Assault Rifles, FN Herstal factory, 2022

rifling equipment, enabling increased production capacity of the FN M2 HB-QCB and FN M3M machine guns. At that time, these guns were a major workload for the plant.[272] Production was boosted, as well as the quality aspect of the rifling and chamber. In the machining workshop, in 2015, the company added machinery for three-dimensional work and brought in more cylindrical grinding machines. The new investments also support greater automation and milling capacity. From 2017 to 2020, new lines of Heller 5-axis CNC machines and Gröb machining centres were added to replace the last machines dating back to the 1990s. Robotics are needed to lighten the workload and enable operators to focus on high value-added tasks and quality control. Automation of industrial tools is controlled in real time via RFID tags. This is one more step on the long road towards the fourth industrial revolution (Industry 4.0), a new technical system based on computerisation of the industry and bringing together a new generation of connected and intelligent factories. The aim is to pursue this evolution by applying an artificial intelligence system capable of performing predictive maintenance.

In England, the FNH UK factory is adapted to FN Herstal's organisational standards. It has integrated the concepts of synergy, IT support and Technical Data Package, which aims to develop and integrate technologies to improve

Assembly bench for M240 (FN MAG®) Machine Guns, FN America factory, in Columbia, South Carolina, 2022
For FN America to continue being a major supplier of small arms to the world's largest army, it has had to meet strict technical specifications and to ensure the quality of its products.

materials, processes and techniques. Investments are undertaken to bring the workshops up to standard, by introducing new machines and adapting them to safety standards. The manufacturing processes targeting the design and sturdiness of the FN MAG and .50 machine guns have been upgraded.

Across the Atlantic, the Columbia plant has also been continuously adapted to shift from a production source focused on the M240 and M249 machine guns, followed by the M16 rifles, to a diversified production of weapons driven by Law Enforcement activities. For example, the increased production of pistols has required the installation of new and more compact machines in the barrels

manufacturing and machining centre. Simultaneously, it proved necessary to overhaul the organisation of the production lines, and the manufacturing process, as well as to introduce the Supply Chain concept – under FN Herstal's guidance. So the last decade has seen continuous improvement of industrial tools and the way they are managed, all influenced by the 'lean' principle, which aims to reduce costs and improve production capacity.

B25 Over and Under Shotgun with side-plates inlaid with fine gold, posed on an FN car, 2022
A shotgun and a car, two high-end objects that symbolise the company's rich history. The Ars Mechanica Foundation has been entrusted with preserving and showcasing this history since 2008.

More than 130 years of heritage

Throughout its history, the Herstal Group has highlighted its prowess in a vast variety of civilian and military products. These cover everything from hunting rifles to machine guns, as well as cars and turbojet engines. Thanks to its ingenuity and expertise, the company has pioneered many changes in weapons manufacturing and more broadly in mechanical engineering products, not to mention the modernisation of industrial processes. These methods and products, today part of the annals of science and technology, have forged an invaluable industrial heritage. They shine a light on our societies' technological evolutions and know-how, to which the Herstal Group has contributed enormously.

Fabrique Nationale quickly became aware of its rich heritage and started to actively preserve it after the Second World War, amid growing and renewed interest in industrial heritage. In 1964, based on its 75 years of experience, FN published a first monograph on its history. This was followed by two other books, in 1989 and 2008. Each time, the aim was to mark key anniversaries. Because preserving the company's memory means shining a light on the path taken and looking to the future while still mindful of the past.

The company also recognised the importance of keeping a record of techniques that would soon disappear as industrial modernisation grew rapidly. The FN Museum of Industrial Archaeology was therefore established in 1977. This initiative was recognised by the Council of Europe, which included it in a shortlist of the year's top nine museums.

An old system made way for a new one, upending reference points and raising awareness. This happened again in the early 2000s when the company, which was celebrating its 120th anniversary and entering a new technological phase, resolved to equip itself with a strategic and permanent tool to manage its heritage.

Thus the Ars Mechanica Foundation was set up in 2008. This is a heritage foundation serving the public interest, both recognised and supported by the King Baudouin Foundation. The mission of this foundation – a genuine conservatory of mechanical arts – is to centralise, safeguard, study and showcase the physical and intellectual heritage of the Herstal Group in Belgium and worldwide. Heritage is a common denominator, simply because it covers all the Herstal Group's activities and entities. Consequently, heritage is an ideal space for communication, turning the Ars Mechanica Foundation into an ambassador for the company's brand image. Calling on digital resources, publication and exhibitions, this

Exhibition space of Ars Mechanica, 2022
Ars Mechanica Foundation's museum work led to the opening, at the heart of the Herstal site, of a private exhibition place – which shows some of the major products behind the success of the FN Herstal, Browning and Winchester brands. Through the heritage on display, this showcase is a useful tool to promote the company's brand image.

heritage and diplomatic tool is harnessed to disseminate all the know-how and talent of the Herstal Group's employees. It cultivates the company's culture, unites the large community of aficionados of the FN Herstal, Browning and Winchester brands, creates strong bonds with customers and stimulates dialogue with civil society.

Over 130 years of excellence
Leading to the future

Pioneer in modern small arms

Backed by over 130 years of history and know-how, the Herstal Group designs, develops and markets a range of innovative, unique and diversified range of products and services based on small-calibre light firearms. A tradition and commitment that the Herstal Group pursues daily to meet the demand for excellence of its customers and partners.

Today, the Herstal Group is:
- A worldwide sales network.
- Fully in control of manufacturing its products, from design and development to marketing and quality assurance.
- Constantly innovating.
- Two fields of activity: Defence and Security - Hunting and Sports Shooting.
- Seven production units in six countries: Belgium, Finland, Portugal, United Kingdom, United States, and Japan*.
- Three internationally renowned brands: FN Herstal, Browning and Winchester Firearms**.
- 3,000 employees worldwide, including 1,500 in Belgium.

* Miroku operates as a partner, not as a subsidiary.
** Winchester Firearms is a trademark registered by the Olin Corporation.

2021, the baton is passed on

In October 2021, Julien Compère, former chief of staff of the Walloon Region's Minister of the Economy, then CEO of Liège University Hospital (CHU de Liège), succeeded Philippe Claessens as CEO of the Herstal Group. Now heading the Herstal Group, Compère is faced with many challenges – not least because this new decade is still marked by the global health crisis and shaken by the war between Russia and Ukraine. The Herstal Group must address the new geopolitical, socio-economic and environmental landscapes. Doing so will ensure that the Group maintains a leading role in its two core businesses, is part of the strengthening of European defence and NATO, diversifies its portfolio of activities, pursues its international development policy, and maintains its role as a key structuring force in Belgium and particularly in Wallonia. Herstal Group will have to address these future strategic challenges.

Postface

Herstal Group, backed by its 3,000 employees and through its three flagship brands – FN Herstal, Browning and Winchester (under licence), has become a global leader in civilian and military light weapons systems. The Group boasts a solid European/American base, and this is reinforced by a special partnership with Japan. FN Herstal, FN America, FNH UK, Noptel, Browning North America, Browning International, Browning Viana, and Miroku make up the Herstal Group. They are an integral part of its history and they underpin its forward-looking vision.

Since the company's founding in 1889, several generations have followed in each other's footsteps to build the Group we know today. All of them, as heirs to centuries-old expertise and open to the innovations of their time, have worked together to elevate the business and its brands to the status of a global benchmark. Ever resilient, the "Fabrique Nationale" as it was called in the 19th century, has succeeded in adapting itself to geopolitical upheavals, economic crisis, and technological revolutions. It has reinvented itself, kept on innovating, invested, until it became an undisputed world leader in its fields of activity.

In 1998, the Walloon Region became the Herstal Group's sole shareholder. Since then, the region has supported the strategy of the company, enabling it to reposition itself as a leading global player in all its sectors.

Strong innovation, plus the drive to expand and diversify within its portfolios of activities and markets, have borne fruit for the company year after year. Thanks to a can-do attitude and the multiple talents of its employees, Herstal Group is today a high-performing 21st century industry. Its annual turnover today is close to a billion euro.

In 2022, the geopolitical situation in Eastern Europe had a huge impact, obliging Europe to strengthen its security and defence policy. Calling on its expertise, as well as its human, industrial and technological assets and an international presence, Herstal Group is determined to play a major role in this new environment.

The decade ahead will undoubtedly trigger profound changes. The Group's flexibility, its capacity to anticipate change plus an ability to remain at the cutting edge will be crucial to sustaining its position. Just as it has since its earliest days, the Herstal Group is ready to face this new challenge.

Julien Compère
Chief Executive Officer of Herstal Group

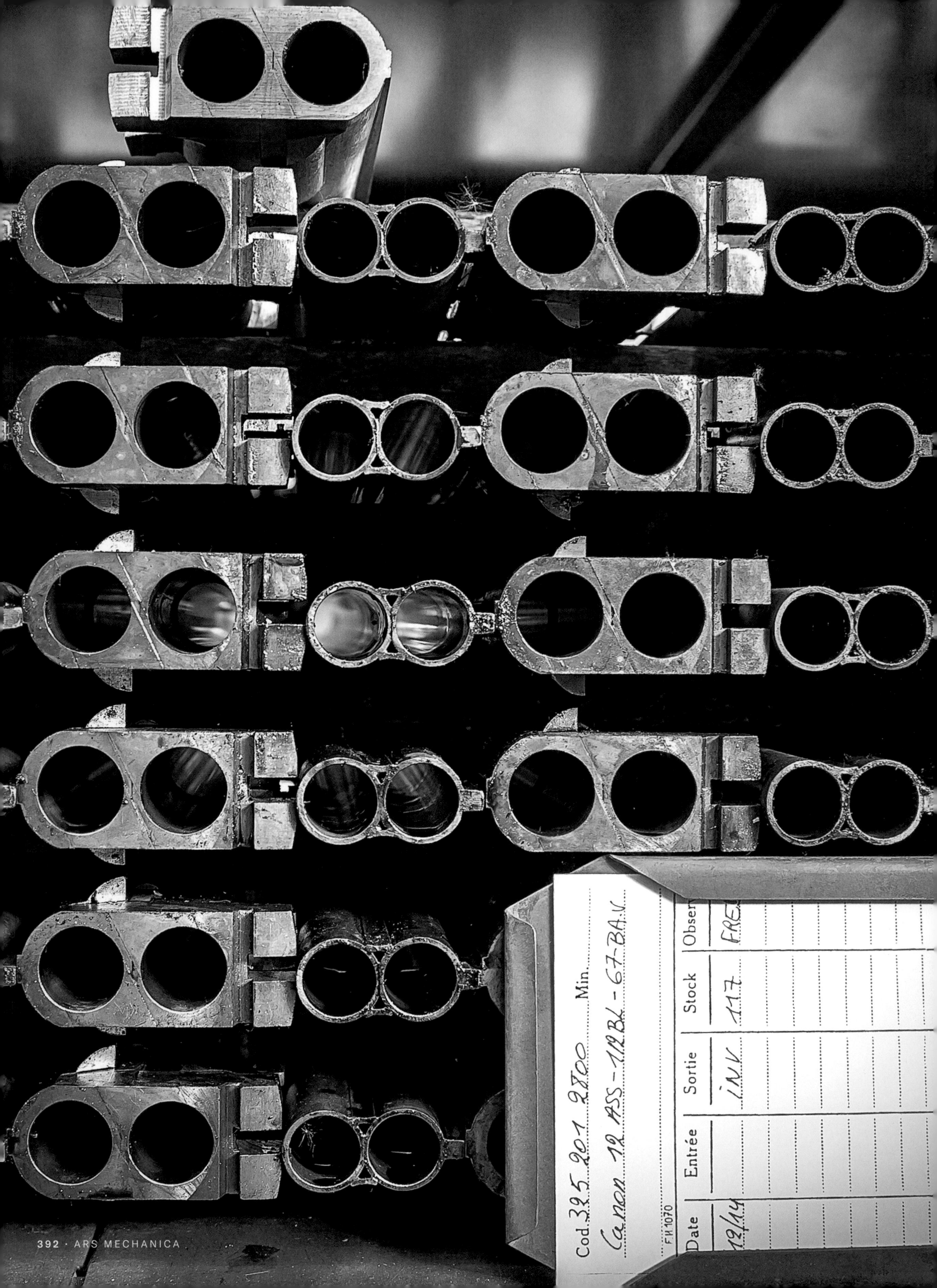

Cod. 325.201.2800   Min.

Canon 12 ASS - 112BL - 67-BAV.

FH 1070

Date	Entrée	Sortie	Stock	Obser
		INV	117	FRE
19/14				

Notes

1 Medical commission of the Province of Liège, rapporteur C. Wessaige, *Mémoire sur la condition des ouvriers et le travail des enfants dans les mines, manufactures et usines* [*Report on the condition of workers and on children working in mines, manufacturing and factories*], Liège, 1847. This Commission noted, for example, 50 cloth factories, 28 lime kilns, four lead and tin refineries, 49 brick factories, 42 juniper distilleries, three powder magazines, two hat factories, 12 salt refineries and 24 soap producers. In the same time period, authorisation was given to place 278 steam engines either in these establishments or in others that existed beforehand and that were using other power sources.

2 Victor Hugo, *Le Rhin, Lettres à un ami*, lettre VII [*The Rhine, Letters to a friend*, letter VII] 1842.

3 This refers to the Universal Exhibition in Paris in 1889, best known for constructing the Eiffel Tower.

4 General Alfred van der Smissen, Léopold II's former aide-de-camp, chaired the committee that was responsible for choosing the rifle for the Belgian Army. He was the one who brutally put down the 1886 riots in Liège and Charleroi.

5 *Le Peuple*, 27 October 1889.

6 *La Réforme,* 27 October 1889.

7 *La Réforme*, 17 November 1889.

8 *La Meuse*, 22 October 1890.

9 We will see later that this figure was greatly underestimated.

10 *La Meuse*, 28 October 1890.

11 *La Meuse*, 18 December 1890.

12 The actual area was of course 10,000 square metres. The figure would be rectified in a later edition of the newspaper.

13 This is in fact Hartford in Connecticut.

14 *La Meuse, 24 June 1891.*

15 The 'restructuring of the company' was a discreet way of referring to the takeover of FN by the German group, Loewe, in 1896.

16 The Deutsche Metallpatronenfabrik was established in Karlsruhe in 1872 and bought by Ludwig Loewe & Co in 1889.

17 Minutes of the Board of Directors meeting of 10 October 1891.

18 This consortium had to undertake not to offer the same facilities to competitors of FN. A practical application of this agreement was the advance granted, for two years, of one million pounds sterling. This was the amount of the Chinese order, for which payments could therefore be spread out (see minutes of the Board of Directors of 15 March 1895).

19 Minutes of the Board of Directors of 28 July 1893. On that day, the Dutch question was raised. The Government of the Netherlands was apparently inclined to select the Mannlicher rifle. So the company Steyr would be more likely to secure the order of 50,000 rifles than FN, which was more specialised in producing the Mauser.

20 The 28 July, 1 and 15 September, 15 December 1893, 16 March, 18 and 23 May, 17 August and 21 September 1894, the date when this proposal's failure was officially noted.

21 The slogan ran: '8 hours of work, 8 hours of leisure and 8 hours of rest.'

22 Universal suffrage (i.e. male universal suffrage) had been adopted two years earlier, but it was weakened by a complex system that allowed the wealthiest and best educated to have additional votes.

23 During the same session, the Board protected itself from political strikes. In this scenario, the management would immediately close the factory…

24 These initiatives came after the partial victory of the Belgian Parti Ouvrier [Workers' Party] in the 1894 elections, where it garnered nearly 20% of the vote, allowing it to secure 28 of the 152 seats in the Chamber of Representatives.

25 *Le Soir*, 23 December 1895.

26 *Le Soir*, 25 December 1895.

27 *Le Peuple*, 26 December 1895.

28 Minutes of the Board of Directors of 17 March 1890.

29 Headline in *L'Express*: 'The Prussians in Liège' or 'Herstal to the Prussians'.

30 On 1 January 1896, *Le Soir* led the charge against Loewe, in the name of economic patriotism, taking care to note its Israelite faith, and citing a vehement German anti-Semite, Hermann Ahlwardt, the author of a brochure with the revealing title: *Les fusils juifs* (1892) [The Jewish rifles].

31 Those resigning were Bormans, Pirlot, Simonis, Dumoulin and Ancion. Francotte and Nagant resigned from the Board in July 1894.

32 On this subject, see Marie-Thérèse Bitsch, *La Belgique entre la France et l'Allemagne. 1905-1914.* Paris, Éditions de la Sorbonne, 1994.

33 'French diplomats based in Brussels keep denouncing the German invasion and supporting the Germanophobic press, which periodically sounds the alarm', in Bitsch, *op. cit.* p.168.

34 Minutes of the Board of Directors of FN meeting of 20 February 1897.

35 Auguste Francotte Jr, *La Fabrique nationale d'Armes de Guerre, 1889–1964*, Liège, Desoer, 1965.

36 Serbia turned out to be very bad at paying what it owed.

37 'Articles of association adopted by the general meeting of shareholders', *Le Moniteur belge*, 20 and 21 June 1896.

38 We could mention the Nagant and Pieper bicycles and cars, which were geographically close by.

39 'Articles of association', minutes of the Board of Directors of 12 October 1897.

40 In 2019, the American reference website *thetruthaboutguns.com* came up with the headline: 'John Moses Browning and Fabrique Nationale d'Herstal – The Greatest Partnership in Firearms History.'

41 Minutes of the Board of Directors, 28 November 1896.

42 Active in Herstal between 1899 and 1946, this cooperative brought together Neumann Frères, Janssen & Fils et Dumoulin fils & C°. It sold weapons using standardised parts produced by FN. The weapons were finished in a wide variety of styles.

43 Minutes of the Board of Directors meeting, 25 June 1903.

44 It turns out that the first meeting between Berg and Browning was in Hartford in 1893, when the inventor was developing his first gas-operated model of machine gun for Colt

45 See Nathan, *The guns of John Moses Browning*, New York, Scribner, 2021, pp. 118-119.

45 *The Salt Lake Herald*, 17 November 1895.

46 The first breech-loading firearms date back to the early 19th century: Jean Samuel Pauly, a Swiss gunsmith based in Paris, filed a patent for this in 1812. The Pauly rifle used the first modern and self-contained cartridge, which he had created in 1808.

47 John Moses Browning adapted Winchester lever-action rifles to new, more powerful calibres: .45-70 for the 1886 Model, .30-30 with smokeless powder for the 1894 Model.

48 'He has not made any inventions in pistols.'

49 In 1897, the pistol was sold for 30 francs. So the royalties paid to the Browning Brothers amounted to nearly 7% of the sales price, which was a lot.

50 *The Salt Lake Herald*, 19 August 1900.

51 La Meuse, 26 May 1899.

52 The journalist made much of the fact that the pistol was light (600 grams), cited the tests that it had undergone and described how it functioned. The Browning pistol was used by famous robbers, including the Bonnot gang during a bank robbery in Paris, rue Ordener, in 1911.

53 *La Meuse* from 13 June 1906 added, during a trial against an anarchist: 'The Browning is easy to handle. The slightest of pressure fires the shots, all seven of which can be fired without interruption.'

54 The newspaper added: 'The newspaper was printed in French and this article was translated by Ms Hawkes, who teaches French at the school in Ogden.'

55 *The Ogden Standard* – or Ms Hawkes – somewhat mixes up the titles of those attending this celebration. Ludwig Hagen is therefore described as Prime Minister whereas he is only Vice-Chairman of the Board of Directors.

56 Minutes of the Board of Directors meeting, 12 August 1895.

57 For comparison, an illustrator was paid 3,600 francs per year and a head accountant 6,000 francs per year. In 1897, a female typist, Ms Tack, was paid 1,200 francs per year and a male typist, Mr Dartois, 2,400 francs per year…

58 This commission only ended up being cut in half, even though Berg played no part in making a sale.

59 Minutes of the Board of Directors meeting, 1 September 1899.

60 Minutes of the Board of Directors meeting, 12 October 1897.

61 Minutes of the Board of Directors meeting, 26 November 1898.

62 Minutes of the Board of Directors meeting, 30 January 1899.

63 *La Meuse*, 27 March 1897.

64 As was then customary, FN only produced the chassis, to which the customer added the bodywork in specialised workshops. FN's first car with bodywork, made of wood, was produced in 1913.

65 Minutes of the Board of Directors meeting, 1 September 1899.

66 Minutes of the Board of Directors meeting, 17 November 1899.

67 Minutes of the Board of Directors meeting, 8 October 1899.

68 This refers to what would later be considered the first automobile fair in Paris. In France, only 2,000 cars were on the roads back then, but the event attracted at least 140,000 visitors.

69 Minutes of the Board of Directors meeting, 26 November 1898.

70 Minutes of the Board of Directors meeting, 9 December 1901.

71 The anecdote is quoted by François-Marie Dumas's excellent website (www.moto-collection.org/moto-collection/fmd-moto-FN-7317.htm).

72 A British advertisement at the time described it as follows: 'She's a beauty, so smooth and flexible'.

73 Minutes of the Board of Directors meeting, 23 July 1913.

74 On 8 October 1907, FN therefore borrowed one million francs from Crédit liègeois. On 26 November that year, FN decided to increase its capital by 1.1 million francs by issuing new shares. On 5 May 1908, it took out loans of two million Belgian francs via 500 Belgian franc bonds, reserved exclusively for the Liège-based bank, which bought them for 475 Belgian francs (see the minutes of the Board of Directors on these dates).

75 FN's forge dated from the very early days of the company and its own foundry from 1906.

76 Minutes of the Board of Directors meeting, 29 February 1912.

77 In 1896, when the first major employment-related conflict broke out at FN, the reaction was fairly sharp. The company even tried to generate highly negative press articles in order to discredit the strikers.

78 *La Meuse*, 16 February 1897.

79 *La Meuse*, 3 May 1897.

80 Minutes of the Board of Directors meeting, 2 September 1909.

81 Minutes of the Board of Directors meeting, 15 February 1910.

82 *La Meuse*, 22 June 1910.

83 Minutes of the Board of Directors meeting, 13 September 1910.

84 Société générale de Belgique, *report presented to the general assembly of shareholders in the year 1917*, Brussels, 1918.

85 Alexandre Galopin and Gustave Joassart were to become Managing Directors of FN, the former in 1919 and the latter in 1923.

86 Minutes of the General Assembly of 28 January 1919, i.e. just over two months after the armistice.

87 Minutes of the Board of Directors meeting of 28 January 1919.

88 The Board of Directors meeting of 9 June 1921 noted, regarding the Vélos, Autos, Motos [Bicycles, Automobiles, Motorcycles] Division: 'Customs barriers are increasingly limiting our activities to the Belgian and Dutch markets.' In 1923, the Belgian market was viewed as 'FN's biggest customer' for these products.

89 The League of Nations met for the first time on 10 January 1920. Created under the Treaty of Versailles, it was intended to be a forum for arbitration between nations that aimed to avoid conflicts.

90 Minutes of the Board of Directors meeting of 9 June 1921.

91 Minutes of the Board of Directors meeting of 14 September 1921.

92 In Belgium, the 'retail price index', which tracks the monthly change in consumer prices, was created in 1920: its base in April 1914 was 100.

93 Minutes of the General Assembly of 9 October 1923.

94 Minutes of the General Assembly of 14 October 1924.

95 Claude Gaier, *Ars Mechanica, op. cit.*

96 Minutes of the Board of Directors meeting of 12 January 1922. The example given is that of the Baltic countries.

97 Minutes of the Board of Directors meeting of 12 January 1922.

98 Yvan Makhonine was a Russian engineer, inventor of the coal liquefaction process, which led to the production of a new fuel.

99 Minutes of the Board of Directors meeting of 7 May 1928.

100 Minutes of the Board of Directors meeting of 3 May 1929. For the staff, FN undertook to continue all the employment contracts, written or verbal, of the workers and employees. But it reserved the right to modify their job assignment within the merged company, whether in Herstal or in Bruges, where the installation of a cartridge factory was then envisaged.

101 This law largely stimulated the consolidation of companies and the streamlining of their production. It led to reduced competition between Belgian companies of the same kind and the formation of groups sufficiently strong to compete on foreign markets. In the banking sector, for example, this law favoured the acquisition by the Société Générale du Crédit général liégeois and the Banque de Bruxelles (and several other establishments): both banks had played a major role in FN's development before the war.

102 Minutes of the Board of Directors meeting in June 1920.

103 The paper mark replaced the Goldmark in 1914. After the war, it continually lost its value against the dollar. In 1920, a dollar was worth 4.20 marks; in 1923, it was worth 4,200 billion.

104 In February 1929, FN, which employed 7,569 workers, announced that it would need 1,000 more.

105 Minutes of the Board of Directors meeting of 18 February 1929.

106 Fafnir was a German automobile manufacturer based in Aachen, producing vehicles from 1903 to 1926.

107 Minutes of the Board of Directors meeting of 19 January 1937.

108 Minutes of the Board of Directors meeting of 13 September 1920.

109 Reports of the Board to the annual general assemblies. See on this subject the minutes of the general assemblies of shareholders from 1919 to 1939.

110 When this quality declined, the Board of Directors became concerned and searched for the reasons, as happened on 21 September 1929: 'The General Manager spoke at length about the decline in quality that had occurred in the car and motorcycle divisions.' He listed all the reasons that had led to this situation and concluded that the company's priority was 'to take the utmost care in performing the work and maintaining the spirit of quality'.

111 The journalist from *La Meuse* quoted for the cars: a new chassis of 24/30 HP; for the motorcycles: two four-cylinder 7 HP, with a more powerful engine than in 1914 and with three speeds and a light single-cylinder of 2¾ HP with two speeds, and finally bicycles without chains or 'chainless' bicycles.

112 H. Houben, 'Les restructurations dans l'industrie automobile en Belgique', *Courrier hebdomadaire* du CRISP, 2016/10, n° 2295-2296, pp. 5-71.

113 In 1919, the Americans chose Antwerp as their commercial base in Europe, because the ease of access to the port outweighed manufacturing tradition. They would later set up their production plants there. In 1922, Ford assembled its first cars in Antwerp (50,000 units were assembled there in 1925); General Motors set up there in 1924; Renault did the same in Vilvoorde in 1925 and Citroën in Forest in 1926, etc.

114 Negotiations had taken place with Chrysler, Fiat and Peugeot.

115 Minutes of the Board of Directors meeting of 18 May 1935.

116 Minutes of the Board of Directors meeting of 19 May 1938.

117 Minutes of the Board of Directors meeting of 8 May 1919.

118 Minutes of the Board of Directors meeting of 21 March 1921.

119 The Citroën-Kégresse was one of the first tracked vehicles whose front wheels pull and rear tracks push, named after its inventor Adolphe Kégresse.

120 At that time, FN referred to an 'all-terrain' staff vehicle for the Department of National Defence, while adding 'colonial car'.

121 *Le Soir*, 11 May 1928.

122 *Idem.*

123 *La traversée du Sahara à moto* is a silent film of 5' 11'', made in 1927 by an unknown producer in Marseille and Liège. It shows the triumphal return of the 'desert motorcycle victors', Royal Belgian Film Archive (www.youtube.com/watch?v=F6K2v3jZPSM).

124 Under French colonial rule, this was the name of the city known today as Béchar, in Algeria.

125 FN won the Coupe du Roi [King's Cup] in 1925, 1926 (with an FN 1300), 1932 (three cars 8 cylinders), 1933 with Prince Baudouin, model 1933.

126 In 1931, the engineer Henry Van Hout had the bronze cylinder head replaced by an aluminium one on the 350 cc; the cylinder was also made of aluminium, with a nitrided cast iron liner, chemically treated with nitrogen. The support from the company's laboratory was obviously vital in these operations.

127 On 8 May 1919, it was announced that an order for 10,000 pistols for the Belgian government would be completed for June and that this would likely include a new order for 5,000 pistols 'as a follow on'.

128 Production of the mounts of these guns helped to reduce the impact of the end of automobile production at Herstal in 1935.

129 Note cited in *SAM40.fr/, Stratégie, Aviation & Guerre mécanique 1940-1914. Le blog d'histoire militaire de Pierre-Yves Hénin.*

130 Minutes of the Board of Directors meeting of 19 March 1940.

131 In Belgium, the Service du Travail obligatoire – or STO – was established in 1942. It covered men aged 18 to 50 and women of 21 to 35. In June 1943, the lower age limit was reduced to 18. The occupiers' law was aimed at requisitioning local workers into German industry.

132 In February 1945, the Managing Director underscored the material and moral damage suffered due to these 'Flying bombs'. He then invited the directors to first assist the bereaved families, by granting them compensation, until the State could take over.

133 The Board of Directors meeting of 7 October 1944.

134 Minutes of the Board of Directors meeting of 7 October 1944.

135 Created by the Treaty of Paris in 1951, the ECSC (European Coal and Steel Community) was designed to coordinate the coal and steel market between six European countries, including Belgium.

136 Minutes of the Board of Directors meeting of 3 February 1945.

137 See on this subject Dirk Luyten, 'L'épuration économique en Belgique', dans *L'épuration économique en France à la Libération* Rennes, Presses universitaires de Rennes, 2008, pp. 203-213 (available online: books.openedition.org/pur/4779).

138 Minutes of the Board of Directors meeting of 26 October 1944.

139 These permanently lost machines were located in the area occupied by the Red Army, which became the German Democratic Republic in 1949.

140 *La Meuse*, 5 December 1944.

141 350 million Belgian francs would then be paid to FN.

142 Minutes of the Board of Directors meeting of 6 June 1945.

143 Minutes of the Board of Directors meeting of 7 October 1944.

144 See www.designmuseumgent.be/fr/object/melk-machine.

145 'Many delicate questions remain unresolved because of the vagueness of the legal texts. New interpretative legislation has been awaited for months and, in the meantime, everything remains up in the air,' declared the General Assembly on 25 October 1945. The company's management repeated this despondent statement many times.

146 Minutes of the General Assembly of 25 October 1945.

147 The Brussels Treaty, signed by the United Kingdom, France, the Netherlands, Belgium and Luxembourg on 17 March 1948, envisaged military assistance if one of the signatories were to be attacked. The treaty also envisaged closer economic and cultural cooperation.

148 The Benelux was an intergovernmental cooperation agreement between Belgium, the Netherlands and Luxembourg. These three neighbouring constitutional monarchies, between France and Germany, came together at the end of the Second World War to form a stronger economic entity to face much bigger neighbouring countries.

149 This relates to the Indonesian war of independence. Indonesia was recognised as independent on 27 December 1949.

150 The SNCFB is the acronym for the Société Nationale des Chemins de Fer Belges [National Company of Belgian Railways], which was later simplified to SNCB.

151 No dividend had been paid out since 1938, 12 years earlier.

152 In accordance with the law FN set up a Works Council during the period from 1948 to 1949. Elections were held without a hitch and without wanting to anticipate the efficiency of this measure, the directors firmly hoped that this new body would develop 'the spirit of cooperation and mutual understanding, for the greater good of the company and its workers at all levels of the hierarchy'. *Minutes of the General Assembly of 26 October 1950.*

153 This was in particular the 'bras-charbor' [labour for coal] agreement signed in 1946, under which 200 kilos of coal were delivered to Italy every day for each of the 50,000 Italian immigrant workers in Belgium.

154 Belgium, Canada, Denmark, the US, France, Iceland, Italy, Luxembourg, Norway, the Netherlands, Portugal and the UK were the first signatories of the treaty.

155 Minutes of the General Assembly of 27 October 1955.

156 Minutes of the General Assembly of 22 October 1959.

157 See above for Gustave Joassart's negotiations with the German authorities to keep two offices in Liège.

158 Minutes of the Board of Directors of 10 December 1956.

159 Minutes of the General Assembly of 24 October 1957.

160 There has always been a real dividend policy at FN, which is based on the simple observation that it is the shareholders, their means and their trust that allow the company to invest and therefore to make progress. Whenever it can, the company remunerates those who show trust in it as generously as possible and, each time that it's impossible, it apologises for that.

161 Minutes of the Board of Directors of 11 December 1950.

162 Minutes of the Board of Directors of 9 July 1962.

163 Claude Gaier, in *Ars Mechanica, op. cit.*

164 Minutes of the Board of Directors of 15 January 1951.

165 John Val Browning returned for good to the US after a twelve-year stay in Liège, in October 1959 (Minutes of the Board of Directors of 22 October 1959).

166 John Moses Browning died on 26 November 1926.

167 Minutes of the Board of Directors of 18 January 1960. Details of the order: 109,400 5-shot automatic shotguns, 2,100 double-automatic shotguns, 9 900 shotguns, 22,400 automatic rifles, 14,900 pistols and 3,000 Mauser rifles.

168 Minutes of the Board of Directors of 12 September 1960.

169 Minutes of the Board of Directors of 11 March 1957.

170 Minutes of the Board of Directors meeting of 11 April 1960.

171 Bell & Howell was a company founded in 1907 in the United States by two projectionists. It is a specialist in the production of photography equipment, cameras and lenses.

172 The directors said: 'the relations between Browning and FN are less trusting than they have been in the past due to the way that the Browning family decided to put some of its shares on the market' (minutes of the Board of Directors of 26 October 1961).

173 Minutes of the Board of Directors meeting of 10 May 1965.

174 In 1960 for example, the 'strike of the century' brought Belgium to a standstill. The strike was the result of opposition to the government bill for a Unitary Law, which raised the retirement age in the public sector, increased indirect taxes and reduced unemployment benefits. In 1966, there was a 'women's strike', as well as multiple work stoppages that were more sectoral.

175 The Anson & Deeley mechanism is named after two English gunsmiths, William Anson & John Deeley. In 1875, they designed a simple and ingenious system which could arm itself when the barrels were pivoted. The mechanism is integrated inside the hollowed-out action frame, unlike the models with side plates, where the same mechanism is installed on plates (side plates) on either side of the action frame.

176 The Ecole d'Armurerie Léon Mignon de Liège was founded in 1897 by the City of Liège at the initiative of the Union des Fabricants d'Armes. Even today, students are trained to become gunsmiths and engravers, among other trades.

177 A scooter is a two- or three-wheeled motorised road vehicle, featuring small-diameter wheels, an open frame that forms a floor (a large space between the wheels to accommodate feet and any luggage) and a fairing.

178 Minutes of the Board of Directors meeting of 10 July 1950.

179 This feat was reported in the 1954 list of FN's greatest victories.

180 Report of the Board of Directors to the General Assembly on 24 October 1968.

181 The General Assembly of 27 October 1961 predicted: 'The Common Market may well result in some growth [in agricultural equipment]'.

182 Minutes of the General Assembly of 24 October 1963.

183 Auguste Francotte, Fabrique Nationale d'Armes de Guerre, 1889-1964, Liège, Mathy, 1965.

184 Report of the Board of Directors to the General Assembly of 1967.

185 In 1936, Bofors Aktiebolaget granted FN the licence to produce a 40 mm calibre automatic anti-aircraft gun. One hundred and fifty pieces of this type were sold to the Belgian Army.

186 According to the Observatoire de la Finance (a Swiss non-profit foundation) 'the process of financialization results from the growth of practices, techniques as well as representations and values inspired by finance. Until now, finance was only a tool at the service of work, the only factor in the creation of value. From now on, it is work that is at the service of financial results.'

187 The Works Council is a regulatory and permanent body of the company. It brings together union and employer representatives in a discussion structure, where problems of all kinds affecting the life of the factory can be raised. It is within this Council that the employer consults and informs the workers.

188 These agreements were reached following the Joint Declaration of Productivity in 1954 (agreement by unions and employers to link the reduction of working time to an increase in productivity).

189 Report of the Board of Directors to the General Assembly of 22 October 1964.

190 ELDO: European Launcher Development Organisation, European centre for the construction of spacecraft launch vehicles.

191 In the minutes of the Board of Directors meeting of 8 February 1960, René Laloux reaffirmed to the Board 'that it was not the role of industry to select equipment, but to build equipment selected by the governing military authorities'.

192 The Laloux family is one of the families that have managed FN since its foundation. René Laloux (1895-1981) was successively Deputy Director (1932), then a partner of Gustave Joassart (1948), before finally becoming Managing Director from 1950 to 1963, then Chairman and Managing Director of FN from 1963 to 1971. His vision of the company (based on a commitment to quality work and paternalism) often clashed with the new management that followed.

193 As early as 1973, the subdivision of the 'armoury department' into three divisions (defence and security; hunting and sports; serial weapons production) was envisaged, as well as the creation of new directorates a few months later: personnel directorate; administrative directorate; financial directorate; planning and equipment directorate.

194 See page 256.

195 Report of the Board of Directors to the General Assembly of shareholders of 24 October 1974.

196 Report of the Board of Directors to the General Assembly of shareholders of 23 October 1975.

197 Report of the Board of Directors to the General Assembly of shareholders of 28 October 1976.

198 SONACA (Société Nationale de Construction Aérospatiale) was the successor of the Société Belge des Avions Fairey, created in 1931 to produce under licence 43 aircraft for the Belgian Air Force.

199 One can get an idea of the power of this aircraft by reporting a counter-example. US President Jimmy Carter ordered a limitation on the proliferation of conventional weapons. An export version of the F-16 with reduced performance was built – the F-16/79 – but its poor performance meant it could not be sold.

200 As part of CFM International, GE Aircraft Engines is tasked with the high-pressure part of the engine, also known as the 'core' (high-pressure compressor: combustion chamber and high-pressure turbine) and SNECMA is responsible for the low-pressure part (fan, low-pressure compressor or booster and low-pressure turbine), as well as the auxiliaries and the exhaust nozzle.

201 Minutes of the Board of Directors meeting of 23 May 1977.

202 Report of the Board of Directors to the General Assembly of shareholders of 27 October 1977.

203 Report from the Board of Directors to the General Assembly of shareholders of 7 June 1979.

204 The company has always wanted, as much as possible, to control the distribution and sale of its products. This has been done by creating subsidiaries – like FN England in the early 20th century – or by acquiring a stake in their capital in order to ensure at least a presence on the board of directors, as was the case for Cartoucherie Française in the 1920s.

205 Minutes of the Board of Directors meeting of 20 December 1976.

206 Minutes of the Board of Directors meeting of 7 March 1977.

207 Minutes of the Board of Directors meeting of 10 December 1973.

208 The Browning company was prosecuted by the American courts for having, in the opinion of the customs service, undercut the price of its arms imports from Belgium by contributing significantly to the cost of equipping the Herstal factories for their manufacture. Five people from FN were invited to testify in this case by the Attorney General, which placed the factory in an 'embarrassing situation' (source: Minutes of the Board of Directors meeting of 22 December 1975). It was added that FN's decision as to whether it should testify or not should 'realistically take into account the very strong bargaining power of the US Government with respect to us (F.100 and MAG)'. This legal dilemma only ended in April 1983, when Browning paid $3.5 million to the US Customs Service.

209 Browning had 10,000 sales outlets in America, which alone accounted for half of the world market for 'sporting guns'.

210 The six subsidiaries were: Browning Manufacturing Co. (Utah); Browning Arms Co. (Utah); Browning Arms Co. of Canada; Caldwell Lace Leather Co. (Kentucky); Jarman Co. (Oregon); and Browning A. (Belgium).

211 Turnover is the sum of the sales of goods or services of a company. It is equal to the amount (excluding taxes) of all transactions carried out by the company with third parties, in carrying out its normal and current business activities.

212 Pascal Deloge, *Une histoire de la Fabrique Nationale de Herstal*, Liège, Éditions du Céfal, 2012.

213 Report of the Board of Directors to the General Assembly of 3 June 1982.

214 Report of the Board of Directors to the General Assembly of 6 June 1985.

215 Report of the Board of Directors to the General Assembly of 6 June 1985.

216 Remark made by the US Secretary of the Treasury, John Connally, to a European delegation in late 1971.

217 Aglietta Michel, Coudert Virginie, 'III. Les fluctuations des cours du dollar : origines et conséquences', in: Michel Aglietta éd., Le dollar et le système monétaire international. Paris, La Découverte, 'Repères', 2014, pp. 65-84.

218 The 'dollar zone' is defined as all the countries whose currencies are pegged to the dollar. Broadly speaking, it covers a very large number of emerging countries, in Asia, Latin America and the Middle East, which seek to stabilise their currency against the dollar.

219 On 12 September 1985, the dollar reached its highest value. On 22 September, the finance ministers of the five most industrialised countries (the US, France, UK, Japan and Germany), who were meeting in New York, agreed to favour a fall in the value of the dollar. Following this press release, the dollar fell markedly.

220 *Le Monde*, 28 October 1980.

221 Report of the Board of Directors to the General Assembly of 3 June 1982.

222 1986 report to the General Assembly of 4 June 1987.

223 Under the new Belgian legislation on Coordination Centres, on 8 November 1985 FN Herstal created the SA FN Coordination Center with a capital of 500 million Belgian francs. Its mission, mainly financial, was to take out the long-term loans necessary to finance the investment programmes and R&D of Belgian companies in the FN Group. The Center's work would, through the fiscal advantages that the law provided, reduce by 25% the cost of the capital borrowed like this. Secondly, the Center would centralise the financial management activities that were carried out for the benefit of the FN Group's different entities.

224 Report of the Board of Directors to the General Assembly of 27 October 1977.

225 Browning America, Browning Sport Ltd. (UK), Browning SA (France), Browning Viana (Portugal), Browning Sports Italia SRL and Browning Canada Sports Ltd-Ltée.

226 Winchester owed a number of its leading products to John Moses Browning, from the one-shot Model 1885 to the pump-action 1893 and 1897 Models, as well as the legendary lever-action 1886, 1892 and 1894 Models.

227 42.5% of shares for FN and 6.5% for the Walloon Region (see *Flash FN*, no. 351, 12 June 1989).

228 Rhéo was tasked with the research and development of machines and processes to exploit industrial waste and urban plastics, by extending the plastification technology developed at the end of the 1970s.

229 Tangible proof that the situation was improving was that FN was able, for the first time in years, to pay a dividend to its shareholders in 1992 and 1993. The results were stable until 1995. They were then less good, but remained at an acceptable level.

230 In other words, clearance to access classified information.

231 See McNAB C., *The FN Mag Machine Gun*, Osprey Publishing, 2018.

232 *Herstal 2000*, n° 1, June 1998.

233 See, in particular, Bernard Adam *et al.*, *La nouvelle architecture de sécurité en Europe*, Brussels, GRIP, 1999.

234 See Stanley R. Sloan, *Defense of the West: NATO, the European Union and the Transatlantic Bargain*, Manchester, Manchester University Press, 2016.

235 *Dépenses militaires et transferts d'armements conventionnels [Military expenditure and transfers of conventional weapons] – Compendium 2006*, Bruxelles, GRIP, 2006.

236 Frédéric Mauro, *Où en est la recherche européenne de défense?*, Brussels, GRIP, 2017.

237 FN Herstal had ventured into this area, together with the Microelectronics Department of Liège University, as part of a European programme to develop electronics technologies.

238 *Made in Belgium*, December 2000.

239 *Herstal 2000*, n° 2, May 1999.

240 Groupe Herstal, *Informations de base*, 2004 version, updated in 2007.

241 *Made in Belgium*, June 2001.

242 *Herstal 2000*, n° 2, May 1999.

243 Groupe Herstal, *Informations de base*, 2000 version, updated in 2003.

244 *Made in Belgium*, January 1999.

245 See Stéphane Vial, *L'être et l'écran. Comment le numérique change la perception*, Paris, PUF, 2013.

246 *Made in Belgium*, March 2003.

247 *Management report of the General Assembly's Board of Administration, 2 June 2008.*

248 Leigh Neville, *Infantry Small Arms of the 21st Century*, Pen & Sword Military, 2019, p. 215.

249 Chris McNab, *Weapons of the US Special Operations Command*, Oxford, Osprey Publishing, 2019, p. 4

250 Pierre Thys (dir.), *Les armes 'à létalité réduite'*, Paris, L'Harmattan, 2010.

251 Luc Mampaey, *Les armes à 'létalité réduite'. Solution ou perversion ?*, Brussels, GRIP, 2009.

252 Gordon L. Rottman, *Browning .50 caliber Machine Guns*, Oxford, Osprey Publishing, 2010, p. 24.

253 Alain De Neve, 'Mutations technologiques et transformations militaires: que reste-t-il du discours de la RMA ?', in *Pyramides*, 21, 2011, pp. 27-52.

254 *Made in Belgium*, April 2002.

255 *Idem.*

256 Minutes of the Board of Directors meeting of 12 June 2007.

257 *Made in Belgium*, August 2004.

258 *Idem*, April 2006.

259 *Idem*, December 2003.

260 *Idem*, October 2006.

261 Yannick Quéau, *Le projet de défense européenne: l'Arlésienne de l'UE*, Bruxelles, GRIP, 2014.

262 See in particular Dominique Foray, *L'économie de la connaissance*, Paris, La Découverte, 2018.

263 In 1994, the White Paper on growth, competitiveness and employment published by the European Community already estimated that 75% to 95% of the wage bill of companies would henceforth be allocated to roles for organisation rather than direct production, i.e. IT, engineering, training, accountancy, sales and research. See *Growth, competitiveness, employment. The challenges and ways forward into the 21st century.* White Paper, Brussels, CECA-CE-CEEA, 1994, p. 80.

264 *L'Écho*, 5 November 2020.

265 Chris McNab, *The FN MAG Machine Gun*, Oxford, Osprey Publishing, 2018.

266 For the different versions, see in particular Chris McNab, *The FN MINIMI Light Machine Gun*, Oxford, Osprey Publishing, 2017.

267 Spiegel and Paul Shipley, 'Lightweight Small Arms Technologies (LSAT) Program Update', in *NDIA Joint Services Small Arms Systems Annual Symposium, May 2006*.

268 Federico Santopinto, *Le financement de la recherche de défense par l'UE*, Brussels, GRIP, 2016.

269 Federico Santopinto, *Le Fonds européen de la Défense ouvre le débat sur les exportations d'armes*, Brussels, GRIP, 2019.

270 Federico Santopinto and Julien Maréchal, *La Facilité européenne pour la Paix: un nouvel outil au service de la politique d'assistance militaire de l'UE*, Brussels, GRIP, 2020.

271 *Made in Belgium*, December 2020.

272 *Idem*, December 2019.

Bibliography

Publications

1889–2014, 125 ans en images, FN Herstal, 2014.

A History of Browning Guns, from 1831, Ogden, J. M. & M.S. Browning Company, 1942.

Mémoires de la FN, vol. 1-12 (2012–2021), Herstal, Ars Mechanica Foundation.

BARTHOLOMÉ P., *Techspace Aero, 50 ans d'histoire et d'aventure*, Herstal, Par amour du ciel, 1999.

BEDU L., *De Mauser à Blaser. Un siècle de carabines de chasse*, Paris, 2012.

BITSCH M.-T., *La Belgique entre la France et l'Allemagne, 1905–1914*, Paris, Publications de la Sorbonne, 1994.

BOORMAN D. K., *The History of Winchester Firearms*, Guilford, The Lyons Presse, 2006.

Browning Firearms Collection, Ogden, The American Society of Mechanical Engineers, 1989.

CASHNER B., *The FN FAL Battle Rifle*, Oxford, Osprey Publishing, 2013.

CHIQUET P., *La gabegie. Le scandale du complexe militaro-industriel français*, Paris, Albin Michel, 1997.

COENEN M-T., *La grève des femmes de la FN en 1966*, Brussels, POL-HIS, 1991.

COPÉ M., *Les travailleurs aussi fabriquent l'histoire de la FN*, Liège, Fondation André Renard et FGTB., 1989.

Croissance, compétitivité, emploi. Les défis et les pistes pour entrer dans le XXIe siècle, Livre blanc, Brussels, CECA-CE-CEEA, 1994.

DE BECKER G., *Quand la FN avait des roues*, Heinstert-Attert, 2013.

DELOGE P., *Une coopération difficile : Belgique et Grande-Bretagne en quête de sécurité à l'aube de la Guerre froide*, Brussels, Musée royal de l'Armée [Royal Military Museum Brussels], 2000.

DELOGE P., *Une histoire de la Fabrique Nationale de Herstal*, Liège, Éditions du Céfal, 2012.

DENAYER J., PACYNA D. et BOULVAIN F., *Le minerai de fer en Wallonie. Cartographie, histoire et géologie*, Jambes, DGO des Ressources naturelles, Service Public de Wallonie [Public Service of Wallonia], 2011.

Dépenses militaires et transferts d'armements conventionnels – Compendium 2006, Brussels, GRIP, 2006; 2019; 2020.

DESTATTE P., *Histoire de la Belgique contemporaine. Société et institutions*, Brussels, Larcier, 2019.

DRECHSEL O., *F.N. Le service poids-lourds*, tomes I à III, Herstal, typed edition, *s.d.*

DRECHSEL O., *FN Turboréacteurs, 25 années de production*, Liège, 1976.

DUCHESNE J.-P., *L'affiche en Belgique. Art et pouvoir*, Brussels, Labor, 1989.

DUCHESNE J.-P., 'Les arts plastiques et graphiques aux XIXe et XXe siècles', in *Histoire culturelle de la Wallonie*, Brussels, Fonds Mercator, 2012.

DUMONT F., 'L'affiche en Wallonie de la fin du XIXe siècle à nos jours', in *La Wallonie, le pays et les hommes*, t. III, La Renaissance du Livre, 1979.

FORAY D., *L'économie de la connaissance*, Paris, La Découverte, 2018.

FRANCOTTE A., *Fabrique Nationale d'Armes de Guerre, 1889–1964*, Liège, Mathy, 1965.

FRANCOTTE A., GAIER C., *FN 100 ans, Histoire d'une grande entreprise liégeoise 1889–1989*, Brussels, Didier Hatier, 1989.

FRANCOTTE A., GAIER C., KARLSHAUSEN R., *Ars Mechanica, le grand livre de la FN. Une aventure industrielle extraordinaire*, Herstal, Herstal Group, 2008.

GAIER C., *Mémoire du capitaine Michel François Dale, 'La fabrication du fusil 'modèle 1777' à la Manufacture impériale d'Armes de Liège en 1810. Contribution à l'histoire industrielle de la Belgique'*, FN Herstal, 1977.

GAIER C., *The American Gunmaker John M. Browning*, FN Herstal, 1978.

GAIER C., *Cinq siècles d'armurerie liégeoise*, Alleur, Perron, 1996.

GAIER C., *Le banc d'épreuves des armes à feu en Belgique*, Liège, 2013.

GASPARD G., *Les dames de la Basse-Meuse. La motocyclette liégeoise de 1940 à 1965*, Liège, Vaillant-Carmanne, 1978.

GASPARD G., *Les demoiselles de Herstal. La motocyclette liégeoise des origines à 1940*, Liège, Vaillant-Carmanne, 1983.

GENTRY G., *John M. Browning, American Gunmaker*, Browning Co., 1994 [1964].

GOLDSMITH D. L., *The Browning Machine Gun*, vol. 1, Collector Grade Publications, 2005.

GORENSTEIN Nathan, *The Guns of John Moses Browning*, New York, Scribner, 2021

GUNSTON B., *Rolls-Royce Aero Engines*, Patrick Stephens Ltd., 1989.

'Herstal-Browning, 125 ans, les racines de l'avenir', in *Jours de Chasse*, special edition n° 5 (2014).

HUON W., *Ariane, une épopée européenne*, Boulogne-Billancourt, ETAI, 2019.

JOHNSON W., *The FN-49, the Last Elegant Old-World Military Rifle*, Greensboro NC, 2004.

KEMPF O., *L'Otan au XXIe siècle, la transformation d'un héritage*, Monaco, Éditions du Rocher, 2014.

KUPELIAN Y., KUPELIAN J., SIRTAINE J., *Histoire de l'automobile belge*, Brussels-Paris, Paul Legrain, 1979.

KUPELIAN Y., KUPELIAN J., SIRTAINE J., *Soixante ans de compétition automobile en Belgique 1896–1956*, Overijse, De Boeck, 1981.

LAUREYS D., *La contribution de la Belgique à l'aventure spatiale européenne, des origines à 1973*, Paris, Beauchesne, 2008.

L'affiche en Wallonie à travers les collections du Musée de la Vie wallonne, Brussels, Ministère de la Communauté française [Ministry of the French-speaking Community], 1980.

La nouvelle architecture de sécurité en Europe, Brussels, GRIP, 1999.

L'économie mondiale 2021, Paris, La Découverte, 2020.

Les 100 ans de la FN à travers l'art publicitaire, Herstal, Musée communal d'Herstal [The Herstal Museum], 1989.

MABILLE X., TULKENS C.X. et VINCENT A., *La Société générale de Belgique 1822–1997. Le pouvoir d'un groupe à travers l'histoire*, Brussels, CRISP, 1997.

MADIS G., *The Winchester Book*, Brownsboro, Art and Reference House, 1977.

MALINGUE B., *Guide pratique FN-FAL*, vol. 1, Chaumont, Crépin-Leblond, 2017.

MAMPAEY L., *Les armes à 'létalité réduite'. Solution ou perversion ?*, Brussels, GRIP, 2009.

MAMPAEY L., *Groupe Herstal, l'heure des décisions*, Brussels, GRIP, 2000.

MAMPAEY L., *Radiographie de l'industrie d'armements en Belgique*, Brussels, GRIP, 2010.

MAURO F., *Où en est la recherche européenne de défense ?*, Brussels, GRIP, 2017.

McNAB C., *The FN Minimi Light Machine Gun*, Oxford, Osprey Publishing, 2017.

McNAB C., *The FN Mag Machine Gun*, Oxford, Osprey Publishing, 2018.

McNAB C., *Weapons of the US Special Operations Command*, Oxford, Osprey Publishing, 2019.

MILLER D., *The History of Browning Firearms*, Guilford, The Lyons Presse, 2006.

NEVILLE L., *Infantry Small Arms of the 21st century*, Pen & Sword Military, 2019.

PASQUASY F., *La sidérurgie au Pays de Liège. Vingt siècles de technologie*, Liège, Société des Bibliophiles liégeois [Society of Bibliophiles in Liège], 2013.

PELLETIER A., *Boeing, géant de l'aéronautique, de 1916 à nos jours*, Boulogne-Billancourt, ETAI, 2008.

PURAYE J., *La gravure sur armes à feu au pays de Liège*, Liège, 1964.

QUÉAU Y., *Le projet de défense européenne : l'Arlésienne de l'UE*, Brussels, GRIP, 2014.

RENARDY C., *Liège et l'Exposition universelle de 1905*, Brussels, Fonds Mercator, 2005.

ROBERT R., HODGES Jr., *The Browning Automatic Rifle*, Oxford, Osprey Publishing, 2012.

ROTTMAN G. L., *Browning .50 Caliber Machine Guns*, Oxford, Osprey Publishing, 2010.

ROTTMAN G. L., *The M16*, Oxford, Osprey Publishing, 2011.

SANTOPINTO F., *Le financement de la recherche de défense par l'UE*, Brussels, GRIP, 2016.

SANTOPINTO F., *Le Fonds européen de la Défense ouvre le débat sur les exportations d'armes*, Brussels, GRIP, 2019.

SANTOPINTO F., MARÉCHAL J., *La Facilité européenne pour la paix : un nouvel outil au service de la politique d'assistance militaire de l'UE*, Brussels, GRIP, 2020.

SHIRLEY H. M. Jr., VANDERLINDEN A., *Browning Auto-5 Shotguns, the Belgian FN production*, Greensboro, Wet Dog Publications, 2003.

SLOAN R. S., *Defense of the West: Nato, the European Union and the Transatlantic Bargain*, Manchester, Manchester University Press, 2016.

SPARACO P., *Snecma, les moteurs du ciel*, P. Galodé Éditeurs, Saint-Malo, 2007.

SPARACO P., *Airbus, la véritable histoire*, Toulouse, Privat, 2005.

THYS P. (dir.), *Les armes à létalité réduite*, Paris, L'Harmattan, 2010.

VANDERLINDEN A., *FN Browning Pistols. Side-Arms that Shaped World History*, Greensboro, Wet Dog Publications, 2009.

VIAL S., *L'être et l'écran. Comment le numérique change la perception*, Paris, PUF, 2013.

C. WASSEIGE, *Mémoire sur la condition des ouvriers et le travail des enfants dans les mines, manufactures et usines de la Province de Liège*, Brussels, Th. Lesigne, 1847.

WILSON R.L., *Winchester, an American Legend*, New York, Random House, 1991.

Winchester. Armes de légende, Liège, Musée d'Armes de Liège [Arms Museum of Liège], 1992.

Articles

ADAM B., ZAKS A., DE VESTEL P., 'Contexte et perspectives de restructuration de l'industrie de l'armement en Wallonie', in *GRIP, Dossier, notes et documents*, n° 161-162 (September) October 1991.

BERKOWITZ H., DUMEZ H., 'Le système Gribeauval ou la question de la standardisation au XVIIIe siècle' [online], in *Annales des Mines*, 2016/3, n° 125, pp. 41-50 (www.cairn.info/revue-gerer-et-comprendre-2016-3-page-41.htm).

BRUN J.-F., 'La mécanisation de l'armurerie militaire (1855-1869)', in *Revue historique des Armées*, n°269 (2012).

CASSIERS I., 'Une statistique des salaires horaires dans l'industrie belge, 1919-1939' [online], in *Recherches économiques de Louvain / Louvain Economic Review*, vol. 46, n° 1 (1980), pp. 57-85 (www.jstor.org/stable/40723610).

CHARBONNEAU J., COUDERC N., 'Globalisation et (in)stabilité financières', in *La Découverte, Regards croisés sur l'économie*, n°3 (2008), pp. 235-241.

CRISP [Centre de Recherche et d'Information socio-politiques], 'Le sport cycliste en Belgique' [online], in *Courrier hebdomadaire du CRISP*, n° 94 (1961). pp. 1-20 (www.cairn.info/revue-courrier-hebdomadaire-du-crisp-1961-4-page-1.htm).

CRISP [Centre de Recherche et d'Information socio-politiques], 'La structure de propriété des holdings belges' [online], in *Courrier hebdomadaire du CRISP*, n° 566 (1972), pp. 1-22 (www.cairn.info/revue-courrier-hebdomadaire-du-crisp-1972-20-page-1.htm).

DEFRAIGNE J.-C., 'La reconfiguration industrielle globale et la crise mondiale', in *Outre-Terre*, n°46 (2016), pp. 149-198.

Diversification et reconversion de l'industrie d'armement, Brussels, GRIP, 1992.

DE BOER M.-G., 'Guillaume Ier et les débuts de l'industrie métallurgique en Belgique' [online], in *Revue belge de Philologie et d'Histoire*, tome III, fasc. 3 (1924), pp. 527-552 (www.persee.fr/doc/rbph_0035-0818_1924_num_3_3_6304).

DE NEVE A., 'Mutations technologiques et transformations militaires : que reste-t-il du discours de la RMA ?', in *Pyramides*, 2011, pp. 27-52.

DESAMA C., 'L'expansion démographique du XIXe siècle', in *La Wallonie, le pays et les hommes*, t. II, Bruxelles/Tournai/Waterloo, La Renaissance du Livre, 1980, pp. 139-158.

DUCHESNE J.-P., 'Les arts plastiques et graphiques aux XIXe et XXe siècles', in Bruno DEMOULIN (dir.), *Histoire culturelle de la Wallonie*, Brussels, Fonds Mercator, 2012.

DUMONT F., 'L'affiche en Wallonie de la fin du XIXe siècle à nos jours', in HASKIN H., LEJEUNE R. et STIENNON J. (dir.), *La Wallonie, le pays et les hommes*, t. III, Bruxelles/Tournai/Waterloo, La Renaissance du Livre, 1979, pp. 341-344.

DUPRIEZ L. H., 'La conjoncture économique de la Belgique de 1919 à 1929', in *Bulletin de l'Institut des Sciences économiques*, 1e année, n°2 (1930), pp. 75-103.

FISCHER W., 'American influence on German manufacturing before World War I: the case of the Ludwig Loewe Company' in *L'américanisation en Europe au XXe siècle : économie, culture, politique*, vol. 1, Lille, Publications de l'Institut de Recherches historiques du Septentrion, 2002.

GAIER C., 'Un grand nom de l'armurerie, René Laloux (1895-1981)', in *Le Musée d'Armes* [Arms Museum of Liège], n° 33, April 1982, pp. 17-20.

GAIER C., 'Le Mauser à Liège', in *Le Musée d'Armes* [Arms Museum of Liège], n° 127 (2013), pp. 19-26.

GAIER C., 'Les graveurs sur armes liégeois', in *Le Musée d'Armes* [Arms Museum of Liège], n° 136 (2018), pp. 3-35.

GEERKENS É., 'La production en série dans 'industrie armurière belge' [online], in *Parlement[s], Revue d'Histoire politique*, n° 33 (2021), pp. 51-60 (www.cairn.info/revue-parlements-2021-1-page-51.htm).

GIRAUD P.-N., 'Comment la globalisation façonne le monde', in *Politique étrangère*, 2006, pp. 927-940.

HOUBEN H., 'Les restructurations dans l'industrie automobile en Belgique' [online], in *Courrier hebdomadaire du CRISP*, n° 2295-2296 (2016), pp. 5-71 (www.cairn.info/revue-courrier-hebdomadaire-du-crisp-2016-10-page-5.htm).

HUCORNE M., 'L'industrie aéronautique en Belgique' [online], in *Courrier hebdomadaire du CRISP*, n° 1059 (1984), pp. 1-30 (www.cairn.info/revue-courrier-hebdomadaire-du-crisp-1984-34-page-1.htm).

KURGAN-VAN HENTENRYK G., 'Le patronat en Belgique (1830-1960)' [online], in *Histoire, économie et société*, n° 1 (1998), p. 189-211 (www.persee.fr/doc/hes_0752-5702_1998_num_17_1_1980).

L'industrie européenne de l'armement – Recherche, développement technologique et reconversion, Brussels, GRIP, 1993.

LOUIS-ANDRÉ V., 'Progrès technique et progrès économique' [online], in *Revue économique*, vol. 12, n° 6 (1961), pp. 876-904 (www.persee.fr/doc/reco_0035-2764_1961_num_12_6_407486).

MABILLE X., 'La Société générale de Belgique. Éléments pour une histoire de la Banque mixte 1822-1934' [online], in *Courrier hebdomadaire du CRISP* 1993/29, n° 1414-1415, pp. 1-67 (www.cairn.info/revue-courrier-hebdomadaire-du-crisp-1993-29-page-1.htm).

MANIGART P., 'L'évolution des dépenses militaires en Belgique depuis 1900' [online], in *Courrier hebdomadaire du CRISP*, n° 1009 (1983), pp. 1-25 (www.cairn.info/revue-courrier-hebdomadaire-du-crisp-1983-24-page-1.htm).

MEUNIER F.-X., 'Construction of an Operational Concept of Technological Military/Civilian Duality', in *Journal of Innovation Economics & Management*, n° 29 (2019), pp. 159-182.

PEAUCELLE J.-L., 'Du concept d'interchangeabilité à sa réalisation, le fusil des XVIIIe et XIXe siècles', in *Annales des Mines*, n° 80 (2005).

PEETERS M., 'L'évolution des salaires en Belgique de 1831 à 1913' [online], in *Bulletin de l'Institut de Recherches économiques*, vol. 10, n° 4 (1939), pp. 389-420 (www.jstor.org/stable/40742716).

ROGER C., 'La concentration des entreprises et la rationalisation au cours de l'année', in *Bulletin de l'Institut des Sciences économiques*, 1e année, n° 2 (1930), pp. 75-103.

SAUVY A., 'Variations des prix de 1810 à nos jours' [online], in *Journal de la Société statistique de Paris*, t. XCIII (1952), p 88-104 (http://www.numdam.org/article/JSFS_1952__93__88_0.pdf).

SPIEGEL K., SHIPLEY P., 'Lightweight Small Arms Technologies (LSAT) Program Update', in *NDIA Joint Services Small Arms Systems Annual Symposium*, May 2006.

SWENNEN M., 'Les mouvements anticommunistes dans les années 1920' [online], in *Courrier hebdomadaire du CRISP*, n° 2059 (2010), pp. 5-51 (www.cairn.info/revue-courrier-hebdomadaire-du-crisp-2010-14-page-5.htm).

TILLY P., DELOGE P., 'Milieux économiques belges et occupation allemande de 1914 à 1918 : une stratégie du moindre mal [online], in *Entreprises et histoire*, n° 68 (2012), pp. 11-27 (www.cairn.info/revue-entreprises-et-histoire-2012-3-page-11.htm).

VANDERMOTTEN C., 'La production de l'espace industriel belge : 1846-1984' [online], in *Hommes et terres du Nord*, 1985/2, pp. 100-109 (www.persee.fr/doc/htn_0018-439x_1985_num_2_1_1985).

VANVELTHEM L., 'Histoire du temps de travail en Belgique. Le temps de travail en Belgique durant le 'long XIXe siècle' (1800-1914)' [online], in *Analyse de l'IHOES*, n° 159 (2016) (www.ihoes.be/PDF/Analyse_159_Temps_travail_1.pdf).

VERLUISE P., 'Après 1914-1918 : une Europe redessinée par les traités' [online], in *Constructif*, n° 52 (2019) (www.cairn.info/revue-constructif-2019-1-page-5.htm).

VINCENT A., WUNDERLE M., 'Le tissu industriel wallon : secteurs et actionnariat' [online], in *Courrier hebdomadaire du CRISP*, n° 1761 (2002), pp. 5-47 (www.cairn.info/revue-courrier-hebdomadaire-du-crisp-2002-16-page-5.htm).

WANTY E., 'L'effort d'armement de l'Occident et le fusil automatique FN' [online], in *Industrie, Revue de la Fédération des Industries de Belgique*, no 4 (1954), pp. 174-178 (https://books.cpenedition.org/septentrion/46623?lang=fr).

Sources

Herstal Group's archives

- Browning and Fabrique Nationale d'Armes de Guerre [National Factory of Weapons of War]. Catalogues, brochures and technical manuals, end of the 19th century to 2021.

- *Rapports du conseil d'administration de la Fabrique Nationale à l'assemblée générale des actionnaires* [Reports of Fabrique Nationale's Board of Directors to the General Assembly of shareholders], from 1895 to 1991.

- *Procès-verbaux des conseils d'administration de la Fabrique Nationale* [Minutes of the Fabrique Nationale Board of Directors meetings], from 1889 to 1998.

- Herstal Group's in-house newspapers, published as these titles: *Revue FN* from 1953 to 1977; *Journal FN* from 1977 to 1989; *Infos GIAT Industries* from 1993 to 1994; *La Lettre de Herstal* in 1995; *Herstal 2000* from 1998 to 1999; *Made in Belgium* since 1998; *Flash FN*.

- Interviews.

Other

- State Archives in Liège, Fonds Fabrique Nationale d'Armes de Guerre [Fund of the National Factory of Weapons of War].

- Archives from the Musée de la Vie wallonne [Museum of Walloon Life].

- Articles from over 100 Belgian newspapers from 1814 to 1970 can be consulted online on the website of the Bibliothèque royale de Belgique [Royal Library of Belgium]: www.belgicapress.be.

- Articles from all the US newspapers from 1777 to 1963 can be accessed online on the website of the US Library of Congress: chroniclingamerica.loc.gov.

- The *familysearch.org* website lists billions of miscellaneous documents that have been scanned (civil status certificates, death certificates, lists of passengers arriving in the USA, etc). The website is run by The Church of Jesus Christ of Latter-day Saints (Mormons), of which John M. Browning was a member.

- The archives of the French newspaper *Le Monde* are accessible to subscribers (from 19 December 1944 to the present day): www.lemonde.fr/recherche/

Edition
Ars Mechanica Foundation /
Herstal Group
Fonds Mercator

Publication
Mercatorfonds
Bernard Steyaert,
Managing Director

Coordination
Ars Mechanica Foundation
Adrien Marnat,
Chief Executive Officer,
Heritage Director Herstal Group

Fawzi Amri,
Head of project - Digital

Isabelle Gaillard,
Head of project - Administration

Geoffrey Schoefs,
Head of project - Historian

Georges Muls,
Restorer

Mercatorfonds
Alice d'Ursel,
Editorial Coordinator

Laetitia d'Oultremont,
Editorial Coordinator

Scientific and artistic management
Adrien Marnat

Graphic design
Ars Mechanica Foundation
Grégoire Romefort
Studio Debie

Editorial
Jean-Marc Gay
Geoffrey Schoefs

Image research
Fawzi Amri
Corina Biris

Translation
Terry Brisco
Julian Hale

Photoengraving, printing and binding
Graphius, Ghent

Proofreading
Bruno Bernaerts

Distribution
Distributed outside Belgium,
The Netherlands and Luxembourg by Yale
University Press, New Haven and London
ISBN YALE: 9780300267006

Library of Congress Control Number:
2022940288

www.yalebooks.com/art
www.yalebooks.co.uk

Distributed in Belgium,
The Netherlands and Luxembourg
by Exhibition International
ISBN: 9789462302891
D/2023/703/05

Paper: Magno volume 150 gr/m²

Copyright 2023 Mercatorfonds, Brussels
and the authors

Copyright 2023
Ars Mechanica Foundation /
Herstal Group, Herstal

Acknowledgements
The Ars Mechanica Foundation would like to warmly thank those members, both past and present, of the Herstal Group's
management and staff – as well as the Group's partners – who have contributed to this book's publication.

We would like to mention in particular Glyn Bottomley, Martin Boucquey, Georges Chauveheid, Mark Cherpes, Philippe Claessens,
Philippe Collette, Julien Compère, Becky Costello, Christian Creuven, Rui Cunha, Guy De Becker, Philippe de Bruyne, Marie-Pierre
Dechêne, Nicolas de Gottal, Anne Devroye, Louis Dillais, Laurent Forget, Charles Guevremont, Travis Hall, Thierry Jacobs,
Dimitri Jerôme, Igor Klapka, Lucien Manfredi, Yoshihiko Miroku, Larissa Moors, Joris Renckens, Arto Sepponen, Kristof Verjans,
Vincent Verleye and Patrick Vogne.